Drainage Basin Form and Process
A geomorphological approach

For Chris and Viv

Drainage Basin Form and Process
A geomorphological approach

K. J. Gregory
*Reader in Physical Geography,
University of Exeter*

D. E. Walling
*Lecturer in Geography,
University of Exeter*

EDWARD ARNOLD

First published 1973 by
Edward Arnold (Publishers) Ltd.,
25 Hill Street, London W1X 8LL

ISBN: 0 7131 5707 0

Printed in Great Britain by
Fletcher & Son Ltd, Norwich

Contents

Acknowledgements

We should like to join the publishers in acknowledging permission given by the following to reproduce copyright material: the Aerofilms Library for Plate 23; Grant Heilman, Lititz, for Plates 5, 16, 17, 19 and 20; J. S. Shelton and Robert C. Frampton, Claremont, California, for Plates 1 and 25; Technology Application Center, University of New Mexico for Plate 4.

Individual acknowledgement to all the persons who have helped during the production of this volume is impossible to make but we particularly wish to thank Mr Rodney Fry who has patiently and imaginatively drawn the diagrams, Mr Andrew Teed who photographed the originals and constructed the plates, and Mrs P. Holmes, Miss L. Parker and Mrs B. Whinnam who resolutely typed portions of the text. In the University of Exeter the Inter-Library Loans section of the Library, and especially Miss H. Eva, were invaluable in securing references, and to the Head of the Geography Department, Professor W. Ravenhill, we express our sincere thanks for the encouragement he provided and for the resources made available. Mr B. Cart, Mr C. Park and Mr B. Webb kindly assisted with the preparation of the index. Many of the Devon examples cited as illustrations in the text were obtained during a research project supported at the University of Exeter by the Natural Environment Research Council.

To these, and to many others, we tender our grateful thanks and not least to our wives for their continued forbearance!

Preface

> The absence of a special book on rivers describing the physical geography of their basins, is surprising, considering the importance of the subject, scientifically and commercially, and also in an agricultural and sanitary point of view.

A. Tylor perceptively wrote these words in the *Geological Magazine* in 1875 (443) and in the subsequent eighty years few books were written by or for physical geographers following the direction which Tylor advocated, although the work of Pardé (1933) stands as a notable exception. The situation was remedied after 1960. Leopold, Wolman and Miller in *Fluvial Processes in Geomorphology* in 1964 and Morisawa in *Streams: Their Dynamics and Morphology* in 1968 provided modern reviews of the geomorphology of rivers, and *Water, Earth and Man* (Chorley, 1969) included some consideration of the physical geography of rivers and drainage basins together with assessment of their significance in socio-economic geography. The trends which have affected the study of rivers in the period since 1875 have been shown in the papers collected in *Rivers and River Terraces* by Dury (1970) who noted that fluvial morphology has already 'embraced numerical morphometry, it has assimilated the statistical treatment of a whole range of data on stream channels and of the inter-relationships of these data, and it is now producing models with a stochastic base'. During the last ten profitable years broader works by several geographers have included much that is relevant to the understanding and appreciation of rivers and of drainage basins; techniques for their study were included in *Techniques in Geomorphology* (King, 1960), models appropriate to their understanding were contained in *Models in Geography* (Chorley and Haggett, 1967), the application of networks was embraced in *Network Analysis in Geography* (Haggett and Chorley, 1969), methods of approach to particular problems were detailed in *Numerical Analysis in Geomorphology* (Doornkamp and King, 1971), and a systems outlook was demonstrated in *Physical Geography: A systems approach* (Chorley and Kennedy, 1971).

Related disciplines, also concerned with rivers and drainage basins, did not wait until 1960 to remedy the deficiency indicated by Tylor. Particularly in the field of hydrology, general works by Meinzer (1942), Foster (1948), Linsley, Kohler and Paulhus (1949), and by Wisler and Brater (1959) were among those existing before 1960, but the number of books in adjacent fields also has increased substantially since then (e.g. Chow, 1964; De Weist, 1965) and geographers have also reviewed the relevant parts of the field (Ward, 1967). This growth in the number of books during the last decade has accompanied a great increase in research expressed by an unprecedented expansion of scientific papers, some of which are contained in relevant journals which have been born in this period, including *Journal of Hydrology* (1964), *Water Resources Research* (1965), *New Zealand Journal of Hydrology* (1961) and *Nordic Hydrology* (1970), or in journals which have expanded substantially, such as the *Bulletin of the International Association of Scientific Hydrology*.

Explanation, therefore, appears necessary prior to adding one more book to an ever-increasing field, and it may be found in the statement by Tylor in that no one work hitherto has concentrated exclusively upon the drainage basin. The drainage basin is visualised increasingly as the fundamental unit of study in fluvial geomorphology and this work attempts, for geographers, to review the methods whereby it may be studied, to describe the results which have been achieved, and to point towards the significance which these results have. Such an approach appears warranted by the dependence of fluvial processes upon the drainage basin unit, and justified because the increasing effects of man have substantial implications for the basin and for the river network which it contains. The task is initiated by considering some of the ideas which have emerged in the study of geomorphology as a whole during the present century and, using this as a basis, outlining the way in which the drainage basin functions (Chapter 1). Broad appreciation of the drainage basin is followed by consideration of the manner in which the form of the drainage basin and its components may be described (2), and by the way in which its processes can be measured (3) and expressed (4). The relationships which exist between drainage basin form and process are examined (5) before considering how these relationships vary in space (6) and in time (7). Drainage basins in general, and rivers and streams in particular, provide numerous opportunities for investigation and each chapter includes several detailed examples and is concluded by a selected list of relevant works for further reading; the complete list of references is provided at the end of the book (pp. 407–447). The result of this approach may be to provide a 'do-it-yourself kit' for drainage basins but we hope that if this is the case the result is at least an indication of opportunities which may be realised and a demonstration of stimulating literature which may be explored within the context of a geomorphological view of the drainage basin.

1 Introduction

> . . . rivers and valleys have a special place, for it is impossible to treat the development of land forms, or to describe existing forms in a rational manner, without constant reference to the valleys that have been worn in them and to the rivers by which the waste is washed along the channel in the valley floor.
>
> *W. M. Davis, 1900*

1.1 Drainage basins in geomorphological perspective

Geomorphology has been characterised as the earth shape science (Brown, 1970). Study of this subject, which is concerned primarily with the form of the earth, is little more than one hundred years old. In the study of the form of the earth attention must necessarily be accorded to the development of land form and information necessary to this end can be obtained from three main sources. Firstly, information can be obtained, by mapping and by measurement, about the form of the land and about the spatial distribution of land forms. Secondly, information may be collected on the processes which fashion the surface of the earth at the present time because these processes are responsible for the production of particular types of land forms. Thirdly, the analysis of deposits can provide considerable information about the processes and about the chronology of events which occurred in the past. These three sources of information are basic to the study of geomorphology but information from the three sources is not always easily related. The most apparent difficulty is the dichotomy between process and form in that in many areas the processes operating at present are not the ones which were responsible for fashioning the landforms of that area, or at least the present rate of operation of geomorphological processes is not the same as the rates which obtained in the past. A further difficulty which arises is that the three lines of information are often not capable of study with equal facility. Geomorphological processes often require instrumentation and their study may be more time-consuming than the study of land form. A further qualification apparent in the progress of geomorphology is that some problems are more susceptible to an approach based dominantly upon one of the three lines of enquiry and in the evolution of the subject different approaches have emerged which are founded primarily upon land form, upon process, or upon deposits and chronology. There can be little doubt that many of the problems posed by the surface of the earth and by its landforms require, for their complete solution, methods of study which embrace all three types of information.

1.1a   The context

Reflection of these difficulties is mirrored in the development of the study of geomorphology within physical geography and over the past century several phases of growth may be discerned. Before 1900 the collection of information on surface form, on processes and on deposits was achieved, largely in the latter part of the nineteenth century (Davies, 1969) against the background of incomplete exploration of the earth's surface and of a lack of a coherent theory and a method of approach. A theory was provided in a second phase, initiated at the turn of the twentieth century, when ideas of evolution found expression in geomorphology in the formulation of the concept of the cycle of erosion by W. M. Davis (1899). The notion of evolution of landforms through cycles with stages of youth, maturity and old age and the approach to landscape as a function of structure, process and stage were the hallmarks of this phase, which occupied much of the first 40 years of the twentieth century. A recurring criticism of this phase of the evolution of geomorphology has been that it concentrated upon morphological evidence from landforms and devoted insufficient attention to process and to the study of deposits.

This period, however, witnessed the crystallisation of the subject of geomorphology and was followed, after 1945, by a third phase of diversification which sprang paradoxically from developments of, reactions to, and alternatives for the 'normal cycle of erosion' founded upon structure process and stage. Developments of the work of W. M. Davis included the extension of Davisian ideas and the refinement of the methods of study which he proposed and they included a body of research devoted to the denudation chronology of specific areas (for example, Johnson, 1931; Wooldridge and Linton, 1955). Reactions to the Davisian approach emerged because this approach led to studies which were evolution- rather than process-oriented and to studies which tended away from, rather than towards, the study of geography as a whole (Russell, 1949). Examples of such reactions include the emergence of process studies which were often quantitative in nature (Strahler, 1952), and the development of land form geography (Hammond, 1964), which focused upon the precise description of land form. Alternatives to the Davisian approach evolved independently and included climatic geomorphology (Budel, 1944; Peltier, 1950; Budel, 1963), which was conceived in areas where intimate relationships were encountered between landforms, climate, soil and vegetation. Thus, by 1960 the consequence of diversification had been to produce a number of geomorphologies which reflected the scientific training of the researchers, their location, their motivation, either academic or applied, all of which conditioned the perception of research problems. The emergence of a number of branches of geomorphology concerned with landforms in glacial, periglacial, coastal, and limestone areas, for example, was paralleled by developments in related disciplines such as geology and hydrology which provided new techniques of study, additional data, and further methods of approach.

Since 1960 a fourth phase of intensification and concentration has emerged during which attempts have been made to place the study of geomorphology in perspective. This has been fostered by increasing use of the systems approach, facilitated by model-building and thinking, and effected by the use of more sophisticated techniques and methods of analysis often quantitative in nature. Certain themes have been detected which are common to many of the branches of geomorphology and notable amongst these is the significance of man as a geomorphological agent. Study of the impact of man, directly and indirectly, upon both surface form and process has provided the study of geomorphology with an orientation which is increasingly relevant within the study of geography as a whole.

A paradox has, therefore, been realised where the outcome of diversification and the emergence of several distinctive approaches in geomorphology have been accompanied by integration within geomorphology, within physical geography, and within geography as a whole. Indices of this integration are illustrated by increasing attention devoted to the spatial distribution of contemporary physical processes in climatic geomorphology, to the consideration of physical landscape for human use in landscape evaluation, to the perception of environmental hazards and their economic significance, and to the effect of man upon environment (Brown, 1970). These apparently diverse approaches are united by focusing attention upon present systems, their content, mechanics and spatial variation, as a basis for studies of the past and of the future. Geomorphology can now be visualised as a land form-process science and it is the purpose of this book to elucidate this theme in the context of drainage basins and their fluvial processes, and the purpose of Section 1.2 to indicate the basic ideas upon which the contemporary study of geomorphology is founded.

1.1b   The need

Rivers warrant geographical study for three main reasons. Firstly, because of their existence in the physical landscape and their significance for producing fluvial landforms, secondly, because of their importance indirectly in relation to many other geomorphological processes in fluvially-dominated landscapes, and, thirdly, because of their significance for human use.

Rivers and streams carry a load of sediment derived by erosion and part of this load can be lost through deposition. Fluvial erosion can give rise to distinctive features such as undercut river cliffs, during transport of material very striking features such as potholes can be produced in the river bed, and, subsequently, deposition can produce flood plains where river-transported material is accumulated. A variety of landforms can, therefore, be produced by the direct action of the river but in addition the river can act in conjunction with other landscape-forming processes during the course of landscape development. The margins of a glacier can provide material which can be transported down valley by a river, aeolian processes can similarly provide material for transport, and processes acting on slopes can operate towards the river at the slope base, which can then remove material and indirectly influence and control the course of slope development. The river can operate in conjunction with marine processes in estuaries and in these various ways fluvial processes are intimately associated with glacial, aeolian, mass movement, and coastal processes. For these reasons, rivers are indirectly important in relation to other geomorphological processes and although the river may not be solely, or even directly, responsible for valley formation it is often important in relation to other processes or it is responsible for modifying valleys produced by other processes, by glacial erosion, for example. For these reasons, the significance of the drainage basin, in which the river is the main artery along which energy is available, has been acknowledged by the description of the drainage basin as the fundamental geomorphic unit (Chorley, 1965).

A final reason for the study of rivers arises from their significance to man. The use of rivers for water supply is increasingly relevant against a background of the increasing demand for water, increased from 18 litres per head per day in 1830 to 225 litres per head per day in 1960 in the U.K. Equally, rivers can be used for boundaries, for flotation of timber, for navigation, for fishing, for recreation and, not least, for power. Rivers can also be used as arteries for the disposal of effluent and concern has sometimes been registered over the levels of pollution reached. For these reasons, the margins of rivers in the past have attracted settlement and they continue to attract it despite the liability of some rivers to flood and to cause damage. Rivers in different parts of the world have, therefore, posed questions in relation to these varied uses. The variability of flow during the year and from one year to the next can be important in indicating how much water may be abstracted for supply, how much effluent can be tolerated and how much power can be assured throughout the year. The effects of direct modification of rivers and their basins by man can produce effects elsewhere along the river course. It is, therefore, necessary to measure the flow of rivers and to this end most countries have networks of river measurement stations, and it is equally necessary to understand river activity. Understanding of how rivers have developed in the past and are developing and changing at the present time is essential as a prerequisite to the appreciation of how rivers may change in the future—something which may be largely inspired by, and exceedingly relevant to, the activities of man.

1.1c   The scale

Of the land surface of the earth, 68·7 per cent is drained by rivers to lakes and oceans. Within this area at any one instant in time the world's rivers hold a mere 0·03 per cent of the world's fresh water but the amount of water which flows along the rivers of the world each year is equivalent to a layer of water 28·2 cm (11 inches) deep from the entire land surface of the earth. This water is equivalent to some 9500 litres of water annually for each of the earth's inhabitants. Averages, however, are misleading in that the extremes are often much more pertinent. The number of rivers and streams in the world have not been estimated but the United States may contain as many as 1½ mil-

Plate 1   River flood
Flooding of Los Alamitos, California by Coyote Creek (*top left*) in 1952. River floods can be the result of high intensity rainfall and therefore flooding is an annual occurrence in some climatic zones. High discharges leading to flooding can also be occasioned by snow melt, by storms following a period of prolonged rainfall and by specific causes including dam failure, drainage of subglacial lakes, volcanic eruptions beneath glaciers causing glacier busts. The frequency of river floods may have been influenced by human activity (see pp. 364).

lion unbranched streams (Leopold, 1962). The smallest river is a few metres in length but the Amazon has a length of nearly 6300 km and the eight largest rivers of the world have a total length greater than the equatorial circumference of the earth. Sediment carried by the world's rivers each year is sufficient to coat the British Isles to a depth of 6 cm with a layer of mud and the world average yield of sediment and solutes by rivers, estimated to be between 14 and 64 billion tonnes and 3·96 billion tonnes of dissolved material (Livingstone, 1963), is equivalent to a lowering of the

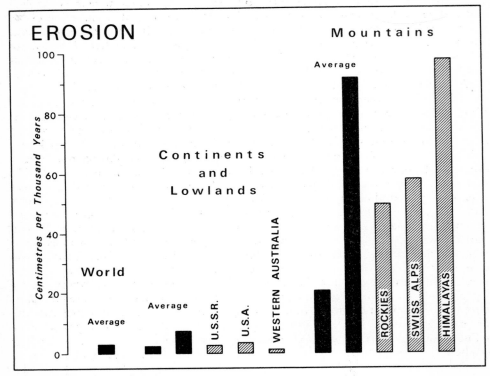

Figure 1.1   Some rates of erosion
Average world rates shown in black are after Stoddart (1969), Corbel (1964) and Schumm (1963). Rates for particular areas are shaded and are based upon Holeman (1968), Stoddart (1969), Douglas (1964), McPherson (1971), and Schumm (1963).

Denudation of the continents	Chemical denudation (Livingstone, 1963)	Suspended sediment discharge (Holeman, 1968)
North America	33·0 tonnes/km²/year	96 tonnes/km²/year
Europe	42·6 (due to moist climate plus fine-grained deposits?)	35
Asia	32·2	600
Africa	24·4	27
Australia	2·3	45
South America	28·3	63

surface of the earth by 3 cm every thousand years. Variations occur from one area to another (Figure 1.1) and the rivers of Asia carry some 80 per cent of the world's sediment to the seas and oceans each year (Holeman, 1968). Rivers provide opportunity to estimate the amount of erosion from the land at the present time because measurements of sediment concentration can be combined with measurements of river flow to give weights of sediment over time, and by knowing the drainage area and assuming a density of material, rates of erosion over the whole catchment have been deduced. Alternatively, it has sometimes been possible to estimate the volume of material which has accumulated in reservoirs and if the date of construction of the reservoir is known this can also provide the basis for an estimate of the rate of land erosion. Such rates are necessarily approximate but afford an indication of the scale of direct and indirect fluvial erosion. The average rate of denudation evaluated in this way is given as 3 cm per 1000 years by Stoddart (1969), as 2·2 cm or 7·2 cm for lowlands by Corbel (1964) and by Schumm (1963) respectively, and as 20·6 cm and 91·5 cm per 1000 years for mountains by the same two authorities. These rates are compared with several rates evaluated for specific areas in Figure 1.1. Average rates for lowlands are usually between 1 and 3–4 cm per 1000 years and these contrast sharply with average rates deduced for mountain areas (Figure 1.1). Many factors influence the rates of denudation in a particular area and, in addition to relief, basin area and the effect of man are amongst the factors which contribute to world-wide variations. Maps showing the pattern of world denudation at the present time have been attempted and in Figure 1.2 (Strakhov, 1967) one view, based upon suspended sediment data, suggests that the highest annual yields of above 240 tonnes/km² occur in south-east Asia where the influence of man is especially notable, the next highest in mountains and in the south-eastern part of the United States, followed by rates in the tropics which are higher than those of temperate latitudes, in turn higher than those of the Arctic. Such calculations have been based upon the drainage basin, a unit employed increasingly in fluvial studies during the twentieth century, and some attempt has been made to extrapolate present rates of erosion back into the past. Based upon erosion rates from large drainage basins it has been estimated that the drainage basin could be reduced to 10 per cent of its initial relief in a period of 11 million years (Ahnert, 1969), and that peneplanation could take about 11 million (Linton, 1957) or between 10 and 110,000 million years (Schumm, 1963).

1.2 The basis

Perhaps the greatest difficulty confronting the geomorphological study of the drainage basin is the discrepancy between process and form. The form of the drainage basin is the product of the processes which have operated in the past on the material locally available to produce a particular drainage basin form but these land-forming processes may not be the same in relative importance or, indeed, in kind as the ones which operate in the drainage basin at the present time. Furthermore, the form of a particular drainage basin, inherited from the past, influences the processes which operate at present and the influence of form over process provides a number of potentially rewarding research problems. This dichotomy between process and form has produced a temporary geomorphological dilemma (Gregory and Walling, 1971)—to study process or to study form. In the past two decades this choice has been answered

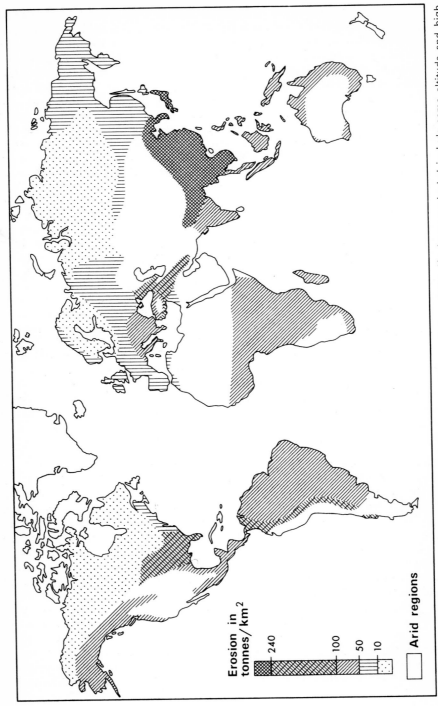

Erosion in tonnes / km²

240
100
50
10

Arid regions

Based upon Strakhov (1967). Other views are discussed in Chapter 6 (pp. 3294–2). Areas unshaded include some altitude and high altitude regions.

by two types of geomorphologist: those primarily concerned with process, and those primarily concerned with the historical evolution of land form and often with the chronology of events. The dilemma has, therefore, seen the development of two types of approach, and although both types are equally related to and dependent upon other disciplines, the form/historical group to Geology and Quaternary studies, the process group to hydrology and sedimentology, for example, it is information from the process school which is increasingly the most directly relevant to much of physical geography and to human geography as well, whereas geological/chronologically orientated geomorphology becomes increasingly detached from the more recent developments in human geography. This difference is producing a dilemma for physical geographers as a whole. Chorley (1971) has drawn the analogy of a tight-rope walker attempting to walk simultaneously on two tight ropes which are becoming more and more separated and posed the question whether the walker would commit himself entirely to one rope or the other or would attempt to continue along both. Until the last decade the research frontier was placed firmly in the very specialised branches of physical geography, whereas at the present time some of the most promising fields of research are those which require consideration of form as well as of process, of geomorphology and of climatology, of physical and of human geography. Thus the dilemma, to study form or process, is only a temporary one: some of the most rewarding research will arise from the investigation of land form-process relationships because these provide results for understanding the past, for estimating the future, and for application to other fields of geography (cf. Carson and Kirkby, 1971). This trend, typical of much of geography, has been recognised by Mead (1969) as a paradox in that increasingly unity of approach is accompanied by an unparalleled diversification of activity within the subject. This paradox has emerged in the study of geomorphology over the last twenty years and in that time certain basic concepts have evolved which are fundamental to the understanding of past and to the progress of future investigations.

1.2a   The systems approach

The drainage basin is an excellent example of a geomorphological system (Plate 1). Chorley (1962) advocated the use of systems thinking in geomorphology, whereby one considers the system as a set of objects together with the relationships between the objects and between their attributes. Closed systems are those which possess clearly defined boundaries, across which no import or export of materials or energy takes place and the Davisian cycle of erosion has been cited as an instance of a closed system. Whereas the closed system may develop after an initial supply of energy, such as initial uplift in the Davisian cycle of erosion, the open system requires a con-tinuing energy supply and is, in effect, maintained by constant supply and removal of energy. Therefore, the drainage basin can be envisaged as receiving energy or input from the climate over the basin, and it is losing energy or output through the water and sediment lost to the basin, largely through the basin mouth. The advantages of the open system approach arise from the facts that it places emphasis upon adjust-ment and upon relationships between form and process, upon the multivariate character of the many geomorphic phenomena, and upon the total physical environ-ment which includes consideration of the influence of man. The open system approach is one which, therefore, focuses upon the spatial distribution, upon the total environment, and upon the relationships between the forms and the processes within a particular environment. The use of systems thinking in geomorphology has been paralleled by its use in other parts of geography (Stoddart, 1965) and it has been accompanied by the emergence of other related ideas as discussed below (p. 13). The systems approach is not without its critics, particularly concerning the precise definition of terms (for example, Chisholm, 1965; Smalley and Vita Finzi, 1969) but it provides a very useful conceptual vehicle for study of the drainage basin.

In the drainage basin open system input of energy occurs from the climate over the basin and from endogenetic forces under the basin, transport of water and sediment takes place within the system over slopes and in stream channels and below the surface, and loss takes place principally by evaporation and transpiration to the atmosphere and by outflow of water and sediment from the mouth of the basin (Figure 1.3). The drainage basin system is influenced very much by scale and the components of the basin identified at one scale (for example, the Amazon basin) may not be identical with those recognised at a different scale (for example, a few square

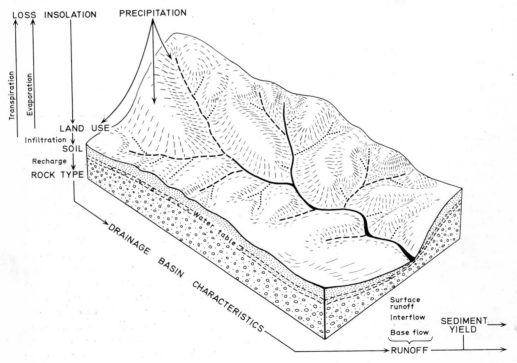

Figure 1.3   A geomorphological view of the drainage basin
Depicting the way in which drainage basin characteristics (form) influence the transformation of input (precipitation-losses) into output of runoff and sediment yield. The dynamic nature of the drainage network is incorporated by representing perennial (solid), intermittent (dashed), and ephemeral (dotted) streams. Based upon Gregory and Walling (1971).

kilometres). The basin contains many subsystems each at a variety of scales such as the hillside slope, the fragment of a slope such as the morphological unit (Gregory and Brown, 1966), or the stream channel subsystem. The drainage basin is a fundamental unit because of its functional significance for fluvial processes and indirectly for other geomorphological processes, it is one which is increasingly utilised as a functional study unit by other disciplines (for example, Bormann and Likens, 1969) and it is one with importance as a unit in human and historical geography (for example, Smith, 1969).

1.2b   Factors influencing systems

According to Budel (1969) the earth consists of seven great spheres (Table 1.1), and although the land surface is the domain of the geomorphologist his sphere is inevitably

Table 1.1   Earth spheres

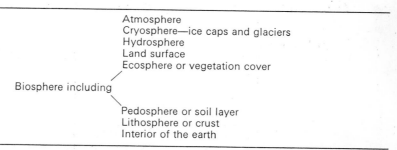

influenced by, and combined with, other spheres in open systems. The land surface and the drainage basins upon it may be visualised as influenced by certain factors. Some of these factors are inspired by the interior of the earth, the lithosphere, or the pedosphere and include those factors which are inherited from the past, such as rock lithology, rock structure, soil character and the broad pattern of topography, together with those endogenetic processes of the present time. It has been estimated (Schumm, 1963) that mean rates of orogeny can be about eight times greater than maximum rates of denudation. Other factors are inspired by the hydrosphere, the cryosphere, and the atmosphere and notably include the effect of climate. Climate has been regarded as of fundamental significance by many workers and this arises from the receipt of radiation and precipitation which give rise to a particular pattern of climate and to a particular assemblage of geomorphological processes in a specific area. Acknowledgement of the importance of climate is expressed in the morphoclimatic zones of French workers (Tricart and Cailleux, 1965), in the morphogenetic zones of German researchers (Budel, 1944, 1963, 1969), by the theory of zonality formulated in the U.S.S.R. by Grigoriyev and developed by later workers (Gerasimov, 1967), and in America by the work of Peltier (1950) who proposed a series of nine morphogenetic zones. Such physical zones have been proposed as a reflection of the effect of climate upon geomorphological processes over the surface of the earth and they can be adopted as a useful vehicle for approaching the study of landscape. The zones which

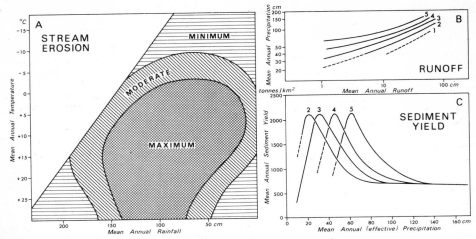

Figure 1.4 Climate and drainage basin processes
A shows an interpretation of three degrees of activity of stream erosion according to mean annual precipitation and mean annual temperature (Peltier, 1950). B indicates the trend of runoff with mean annual precipitation and C indicates the pattern of mean annual sediment yield (Schumm, 1965). In each case mean annual temperatures are approximately −1 °C (1), 4·5 °C (2), 10 °C (3), 15 °C (4), 20 °C (5).

have been recognised were first distinguished according to a single criterion such as climate, soil, or vegetation but more recent attempts have endeavoured to distinguish landscape complexes in which 'natural landscapes' (Passarge, 1929) are arranged in taxonomic categories. Peltier (1950) viewed landscape as a result of weathering by mechanical and by chemical means, followed by transport of the weathering products during which time erosion could occur and be followed by deposition. Consideration of the climatic limits of each weathering and transportation process allowed Peltier to produce a series of graphs showing the likely occurrence of zones of maximum, moderate and minimum activity for each process (see, Figure 1.4A). Consideration of the likely distribution of each process led to the proposal of nine morphogenetic regions of the world, each characterised by an estimated range of average annual temperature, an estimated range of average annual rainfall, and by certain morphological characteristics (Table 1.2).

The significance of climate-determined processes over the earth's surface has been illustrated by considerations of the world pattern of runoff and sediment yield. Work developed from suggestions of Langbein and others (1949) has shown how annual runoff increases as annual precipitation increases, and that runoff decreases as temperature increases with constant precipitation (Figure 1.4B). The relationship between annual sediment yield and annual precipitation for basins of the order of 3500 km² in the United States has been evaluated by Langbein and Schumm (1958) for a mean annual temperature of 10 °C (50 °F) (Figure 1.4C). The curve (Figure 1.4C, No. 2) for 10 °C shows that sediment yield is at a maximum at about 30 cm of annual precipitation but then reduces to lower values with greater amounts of precipitation. Variation in sediment yield with precipitation can be explained by the

Table 1.2 Morphogenetic regions (*after Peltier, 1950*)

Region	Range of average annual Temperature	Rainfall	Action of running water
Glacial	−18° to −6·5 °C	0 to 114 cm	
Periglacial	−15° to −1 °C	12·5 to 140 cm	Weak
Boreal	−8° to 3 °C	25 to 150 cm	Moderate
Maritime	1·5° to 21 °C	125 to 190 cm	Moderate to strong
Selva	15·5° to 29·5 °C	140 to 230 cm	
Moderate	3° to 29·5 °C	90 to 150 cm	Maximum
Savanna	4·5° to 29·5 °C	63 to 125 cm	Strong to weak
Semi-arid	1·5° to 29·5 °C	25 to 63 cm	Moderate to strong
Arid	12·5° to 29·5 °C	0 to 40 cm	Slight

interaction of precipitation and vegetation on runoff and erosion. There is, therefore, a difference of opinion between the views of Peltier (Figure 1.4A) and of Langbein and Schumm (Figure 1.4C) although both acknowledge the importance of the vegetation cover. The effect of climate is intimately bound up with the pattern of vegetation and with major soil types and so world divisions, subsequent to those proposed by Peltier, have focused upon the total physical complex. Tricart (1957) and Tricart and Cailleux (1965) proposed thirteen morphoclimatic zones and drew attention to the zonality of geomorphic phenomena. Zonal phenomena were the direct results of latitudinal climatic belts; azonal phenomena were the result of non-climatic control including endogenetic effects; extrazonal phenomena occurred beyond their normal range of occurrence as sand dunes occur on coasts; and polyzonal phenomena included those which operate in all regions of the globe subject to the same basic laws, such as water flow.

Such regionalisation of physical processes, albeit as 'a good servant but a bad master', is confronted by the difficulties of scale, of criteria for definition, of lack of information in equal detail over the surface of the earth, and of whether regions sufficiently homogeneous exist to be distinguished adequately. The two most prominent obstacles to the successful isolation of morphoclimatic or morphogenetic zones have been changes of climate and changes wrought by man—the former complicating the analysis of landscapes and land systems, the latter restricting the application of contemporary process studies to the past and, indeed, to the future. These two difficulties have been appreciated by Budel (1963, 1969), who distinguished three generations of geomorphological study. *Dynamic geomorphology* concerns the study of particular processes; *climatic geomorphology* considers the total complex of currently active processes in their climatic framework; and *climato-genetic geomorphology* involves the analysis of the entire relief including features adjusted to the present climate and features produced by former climates. Implicit in climato-genetic geomorphology, therefore, are the notions that the present is to some extent governed by the past, that in many areas fossil landforms may be more prominent in a landscape than landforms in harmony with the present processes, and that several generations of climato-genetic regions can be recognised between the equator and the poles. In 1963 Budel recognised five such zones but in 1969 this was increased to seven (Figure 1.5). In each zone a distinctive pattern of processes should operate at the present time reflecting the macro-climate, as in the case of river regime, but variations within the zone can arise from the effects of factors other than climate, notably the effect of rock type, relief, variations in soil or vegetation cover and, of course, the influence of man. Furthermore, the pattern of world zones in the past has not always been a simple derivative of the present picture. During the Quaternary, changes of climate led to adjustment of the pattern of world zones accompanied by fluctuations of sea level, and during the Tertiary the tropical zones were probably much more extensive. Thus in the extra-tropical zone of valley formation (Figure 1.5, No. C), which extends over

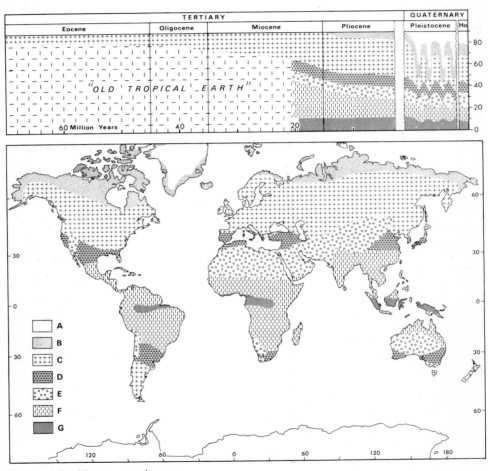

Figure 1.5 Climato-genetic zones
Based upon Budel (1963, 1969). A = Glaciated areas, B = Zone of pronounced valley formation, C = Extra-tropical zone of valley formation, D = Subtropical zone, E = Arid zone, F = Circum-tropical zone of excessive plain formation, G = Intertropical zone of partial plain formation. The upper diagram indicates the way in which the pattern of zones may have changed over time. Three horizontal scales are used to represent the duration of the Holocene (Ho), the Pleistocene and the Tertiary respectively.

much of the middle latitude regions, the relief outlines may be composed of remnants of planation surfaces produced in the Tertiary period, under sub-tropical conditions; the major landforms may reflect the alternation of cold phases of glaciation with interglacials of the Quaternary; and the present processes reflect the impact of the present climate upon a landscape which already possesses an inherited relief, is underlain by specific rock types, and is clothed by soil and vegetation, all of which have been modified by the hand of man.

Uniformitarianism, therefore, does not offer an easy solution to the problem of reconciling past and present processes. Founded upon the ideas of James Hutton (1785) and Playfair (1802), Lyell (1830) advocated the use of 'the present as the key to the past'. This dictum is a useful one if interpreted to signify that the processes operating at present throughout the world are the ones that have operated in the recent and geologic past. However, we know that the distribution of geomorphological processes, and of climato-genetic zones, has changed during the past (Figure 1.5) and we know also that, particularly as a result of the influence of man, the rate at which processes operate at the present time may not be typical of the rates at which the same processes operated previously.

In 1922, Sherlock wrote that 'there are indications that the doctrine of uniformitarianism has been taken too far' and he was drawing attention to the extent of man's influence. Although his work attracted comparatively little following for thirty years, man as a geomorphological agent has received an increasing amount of attention since 1960. In the field of physical geography, the works of George Perkins Marsh (1864) in his volume *The earth as modified by human action*, of R. L. Sherlock (1922) in his analysis of *Man as a Geological Agent* and, more recently, of Thomas in *Man's role in changing the face of the earth* (1956) are three milestones along the road towards understanding the nature of man's impact upon his environment. During the 1960s Fels (1965) proposed the term anthropogeomorphology to connote the geomorphological study of the effects of human action, a presidential address by Jennings (1965) took the title of Sherlock's book of 1922, Brown (1970) devoted an inaugural lecture to the subject of 'Man shapes the Earth' and books and collections of studies (for example, Detwyler, 1970) have appeared. These general statements of the effects of man have been accompanied by an increasing number of research programmes devoted to the investigation of the impact of man (see Walling and Gregory, 1970). This impact has been resolved by Brown (1970) into three categories. *Direct, purposeful human action* includes the positive and negative landforms that man may produce, in altering the pattern of relief, for example, during building construction. *Incidental but direct effects* include the incidental modifications of the surface which arise indirectly from extractive industries such as opencast quarrying or from surface subsidence following underground mining. *Indirect influence* of man is expressed in the numerous ways in which man has altered the rate of operation of geomorphological processes in an area and in some cases has affected the relative importance of the processes operating. The relevance of the study of man's effects on land form and on land process is emphasised by the need to understand the extent to which man has modified the processes operating at the present time. This is necessary so that knowledge of these processes may be applied to past situations before man's effects were so substantial. The study of anthropogeomorphology is encouraged by the relevance of this branch of physical geography to the needs of human geography. Understanding of

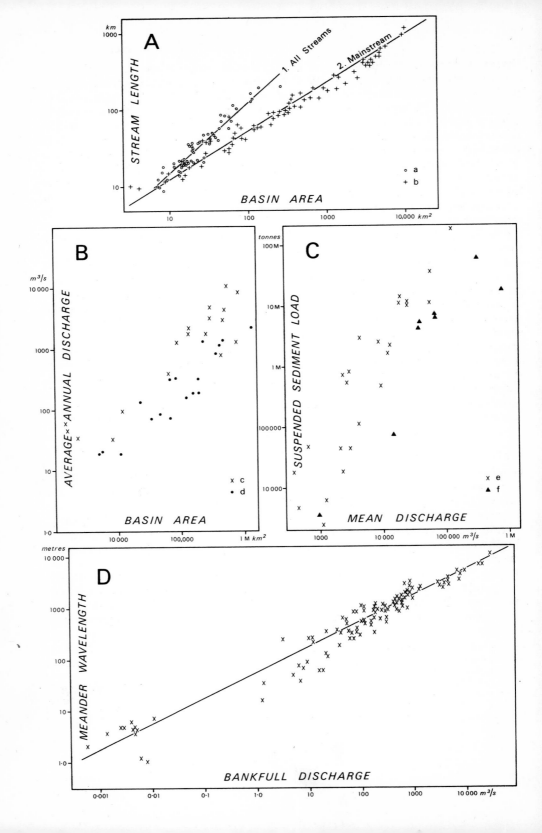

the extent of man's influence upon processes at the present time also affords an opportunity to apply this understanding to the future, to anticipate problems and to devise potential methods of control of future situations.

The geomorphological study of man has arisen in the second part of the twentieth century as a natural development from the studies of the first half of the century. In this development are, therefore, numerous opportunities for physical geography to become more closely integrated, particularly in the light of the systems approach (p. 27), for physical geography to consider present processes in relation to the future as well as the past, and for physical and human geography to become more closely allied as expressed in some recent books (for example, Chorley, 1969) and in the 'constructive geography' of Gerasimov (1967). Vernadsky (1945) advocated the use of the term Noosphere for a new geological epoch inaugurated by man and although this suggestion sprang primarily from the effects of man upon vegetation, it is equally justifiable in the study of rivers and drainage basins. Man has affected input into the drainage basin system, he has influenced the character of the basin in its relief, drainage, soil and vegetational characteristics, he has influenced the character of, and the rate of operation of, processes in the drainage basin, and he has, therefore, influenced the rate, and sometimes the manner, in which water and sediment are removed from the drainage basin system. Rates of erosion consequent upon man's activities can be greatly increased and in Java a rate two-and-a-half times greater than the highest in Figure 1.1 was recorded from an area of cultivation over steep slopes (Douglas, 1967). The International Hydrological Decade (1965–74) includes the theme of the effects of man upon the hydrological cycle (Keller, 1968) and this is one field where the geographer, concerned particularly with the form of the earth's surface, its precise description and its relationship to the processes operating upon it, can make a significant contribution (p. 342).

1.2c   Adjustments in nature

Dynamic equilibrium existing between elements in the physical landscape was perceived by G. K. Gilbert (1877) in his studies of the Henry Mountains, Utah. He referred to the adjustment which obtained between the processes of erosion and the resistance of bedrock in a river channel. A century of investigation has subsequently revealed numerous instances of adjustment between the components and processes of the physical landscape and particularly within the drainage basin. Four examples are illustrated in Figure 1.6. Such relationships, which are discussed in detail in later chapters (2, 4, 5 and 6) have included relations established between measures of basin form, and a particular relationship exists between basin area and mainstream length and between basin area and total length of streams in the basin (Figure 1.6A). Interrelationships have also been demonstrated between basin form and basin process and

Figure 1.6   Adjustments in the drainage basin
A Form-Form relation illustrated by data for south-west England (a) and eastern U.S.A. (b) (based upon Hack, 1960).
B Form-Process relation illustrated for two climato-genetic zones, tropical grassland (c), temperate midlatitude grasslands (d).
C Process-Process relation illustrated for annual data from California (e) and Alaska (f).
D Process-Form relation based upon Dury (1964).

a general relation may be detected, for example, between average annual discharge and basin area (Figure 1.6B). Such a relationship for large rivers in the world can be subdivided according to climato-genetic zones and the relationship is illustrated for two zones (Figure 1.6B). Relationships have been established not only between process and overall basin form but also between the form characteristics of the stream network and basin process, and a relationship between bankfull stream discharge and meander wavelength (Figure 1.6D) has been demonstrated. Relationships between measures of process have also been derived and mean discharge can be related to suspended sediment yield which is illustrated for a particular year comparing two contrasted climato-genetic zones in Figure 1.6C.

Knowledge of such interrelations is important because it reflects the way in which various form and process elements in landscape and in the drainage basin are adjusted, often in a multivariate manner, it demonstrates the fact that physical processes are subject to certain physical laws, and it affords a means whereby the nature of the relationships may be compared from one situation to another. Thus, in all parts of basin systems the continuity equation must be satisfied (p. 235), and theoretical relationships may be established. As examples of such relations, flow equations have been available since the eighteenth century expressing stream discharge in terms of hydraulic radius, channel slope and roughness (p. 236). There are many instances of deviations from the ideal or average conditions—a contingency which has been approached by probability concepts in geomorphology—and there are also numerous instances where some form of balance and adjustment exists between form and process —a feature which has been considered in terms of general systems theory and entropy and expressed by various equilibrium terms (Table 1.3). The open system view of the drainage basin (Chorley, 1962) entails the perception of the basin as a system which is maintained by a constant supply of, and a constant removal of, energy. A characteristic of such an open system is that it can operate in a steady state and the inflow of energy, the system and the outflow of energy exist in a delicate balance. At any one time there is thus a balance between form and form, between form and process, and between process and process. If a change takes place in the input or output of energy or in the form of the system a compensating change will take place to minimise the effect of the change and to restore a state of balance. Such a state of balance is referred to as a steady state and implies a dynamic balance existing between elements in the system at any one time. Leopold and Langbein (1964) have applied the term quasi-equilibrium to the steady state which can exist without involving any regularity of form (Table 1.3).

Leopold and Langbein (1962) reasoned by analogy with thermodynamic entropy and suggested that the most probable distribution of energy in certain geomorphic systems can be derived by considering the system in a steady state. Entropy is essentially the total available energy and its distribution and the distribution of energy in a river system tends towards the most probable state. Two generalisations are, firstly, that there is a uniformly distributed rate of energy expenditure and, secondly, that the minimum total work is expended in the system. The most probable condition would probably be a compromise or intermediate state between these two generalisations. Leopold and Langbein (1963) have also proposed a Principle of Indeterminacy with respect to landscape development and this considers that rivers tend towards a state where total energy expenditure is minimised and energy distribution is uniformly

Table 1.3 Some concepts applicable to the drainage basin

Term	Usage	Example of application
Dynamic equilibrium	An equilibrium state can be maintained as fluctuations are balanced around a constantly changing system condition	To distribution of surface heights and remnants of planation surfaces (e.g. Hack, 1960). These can be regarded as in dynamic equilibrium with form and process variables
Quasi-equilibrium	A state of near equilibrium LACKS REGULARITY OF FORM	To equilibrium between channel form or channel pattern with stream channel processes (e.g. Leopold and Langein, 1964)
Equifinality	Whereby similar final states may be derived in different ways and from diverse origins	To component systems of drainage basins and to deposits. Thus shape of bedload may achieve similar values despite contrasting original rock types and different types of energy environment (see pp. 83–4)
Feedback	Part of the output of a system may act as input to another system and regulate the system either by intensifying (positive feedback) or opposing (negative feedback) the direction of the system	Increased erosion → gullying → channel widening → meandering → decreased slope → decreased velocities → decreased erosion (Negative feedback, see pp. 345–6)
Relaxation time	Time taken to realise equilibrium in a system during change from one equilibrium condition to another one	Varying amounts of time are required to realise equilibrium after changes in network, channel pattern and channel form. A change in climate or in runoff rate could lead to a modified channel pattern over a number of years (e.g. pp 365–7 and Plate 21)
Threshold	A condition characterising the transition from one system state to another	Threshold values of the controls of channel pattern may be identified. These values characterise the change from one pattern to another (e.g. pp. 247–57)

A more comprehensive definition of appropriate systems terms is found in Chorley and Kennedy (1971), pp. 346–59.

distributed, through a channel reach, for example. Thus, some variation in landscape forms arises as a result of incomplete dynamical indeterminacy. The relevance of probability has been further developed by Scheidegger and Langbein (1966) who proposed that rivers and landforms produced by running water may be dominated by random processes, that the assumption of randomness appears to offer a direct approach to landforms and to the hydraulic geometry of rivers, and that the open system steady state of landscape can be described statistically. Scheidegger (1969) has demonstrated that a statistical—mechanical approach can be made to landscape, in a way analogous to the use of the approach in the study of the dynamics of gases, leading towards a 'stochastic' or statistical approach to landscape function and evolution. Rather than a particular landscape this entails conception of a whole series of landscapes which are broadly indistinguishable one from another and, according to Scheidegger, 'the evolution of a particular observable is given by the evolution of its expectation value, i.e. of its average over the ensemble'.

The development of these ideas has been towards the view that the drainage basin is an open system in a steady state, that several elements within the basin are governed

Plate 2   A micro example of adjustment in a drainage basin system
A small catchment in Devon was affected by road construction (cross hatched on map) in 1971 at the head of the catchment. The removal of the grass cover led to increased sediment supply and to greater runoff which led to the replacement of a straight and meandering stream channel by a braided one (A).

by the laws of probability and stochastic processes, and that processes and adjustment of the form of the basin may be dominated in detail by random processes (Mann, 1970). These concepts have been applied particularly to the study of contemporary processes but do not easily apply in exactly the same way when applied to evolutionary studies. Chorley (1964) has distinguished the timeless and the timebound aspects of geomorphology—the former concerned particularly with the study of processes and the latter especially with the evolution of landforms and their chronology. Several authorities have called for concepts which can be applied to both timeless and time-bound studies. Tricart (1966) distinguished temporal factors, associated with evolution and palaeoforms, from dynamic factors which are concerned with action, reaction and interaction. J. T. Hack (1960) proposed a dynamic equilibrium theory of land-scape development which is independent of time and which views the physical land-scape as composed of dynamic equilibrium conditions between form and process. This concept has later been illustrated and developed by Hack in studies of the Shenandoah valley (Hack, 1965) and of the Cumberland escarpment (Hack, 1966), and tends to dispense with the timebound aspects of geomorphology and is based upon a view expressed as, 'it is assumed that the erosion and down-wasting of the central Appalachians were continuous and uninterrupted by periods of base-levelling' (Hack, 1965). However, several other workers (for example, Howard, 1965; Ollier, 1968) have contended that the concept of dynamic equilibrium is not exclusive and that studies involving evolution over time can be accommodated.

1.2d   Geomorphological time

Such concepts (1.2c) must be visualised within the conceptual framework of geo-morphological time and a framework which reconciles several apparently conflicting approaches has been provided by Schumm and Lichty (1965). They drew attention to the fact that, in land form development, the distinction between cause and effect is a function not only of spatial distribution but also of time. Thus the distinction between cause and effect in the shaping of landforms depends upon the size of the system in space and upon the length of time being considered. Three time spans, designated cyclic, graded and steady time, were distinguished and although these do not correspond to specific absolute values they are significantly different in duration. The factors which influence landforms can be either dependent or independent variables according to the time span relevant to a particular study (Table 1.4). At the scale of *cyclic time*, which corresponds approximately to the length of time required for an erosion cycle, only geology, initial relief and climate are independent variables. *Graded time*, which refers to a short span of cyclic time, sufficient perhaps to develop a graded river profile, sees the introduction of vegetation, relief, and hydrology as additional independent variables. *Steady time* applies to portions of the drainage basin such as part of the stream channel over very short periods of time. The advantage of this conceptual framework is that it points to the fact that the drainage basin may present problems which require analysis and solutions conceived at different time scales, it reconciles the historical approach to drainage basin landforms with the dynamic equilibrium approach, and it indicates several growth areas of geomorphology which direct attention to the problems of transferring from one time scale to another.

Consideration of time in the drainage basin necessarily introduces the significance

Table 1.4 A The status of river variables during time spans of decreasing duration (*after Schumm and Lichty, 1965*)

River variables	Status of variables during designated time spans		
	Geologic	Modern	Present
1 Time	Independent	Not relevant	Not relevant
2 Geology	Independent	Independent	Independent
3 Climate	Independent	Independent	Independent
4 Vegetation (type and density)	Dependent	Independent	Independent
5 Relief	Dependent	Independent	Independent
6 Palaeohydrology (long-term discharge of water and sediment)	Dependent	Independent	Independent
7 Valley dimensions (width, depth, slope)	Dependent	Independent	Independent
8 Mean discharge of water and and sediment	Indeterminate	Independent	Independent
9 Channel morphology (width, depth, slope, shape, pattern)	Indeterminate	Dependent	Independent
10 Observed discharge of water and sediment	Indeterminate	Indeterminate	Dependent
11 Observed flow characteristics (depth, velocity, turbulence, etc.)	Indeterminate	Indeterminate	Dependent

B The status of drainage basin variables during time spans of decreasing duration

Drainage basin variables	Status of variables during designated time spans		
	Cyclic	Graded	Steady
1 Time	Independent	Not relevant	Not relevant
2 Initial relief	Independent	Not relevant	Not relevant
3 Geology	Independent	Independent	Independent
4 Climate	Independent	Independent	Independent
5 Vegetation (type and density)	Dependent	Independent	Independent
6 Relief of volume or system above base level	Dependent	Independent	Independent
7 Hydrology (runoff and sediment yield per unit area within the system)	Dependent	Independent	Independent
8 Drainage network morphology	Dependent	Dependent	Independent
9 Hillslope morphology	Dependent	Dependent	Independent
10 Hydrology (discharge of water and sediment from system	Dependent	Dependent	Dependent

of the magnitude and frequency of the geomorphic processes operating. The drainage basin system (Figure 1.3) does not function at a constant rate over short periods because a sudden input, of rainfall for example, can produce a response in the form of a steamflow hydrograph. Over a period of time, such as a year, a series of input events occurs and each one will give rise to a particular sequence of responses. Therefore, not only does variation occur during a time period such as a year but events of a particular magnitude will occur during that time at a particular frequency. It has been shown that in many drainage basins the majority of the work can be accomplished by events which occur on average once or twice each year and also that certain characteristics of the form of the basin and stream channels can be adjusted to events

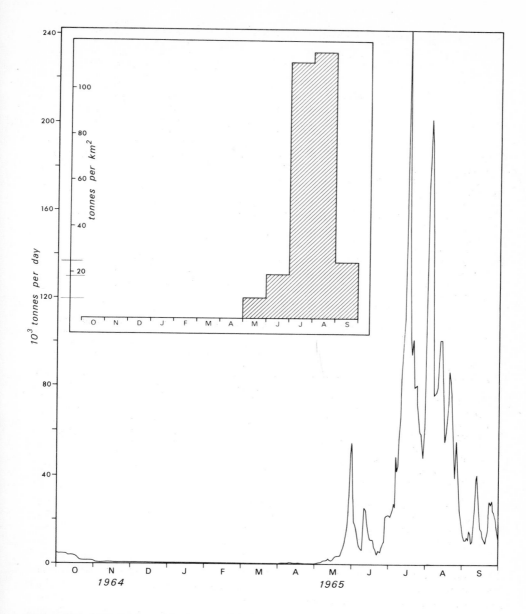

Figure 1.7 Magnitude and frequency of drainage basin processes

Illustrated by daily and monthly (inset) data for suspended sediment output during one water year from the Tanana River, Alaska, catchment area 22 059 km². Based upon data from *U.S. Geol. Surv. Water Supply Paper 1966* (1970).

of a particular frequency. The importance of magnitude and frequency of drainage basin processes has been discussed generally by Wolman and Miller (1960) who have shown, for instance that the Colorado river at Grand Canyon, Arizona (area 353,460 km²) takes a mere 31 days of each year to transport half of the annual suspended sediment load. This concept is illustrated by the suspended sediment record of the Tanana river, Alaska (Figure 1.7). Throughout the water year (1964–5) very small amounts of suspended sediment are transported until the break-up period of May. The maximum load of a single day represents 3·75 per cent of the total load (in 0·27 per cent of the year), over 20 per cent is carried in 6 days (1·92 per cent of the year) and nearly half (47·9 per cent) is conveyed in 26 days (7·12 per cent of the year). Although the climato-genetic zone of active valley formation (Figure 1.5B) has a markedly seasonal regime, water and sediment yield from drainage basins in other zones is characterised equally by events of moderate and high magnitude distributed unevenly through time and, therefore, the size of events in relation to their frequency must constantly be appreciated. The distribution of events in time has prompted analyses of the probability of the frequency of events such as floods and droughts occurring, consideration of the magnitude of the events responsible for influencing particular aspects of the drainage basin, and time series analyses of the sequences of events over time.

1.2e Information and methods of study

The view of the drainage basin which has emerged is that it functions as an open system, that its function changes over short periods of time, that several time scales may be adopted for its study, that there are numerous relationships between the elements of the form and the processes operating in the drainage basin, and that these relationships are very complex and multivariate in character. To understand the drainage basin, measurements are required, relationships must be established, and analysis attempted. An ever-increasing range of techniques is available for these purposes. Information may be obtained from actual drainage basins by field examination, from laboratory models by artificial simulation of drainage basin conditions in a carefully controlled experiment, or from knowledge of physical or statistical laws by analogy. Whichever of these three approaches is used, measurements must be obtained of the drainage basin form, which includes the overall character of the basin as well as of the component parts such as the stream network and the stream channel, and these must be supplemented by measurements of the processes operating as input to, as processes within, and as output from, the basin.

Some information on form may be readily available from topographical map and aerial photograph coverage, and equally records of streamflow and sediment yield may exist for certain stations on major rivers. In the U.S.A., 6283 stations were operating in 1955, in Great Britain 437 stations operated during 1965–6, and Unesco under the I.H.D. has collected and published stream discharge records from 1000 stations on world rivers (Unesco, 1970). Existing data and measurements frequently have to be further developed and new measurements made. Relationships between the measurements may be established using statistical methods (Krumbein and Graybill, 1968; Riggs, 1967), sometimes explored by probability methods, and all facilitated by the use of high speed electronic computers. As Leopold and Langbein (1963) have commented, 'Geomorphology is a field of enquiry rejuvenated not so much by new methods as by realisation of the great and interesting questions that confront the geomorphologist'. The ideas which have emerged during a century of investigation have provided the questions to which answers can now be attempted using the techniques more recently available.

1.3 Approaches to the drainage basin

The drainage basin may be approached in two principal ways (Amorocho, 1967). Attention may be directed towards the component phenomena and their relationships, a method of study in hydrology described as physical hydrology. Alternatively, the drainage basin system may be investigated either by parametric methods, which include the development of relationships among physical parameters involved in hydrological events, or by stochastic methods, which require the use of statistical characteristics of hydrologic variables to solve hydrologic problems. The overall system consists of a number of subsystems which interact with one another and Vemuri and Vemuri (1970) have distinguished the hydrologic processes subsystem from the behavioural subsystem, resulting from human intervention. The conceptions discussed above and the view of the drainage basin have to be assimilated, first, into an overall appreciation of the world setting, second, into the pattern of research activity and, third, into the context of the several methods of approach to the basin which have been advocated.

1.3a The world setting—the hydrological cycle

A simplified version of the components of the hydrological cycle is shown in Figure 1.8. The main locations in which world water is stored include the ocean (94 per cent), subterranean reserves (4·12 per cent) and glaciers (1·65 per cent), and other locations shown in Table 1.5. Movement of water is indicated by the shaded portions (Figure 1.8) incorporating the way in which water is evaporated to the atmosphere from continents and oceans, and is then distributed over the oceans and over the land surface as precipitation. Of the world's fresh water, 75 per cent is held within ice sheets and glaciers, nearly 25 per cent is ground water and the remainder occurs in rivers (0·03 per cent), in lakes (0·3 per cent) and in the soil (0·06 per cent) (More, 1967). Movement of fresh water takes place; as precipitation, part of which may be received by the stream channel; as water flowing over the surface to streams and rivers; as water flowing through the soil and rocks; and water can also be stored in the soil and in the rocks. River flow, which transfers water from the continents to the oceans, can thus be supplied by flow from the surface, from the soil or from the rocks. Precipitation may not be directly responsible for river flow because there can be a delay between rainfall and the recharge of the ground water which later contributes to streamflow, and in some areas rivers can be supplied by meltwater from ice sheets and glaciers or from lakes. There are numerous mechanisms within the hydrological cycle and this is indicated by the large number of subsystems which have been analysed by various workers. The overall cycle cannot be regarded entirely as a closed system because the atmosphere is affected by incoming radiation and small losses or receipts of water can take place beneath the oceans to or from the lithosphere (Dooge, 1968).

1.3b   The pattern of research activity

Study of water bodies, of water movements, and of the systems and subsystems within the general framework of the hydrological cycle is the responsibility of a number of disciplines. Each of the spheres (Table 1.1) is the subject of study by a particular earth science and each of these can include the study of present processes and of change over time. Each discipline has a two-sided personality and this gives the

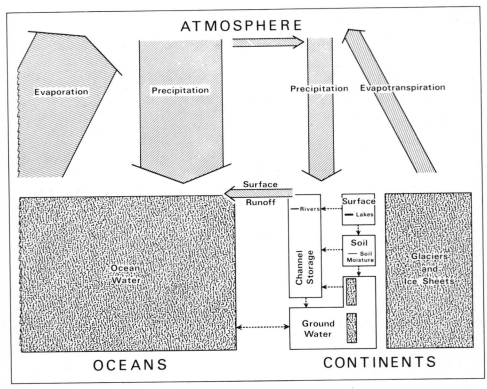

Figure 1.8 The world hydrological cycle
Relative storage amounts are shown by black (as in lakes) and by shaded areas (as ocean water). Diagonally shaded areas represent the annual amount of movement.

dichotomy in fluvial geomorphology, noted above, between the timeless studies of process and the timebound studies of the evolution of land form, echoed by the duality between meteorology and climatology. Although primarily focused upon one of the spheres, as geomorphology is focused upon the form of the land surface and hydrology is concentrated upon the processes involving water, each discipline must be aware of the results and the methods developed by other disciplines. Particularly with the advent of the systems approach in many of the earth sciences the existence of discrete disciplines is now very difficult to envisage and the geomorphologist interested in the drainage basin must take account of other branches of physical geography concerned with the atmosphere and the biosphere. However, the geomorphologist's approach to the drainage basin, although cognisant of the results of work in other parts of physical geography and in other disciplines including sedimentary geology, hydrology and pedology, is primarily directed towards an understanding of the relationships between form and process in drainage basins on the earth's surface and this interaction is not the prime focus of attention in any other discipline. As part of geography it is often

Table 1.5 World water distribution (*after Lvovitch, 1970*)

	Per cent of total volume	Number of years to renew reserves
Ocean	93·93	3000
Subterranean water	4·12	(5000 ?)
Zones of exchange of active water	0·27	(330 ?)
Glaciers	1·65	12000
Lakes	0·016	0·9
Soil moisture	0·005	10
Atmospheric water vapour	0·001	0·033
River water	0·0001	0·027

Annual water balance of the continents in cm (*after Chang and Okimoto, 1970*)

	Rainfall	Potential	Evapotranspiration Actual	Deficit	Surplus
Africa	74	188	65	123	9
Asia	54	104	43	61	11
Australia	49	167	41	126	8
Europe	60	70	47	23	13
North and Central America	70	120	63	57	7
U.S.A.	76	134	71	63	5
South America	151	146	111	35	40

stressed that such study must be in a manner which is conscious of the changing aims and objectives of geography as a whole (Chorley, 1971). Drainage basin systems studied by the geomorphologist should, therefore, portray the drainage basin as the basis for human activity and also embrace the effect of man's actions. The two basic types of system which can be visualised are *morphological* ones, which are concerned with the relationships of properties of form, and *process-response* systems, which include those situations which relate form and process and can also include those relating process with process. In addition to these two types, Chorley and Kennedy (1971) have also distinguished *cascading* systems, which are linked by a cascade or a mass of energy, so that the output from one system may be the input to another subsystem, and *control* systems which consider the significance of prominent variables, especially man, which can promote changes in the distribution of mass and energy in cascading systems and also instigate changes in the morphological system and in process-response systems.

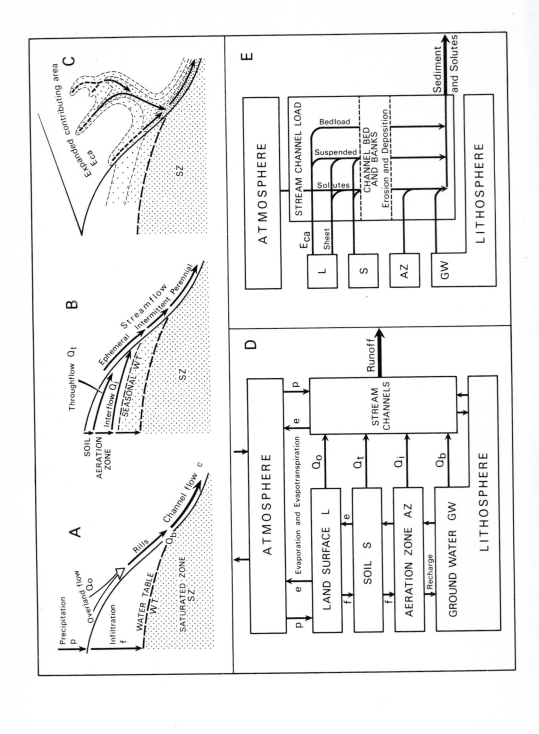

1.3c   Methods of approach to the drainage basin

The geomorphologist, therefore, is provided with a systems approach which can be applied to the drainage basin. Application of the approach, however, depends upon knowledge of the mechanics of the drainage basin and for over half a century the basin has been perceived in several different ways.

An important conceptual view of the drainage basin was provided by R. E. Horton in 1945 and this later came to be regarded as the overland flow model (Figure 1.9A). Basically, this visualised two main types of supply of water to the stream in the drainage basin, namely, overland flow (Q_o) and ground water flow (Q_b). Rain falling on the ground surface would infiltrate into the soil and gradually reach the ground water table below which ground water would be stored and would supply the baseflow component of rivers and streams. The maximum rate at which water could enter the soil was designated the infiltration capacity. If this value was exceeded by the precipitation intensity, water would first be retained on the surface, as surface detention, and subsequently flow would take place on the surface, first as sheets of water in a belt of no erosion, and subsequently concentrated into rills and stream channels to provide the surface runoff component of streamflow. This model was the basis for many attempts to separate stream hydrographs into the two basic contributing components, of surface runoff and base flow, and it envisaged that much of the basin would contribute water directly to the streamflow hydrograph. Difficulties encountered in separating streamflow hydrographs into two components foreshadowed the realisation that overland flow was not often experienced in many areas of the world, the observation that precipitation intensities in many areas seldom exceed the infiltration capacity, and the appreciation that water flow can occur above the saturated zone but beneath the land surface (for example, Whipkey, 1965). Observation of such sub-surface flow led to the suggestion of the throughflow model (Figure 1.9B) and it has been proposed that throughflow is probably the most important mode of water flow on hillsides in humid and in humid temperate areas (for example, Kirkby, 1969). Subsequently, several varieties of flow have been distinguished. Some authorities have referred to throughflow as the flow which occurs in the soil horizons, especially above relatively impermeable layers such as at the junction of the A and B horizons; and to interflow as the lateral flow which occurs in the aeration zone above the level of permanent saturation but below the A and B soil horizons. Others (for

Figure 1.9   Views of drainage basin dynamics
A representing the Horton overland flow model, B depicting the throughflow model and C showing some characteristics of contributing area are explained in the text (pp. 28 and 30). The major components of water movement and sediment movement in the drainage basin are shown in D and E respectively. Runoff (in D) can derive from four main sources (Q_o, Q_t, Q_i, Q_b) each illustrated in A and B and this can be supplemented by precipitation falling directly on to the stream surface and reduced by evaporation and by percolation. Production of sediment and solutes (in E) is visualised from the same four main sources (L, S, AZ and GW) explained in D and can be derived from the land surface by expansion of the channel network (E_{Ca}) or by sheet flow. The land surface and the channel bed and banks can supply all three components to the stream channel load, whereas the other sources provide two or one components. The stream channel load can be supplemented by solutes from the atmosphere and additional changes occur during erosion and deposition along the channel bed and banks.

example, Jamieson and Amerman, 1969) referred to quick return flow in the soil layers, delayed return flow in the aeration zone, and prolonged return flow from the saturated zone (see Figure 7.9). Necessarily, the location and extent of subsurface flow will reflect local conditions and particularly the presence of impeding layers in the profile below the surface. In practice a distinction between the several types is a convenient simplification of the complex continuum which exists in reality.

Although overland flow may occur in semi-arid regions, with the combination of high precipitation intensities and sparse vegetation, and also in arctic areas over permafrost, a different model is more appropriate for other areas. More recently, awareness of the dynamic character of the drainage basin and its included stream network, together with the fact that overland flow velocities are much higher than those of throughflow (perhaps 200m per hour compared with 20 cm per hour), have prompted visualisation of the basin in plan rather than simply in section (Figure 1.9C). Thus the partial area and variable source area models of streamflow production have been developed, which see areas proximate to the watercourses and other hollows as being the areas likely to contribute to runoff formation. In forested watersheds, Hewlett and Hibbert (1965) suggested that runoff production was achieved by the expansion of saturated zones along the valley floors and over the lower portions of the adjacent slopes. In some watersheds the areas which provide runoff on one occasion may not be identical to those which produce runoff under different conditions at another time.

These interpretations emphasise the dynamic character of the drainage basin because the same basin could respond differently, simply according to the prevailing moisture conditions reflecting the antecedent conditions. In general the basin system, represented diagrammatically in Figure 1.9D, can be visualised in terms of the water balance equation, which may be stated as:

Runoff (output of water) = Precipitation (input) —losses (evaporation) ±changes in storage.

Precipitation (p) may fall on water bodies including streams in the basin or it may fall on exposed rock, on bare soil or on vegetation. Depending upon local conditions, some may be detained on the ground or be intercepted on vegetation surfaces, some may flow over the surface as overland flow, and some may infiltrate into the soil. Loss of water to the atmosphere will take place by direct evaporation from water surfaces and by evapotranspiration from soil and plants. At depth there will be a saturated zone below the water table and flow from the ground water reservoir may reach stream channels as groundwater outflow (Q_b). In addition quick return flow (Q_t) and delayed return flow (Q_i) may occur through the zone above the water table. The drainage basin characteristics will obviously influence the manner and the rate by which water is transmitted through the system and the local characteristics will determine exactly how a basin responds to a particular climatic input to produce output in streamflow. Sediment yield from the basin is equally influenced by the drainage basin characteristics and by the way in which the basin generates streamflow (Figure 1.9E). Supply of solutes can come from soil and rock, solutes and suspended material are produced by surface sheet or rill flow, and solutes, suspended sediment and bedload derive from erosion of channel banks and from erosion occurring as the stream network expands and contracts. Thus the sediment system can be visualised as composed of a number of subsystems, particularly the channel subsystem, encompassing erosion and deposition (Figure 1.9E).

The drainage basin geomorphologist, has to describe the characteristics of the basin as exactly as possible, to document the processes operating, and to establish relationships and approach an understanding of the relations between these two as a basis for the elucidation of variations encountered in space and experienced over time. Because the geomorphologist is interested primarily in land form and character, in the way in which it influences process and is in turn fashioned and modified by processes, not all the basin characteristics and basin processes will be his prime concern. In subsequent chapters an attempt is made to focus attention upon the primary subjects while introducing those of a secondary nature to provide a more complete understanding. This aim may be facilitated by a simple example.

1.4 Illustration of an approach

The natural unit of study is the drainage basin and a very small basin is illustrated in Figure 1.10. This basin is underlain by shales, it has a particular relief inherited from the late Quaternary, and, accordingly, its slopes are mantled by periglacial slope deposits. Although the overall form of the basin is, therefore, in part the product of past processes, its form will influence the rate at and way in which processes operate at the present time. In the mid-latitude morphogenetic zone (Figure 1.5C) climatic inputs into this basin will be transformed by the basin characteristics at any time to outputs at X (Figure 1.10) and the basin could be studied as a single unit in terms of the water balance equation as a process-response system. Despite its small size the basin can also be subdivided into several subsystems, principally slopes, network, stream channel reach, and stream channel cross section. In the slope subsystem weathering slowly produces material which is moved, together with the periglacial cover, down the slopes towards the stream channels, and water moves over and beneath these slopes. The slope subsystem could, therefore, be analysed as a morphological system relating angle of slope to variables including depth of superficial material, distance from the stream channel, position in the slope profile. The total network could be regarded as a further subsystem and could be analysed as a process-response system either to relate input of precipitation with output of changed stream network, or to see how the variation of stream network produced a response in varying streamflow from the basin. It would also be possible to analyse a particular channel reach in order to discover how the stream pattern was related to stream processes of a particular magnitude. Channel cross sections could be analysed as subsystems to see, at the level of the morphological system, how channel form was related to locational parameters and to the character of sediments and vegetation; or at the level of the process-response system, to determine how channel form was a response to particular stream channel processes. These subsystems could further be regarded as a cascading system as output from the slopes would provide input to the network and output from particular channel reaches would afford input to channel cross sections.

The recognition of such subsystems underlines the multivariate nature of drainage basins, the scale problem, and the significance of man. It is obviously as arbitrary to separate these subsystems one from another as it is impractical to separate different types of water flow or to isolate individual basin characteristics such as angle of slope

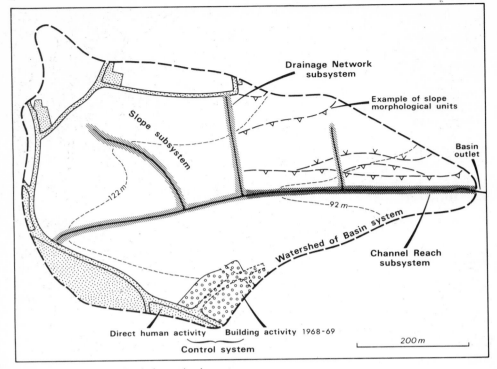

Figure 1.10   A specific drainage basin system
Employed to illustrate the geomorphological approach to the drainage basin this small basin on
the margin of Exeter, Devon is represented here in 1969 and the same basin in 1972 is illustrated
in Figure 8.1. The basin may be treated as a single unit or viewed as composed of slope, network,
channel reach and channel cross-section subsystems. The way in which each of these can be
further subdivided is illustrated by slope morphological units which separate one slope subsystem
into its slope components. The significance of man is represented by the existing built-up area and
by the building activity taking place in 1968-9, both of which indicate the significance of human
intervention as a regulator in a control system.

from depth of material above unweathered bedrock. Subsystems distinguished for
study are influenced by scale and whereas a number of individual point locations
could be studied at points in time it is necessary to group time and space sequences
together for the purposes of study. Different spatial and time scales will necessarily
require distinctive methods of measurement, of analysis, and of interpretation. The
influence of man is particularly notable in this basin because, in addition to the
changes already wrought (Figure 1.10), including enclosure, with its attendant effects
on the drainage network, and road- and house-building which have all modified the
inputs, affected the response of the basin, and conditioned the output, the basin is
being further urbanised (Figure 8.1) and so over time it will be possible to measure the
effects of man regulating this system as a control system (Walling and Gregory, 1970).

Plate 3 Small watershed stream measurement station
A sharp-crested V notch weir installed in a compound section, with weir pool behind, and stage monitored by recorder in wooden box. The station is located at X on Figure 8.1 and at basin outlet on Figure 1.10.

In the final analysis, therefore, the geomorphologist's perception of the drainage basin system should not only embrace the nature of form and process relations but should also indicate how these have been modified by man's influence and how they provide a framework for his actions in the future. In addition, geomorphology, employing a systems approach to the drainage basin, must inevitably rely upon an appreciation of the total physical geography of the basin.

Selected Reading

The systems approach in physical geography is reviewed in:

R. J. CHORLEY and B. KENNEDY 1971: *Physical Geography: A systems approach*, London. (370 pp.)

A field approach to the drainage basin is sketched by:

K. J. GREGORY and D. E. WALLING 1971: Field measurements in the drainage basin: *Geography*, **56**, 277–92.

Specific themes are developed in:

E. H. BROWN 1970: Man shapes the earth: *Geog. J.*, **136**, 74–85.

J. BUDEL 1969: Das system der klima-genetischen geomorphology: *Erdkunde,* **23**, 165–82.

S. A. SCHUMM and R. W. LICHTY 1965: Time, space and causality in geomorphology: *Amer. J. Sci.*, **263**, 110–19.

M. G. WOLMAN and J. P. MILLER 1960: Magnitude and frequency of forces in geomorphic processes: *J. of Geology*, **68**, 54–74.

A useful source of reference is:

D. K. TODD (ed.) 1970: *The Water Encyclopedia* (New York: Water Information Center). (599 pp.)

Part A: Drainage Basin Measurements

2 Drainage basin characteristics

In geomorphologic systems the ability to measure may always exceed ability to forecast or explain. *L. B. Leopold and W. B. Langbein, 1963*

2.1 **Topographic characteristics:** *2.1a Area and order; 2.1b Density of stream network; 2.1c Basin and channel length; 2.1d Basin, network and channel shape; 2.1e Channel, network and basin relief*

2.2 **Rock and sediment characteristics:** *2.2a Rock characteristics; 2.2b Superficial deposits and soils; 2.2c Characteristics of fluvial sediments*

2.3 **Vegetation characteristics**

2.4 **Landscape evaluation and response units**

2.5 **Association between drainage basin characteristics:** *2.5a Drainage basin relations; 2.5b Channel reach relations; 2.5c Dynamic drainage basin characteristics*

2.6 **Example**

A drainage basin is the entire area providing runoff to, and sustaining part or all of the streamflow of, the main stream and its tributaries. The function of the drainage basin and its significance is hinted in the synonyms which have gradually been adopted including *drainage area, catchment area* especially employed in river control engineering, and *watershed*, utilised especially in water supply engineering. Appreciation of the significance of the drainage basin unit arose with the gradual understanding of the mechanism of the hydrological cycle and of the function of the basin in conveying water from precipitation to the river. The need to study the form of the drainage basin derives from two main sources; firstly, to describe the form-form relationships or morphological systems (pp. 26–7), and, secondly, to analyse the form-process relationships. To understand the interrelationships in morphological systems and in process-response systems it is necessary to express the character of the drainage basin in quantitative terms. Numerous methods of describing drainage basins have been proposed; some of these apply to the whole basin, while others apply to a particular characteristic, such as relief or soil. Any single index is inadequate because it attempts to simplify complex reality and to express, often in two dimensions or in a single index, what is in reality three-dimensional and has a time magnitude and a time significance as well.

Methods of describing drainage basin characteristics have arisen in three main ways; firstly, from morphometric attempts to measure the form of the basin, secondly, from expressions of characteristics in a manner relevant to the processes operating in the basin, and thirdly, and incidentally, from other branches of study which have developed methods of landscape description which may be partly applicable to the drainage basin.

Measurement and quantitative expression of the drainage basin perhaps began with the ideas of James Hutton, whose Law of Accordant Tributary Junctions was expressed by Playfair in 1802. During the nineteenth century quantitative measures were proposed for specific areas or for particular problems, but by the early twentieth century methods of stream ordering were being conceived. A great step forward was made by R. E. Horton in 1932 when he crystallised previous work, added new measures, and proposed general methods for the description of drainage basin characteristics. Such characteristics, according to Horton, included morphologic, soil, geologic or structural, and vegetational factors. Horton proposed ways in which examples of each factor could be expressed in a way relevant to the functions of the drainage basin. Many of his ideas were later developed in the hydrological context of 'A hydrophysical approach to drainage basins' (Horton, 1945), were supplemented by W. B. Langbein's contribution on topographic characteristics of drainage basins (Langbein, 1947), and these are basic to the understanding of the methods which developed in the succeeding quarter century. The subdivision proposed by Horton is a meaningful one and so it is expedient to consider first the topographic characteristics of drainage basins, second the rock and soil characteristics, third the vegetation characteristics, and then to proceed to the ways in which these characteristics are interrelated and can be combined in several kinds of drainage basin terrain type.

2.1 Topographic characteristics

The size of the drainage basin influences the amount of water yield; the length, shape and relief affect the rate of water and sediment yield; and the character and extent of the channels affect sediment availability and rate of water yield from the drainage basin. Topographic characteristics need to be visualised in the context of the ways in which they influence drainage basin process but it is equally important to bear in mind the multiform relations which exist between the characteristics; for example, between the shape of the drainage basin and the nature of the drainage pattern. It is unrealistic, but very necessary for the purposes of study, to separate drainage basin characteristics one from another and some ways of classifying topographic characteristics are indicated in Table 2.1. Variety has arisen because the indices have to represent the total basin, the variation within it and the nature of the components, and also because indices have been derived for different purposes. Thus, in Table 2.1, A and B were developed for the basin, whereas G was developed to describe the river and its valley; C and E were developed as a basis for interpretation of drainage basin process, whereas D and F were visualised as a basis for the study of drainage basin process and form. Differing degrees of complexity have also arisen and, whereas a particular problem often merits a specific kind of drainage basin measure, it is desirable that any index can be applied as easily as possible by other workers in contrasted areas using different resources. The principal difficulties which hinder the universal application of all indices surround the limitations and variability of the data available;

Table 2.1   Perception of topographic characteristics of drainage basins

A *Horton (1932)*	B *Langbein (1947)*	C *Johnstone and Cross (1949)*
Form factor	Area	Area
Compactness	Stream density	Overland slope
Mean elevation	Area-distance	Channel slope
General slope	distribution	Size of channel
Mean slope	Length of basin	Condition of channel
Drainage density	Land slope	Stream pattern
Stream number	Channel slope	Stream density
Average fall and	Area-altitude	
slope of streams	distribution	
Direction and length	Area of water	
of overland flow	surfaces	

D *Strahler (1964)*	E *Gray (1965)*
Linear aspects of	Drainage area, size and shape
channel system	Density and distribution of water courses
Areal aspects of	Overland slope or general land slope
drainage basins	Size, length, slope and condition of stream channels
Relief (gradient)	Depressional storage and pondage due to surface channel
aspect of drainage	obstructions forming natural detentions
basin and channel	
networks	

F *Chorley (1967)*	G *Wolman (1967)* (River and valley)
Linear aspects of the	Catchment area
basin—topological	Size of channel
—geometrical	Shape of channel
Areal aspects of basin	Fall
Relief aspects of basin	Rugosity of channel
	Pattern
	Valley form and dimensions—pattern in plan, shape in
	cross-section

Table 2.2   Topographic attributes of drainage basins

Scale	Basin	Network	Channel reach	Channel cross-section
'Dimension'				
Area	Drainage Basin Area	Area tributary to stream channels	Area of channel	Cross-sectional area of channel
	Area of storage, e.g. lakes			
Length	Basin length Basin perimeter	Drainage density Stream length	Channel length ↓ Sinuosity ↑	Width
Shape	Basin shape	Drainage pattern Network shape	Channel shape	Shape
Relief	Basin relief Basin slope	Network relief Network slope	Channel relief Channel slope	Depth

the variations introduced by different operators who may make decisions in different ways; the time-consuming nature of derivation of some indices; the scale factor which dictates that some indices are more appropriate for one scale than another; and the multivariate relationship of drainage basin characteristics which render any single index a poor approximation of complex reality.

Plate 4 River Ganges
Taken from Apollo 7 illustrating the scale of basin and river dimensions and that topographic, rock, soil and land-use characteristics of basins all need to be expressed.

Topographic attributes of drainage basins may be visualised, therefore, for the basin as a whole, for the total channel system or network, for individual parts of channels or reaches, and for the channel cross section. For each of these four categories, or system and three subsystems, indices of area, length, shape and relief are required (Table 2.2). Some of these values are absolute—area, length, perimeter, relief, but are subject to operational definitions; some are obtained by combining two absolute measurements as in density or measures of slope; and others require a definite method for their computation as in basin shape, or drainage network.

2.1a   Area and order

Area was characterised by Anderson (1957) as the 'devil's own variable' because 'almost every watershed characteristic is correlated with area' (Figure 2.1C). The drainage basin must be delimited as precisely as possible. Difficulties arise in defining the basin due to information on which the delimitation is based and due to the significance of basin area for different types of flow. In view of the existence of surface run-off, throughflow and interflow, and baseflow (p. 114) it is possible to have a drainage basin area on the surface that does not correspond to the boundaries of the basin at several levels below the surface. This discrepancy is most evident in areas influenced by a particular pattern of geological structure where the divide on the surface (topographic divide) is not coincident with the groundwater or phreatic divide below the surface (Figure 2.1A).

Definition of the drainage basin is frequently made according to the information shown on topographical maps, on aerial photographs or, alternatively, by field survey. Insertion of the watershed line may be complicated by the problem of discordance of the several divides and in addition by the problems of map or air photo scale and of map convention. This is particularly relevant to contour interval and contour reliability. It is desirable to have the surveyed contours as close as possible and a map scale as large as possible for inserting the watershed line. 1:25,000 maps are frequently suitable and the watershed for a particular basin is drawn as a line which surrounds all the drainage lines and depressions in the basin and passes through the highest points between the stream and adjacent ones (Figure 2.1B). Once the basin area has been determined the area of the basin can be measured by counting squares on squared paper, by planimeter, by cutting out the basin area on high quality paper and weighing to give a value of area, or by using mechanical digitised or scanning techniques.

Because area is so important, and because scale of basins is significant in relation to form characteristics as well as in relation to processes, there has been a search for a method of classifying or ordering drainage basins according to their size. This has usually been based upon the stream network and although a method was proposed by Gravelius in 1914 the ordering system advocated by Horton (1945) was more widely adopted. In this system (Figure 2.2A), each finger-tip tributary was designated a first order stream, two second orders combined to give a third order stream and so on. Once this initial ordering had been completed the highest order stream was projected back to the headwaters along the stream which involved least deviation from the mainstream direction. To order the drainage network in two stages is unnecessarily complicated, it involves an additional subjective decision in projecting the order of streams headwards, and, in addition, all the smallest unbranched tributaries are not of the same order (Figure 2.2A).

These objections were overcome to some extent by the modification proposed by Strahler in 1952. He designated all finger-tip tributaries as the same, first, order, two first orders produce a second order segment, two second orders provide a third order and so on (Figure 2.2B). The advantage of this simple scheme is that it can be derived mathematically from concepts of elementary combinatorial analysis (Melton, 1959), it designates all unbranched segments as the same order, and it gives the highest order to one segment rather than to the whole of the trunk stream. The simplicity and ease

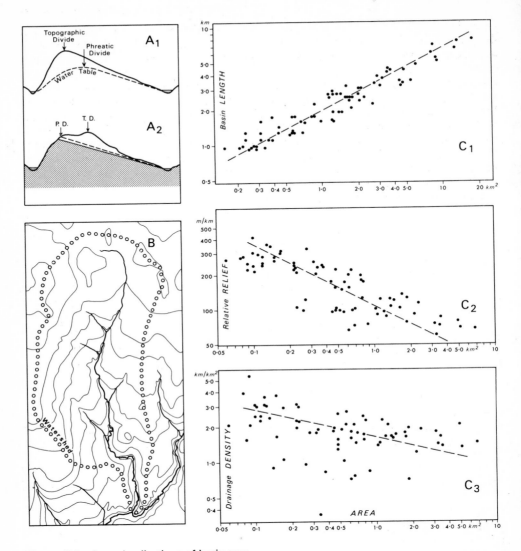

Figure 2.1 Some implications of basin area
A illustrates the possible discrepancy between the phreatic divide and the topographic divide,
B illustrates the way in which a basin watershed is delimited according to contour information on
topographic maps, and C indicates the way in which other topographic characteristics are often
related to basin area. The data from 76 small drainage basins in south-east Devon shows that
the length of the basin, measured parallel to the principal drainage line, is, expectedly, directly
correlated with basin area (C_1) whereas relative relief (C_2) and drainage density (C_3) are, less
expectedly, inversely correlated with basin area.

of application of this method have commended its widespread use in the 1960s. The purpose of stream ordering is not only to index size and scale but also to afford an approximate index of the amount of streamflow which can be produced by a particular network. If all other factors were constant then order of basin should be directly related to size of channel network and increasing order of network should be associated with greater streamflow values. However, the Strahler method of ordering has one limitation for this purpose: the order of the trunk stream (order n) is not changed by the addition of tributary streams of lower order (n−1, n−2, etc.). This is obviously a limitation in that a large number of streams can enter a particular segment without changing its order (Figure 2.3 G2). A further limitation is that small changes of the network can lead to differences in the Strahler order of the trunk stream (Figure 2.3 G3) and in a limiting case the addition of a single first order stream could raise the order of the trunk stream (Figure 2.3 G2).

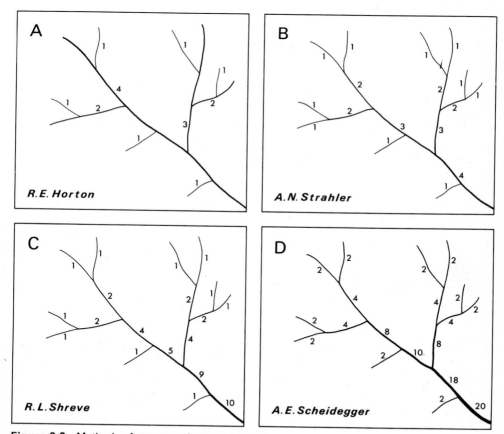

Figure 2.2 Methods of stream and segment ordering
The illustration of the method proposed by A. E. Scheidegger is of the first stage of his consistent ordering. For explanation see text.

Shreve (1957) overcame these obstacles by proposing a method of ordering called segment ordering (Figure 2.2C) in which each outer link or first order segment is designated magnitude 1 and each subsequent link designated as a magnitude equal to the sum of all the first order segments which are tributary to it (Figure 2.2C). A similar method has been evolved by Scheidegger (1965) in a consistent law of stream ordering which involved four postulates for an algebra of stream segment combinations which is commutative as well as associative. These postulates were:

1   when two similar segments are combined the resulting segment has its order increased by an integer,

$$G' * G' = G' + 1$$

2   a combination of two segments of lower order $(G'-1)$ with a given order should increase the order of the latter by one integer;
3   the sequence in which the segments join is immaterial;
4   it does not matter whether a G' segment joins a G'' or vice-versa.

From these four conditions a general law for stream order magnitudes may be expressed as

$$G' * G'' = \log \frac{(2^{G'} + 2^{G''})}{\log 2}$$

Consistent ordering is achieved, firstly, by giving the fingertip tributaries an index of the exponent unity to base 2 and indexing all downstream links by adding index numbers combining upstream at the junction to give a number I (Figure 2.2D). Each link is then given an order magnitude by

$$\text{order magnitude} = G = \log_2 2M$$

where M is the magnitude of exterior links considered as magnitude 1.

In this way 2 becomes 1, 4 becomes 2, 6 becomes 2·59, etc. Woldenberg (1967) used Scheidegger's index (Figure 2.2D) to derive a new order magnitude (W) which, unlike Scheidegger's G, gives an increase in geometrical progression downstream in:

$$W = \frac{\log M}{\log R_b} + 1$$ where M = I/2 and the order magnitude (W) conforms to the geometric progression $Q_u = Q_1(R_b)^{u-1}$

Lewin (1970) has drawn attention to the fact that the Shreve system is based only upon the outer segments and neglects the fact that the inner links gather water as well. Thus a modification of Shreve's method could be proposed and alternative ordering methods can be based either upon junctions (nodes) or upon paths.

The Strahler modification (Figure 2.2B) of Horton's method has been used most extensively, particularly in conjunction with the 'laws of drainage composition' but the Shreve (Figure 2.2C) and the Scheidegger (Figure 2.2D) methods have advantages because they are more descriptive of the total network in relation to streamflow amounts. Ordering is useful because it provides a rapid method of quantitatively designating any stream or stream segment anywhere in the world, but in each case the method of ordering should be specified in conjunction with the scale of map used. A simple method of ordering, useful in relation to process, is to visualise each segment in the network to be of order 1 and to develop ordering additively on this basis (Figure 2.3 G1). Comparable problems are encountered in measuring the surface area of stream channels and in expressing the cross-sectional area of a stream channel at a point (p. 57).

2.1b   Density of the stream network

Density of the stream network has long been recognised as a topographic character-istic of fundamental significance. This arises from the fact that network density is a sensitive parameter which in many ways provides the link between the form attributes of the basin and the processes operating along the stream course. If drainage basins were uniform in every respect streamflow would be proportional to the length of watercourse in a basin, because channel flow is much more rapid than the alternative flow on, or beneath, slopes. As the extent and density of the network reflect topo-graphic, lithological, pedological and vegetational controls, and because they also incorporate the influence of man, network density promises to be a valuable index.

Density of the drainage network was isolated by Neumann in 1900, it was defined by R. E. Horton (1932) as the length of streams per unit of drainage area, and its significance as a factor determining the time of travel by water was appreciated by Langbein (1947). This index has been widely adopted for its ease of comprehension, its simplicity, and its utility. Related measures have been proposed. Stream frequency (F_s) was proposed by Horton (1945) as a measure of number of stream segments per unit area and is, therefore, dependent upon stream order, whereas drainage density is independent of order; the use of the average distance between stream junctions has been suggested (Penck, 1924); and Smith (1950) proposed a texture ratio which was based upon the number of crenulations of the most crenulate contour in the basin. These methods have been complimented by attempts to map drainage density (for ex-ample, Wilgat, 1966) and by an index of drainage intensity (Faniran, 1969) derived by combining drainage density and stream frequency. Horton (1932) considered that the value of drainage density ranged from 1·5 miles per square mile (0·93 km per km²) to 2·00 (1·24) for steep impervious areas in regions of high precipitation, and to nearly zero in permeable basins with high infiltration rates. Langbein (1947) suggested a range of drainage density from 0·89 (0·55) to 3·37 (2·09) in humid regions with an average stream density of 1·65 (1·03). Measurements of drainage density such as these have hitherto been quoted in miles per square mile but henceforth will be given in km per km². More than 20 years of investigation have shown that values of drainage density can be much more variable than these two workers supposed, that problems of definition and of scale are qualifying, and that the task of laborious measurement is sometimes inhibiting.

Measurements are now available for many areas in all continents and a great range of values has been demonstrated, sufficient to underline the usefulness of this para-meter (for example, Figures 5.11, 5.12). With relation to texture ratio, Smith (1950)

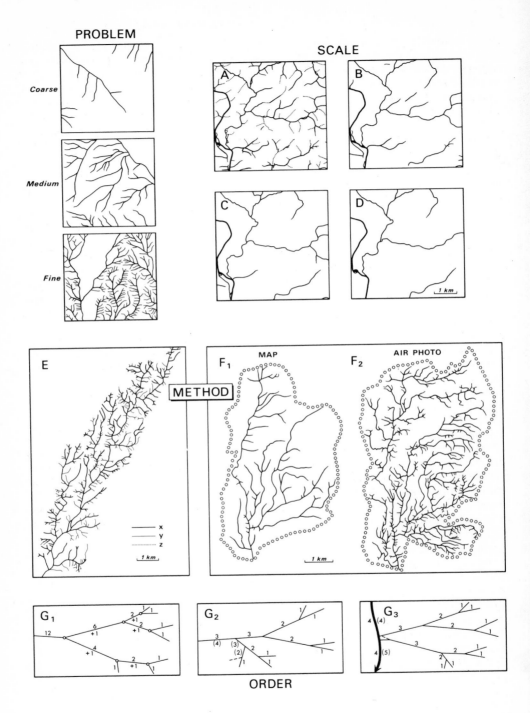

PROBLEM

Coarse

Medium

Fine

SCALE

A

B

C

D

1 km

METHOD

E

x
y
z

1 km

F₁   MAP

F₂   AIR PHOTO

1 km

G₁

G₂

G₃

ORDER

and Strahler (1957) described drainage density values less than 5·00 as coarse, between 5·00 and 13·7 as medium, between 13·7 and 155·3 as fine, and greater than 155·3 as ultra-fine (Figure 2.3). Coarse values are frequent in areas of permeable rocks and of low rainfall intensities. In Britain values are usually well below 5·00 and on Dartmoor average 2·14 (Chorley and Morgan, 1962). Values in the medium category have been recorded from large areas of the humid central and eastern parts of the United States and from New Zealand, whereas fine values have been recorded in Badlands in South Dakota (Smith, 1958) and in badlands on weak clays at Perth Amboy, New Jersey, where Schumm (1956) recorded drainage densities for second order basins from 313 to 820. Although the scope of world-wide variation has been sketched, insufficient measurements are hitherto available for clear relationships to be established with other variables (pp. 271–4).

Values reported must be visualised against the method of delimitation of the drainage network. To obtain drainage density values the network must be determined from maps, air photographs or field survey for the basin, and the area of the basin and the total stream length measured. Horton (1945) originally suggested *composition of the drainage* net to replace the more qualitative term drainage pattern and he constructed a drainage net by using the watercourses shown as blue lines on topographic maps, although he advocated extending these to the watershed to give 'mesh length'. Later workers who have followed Horton's method include government agencies in the United States (Morisawa, 1957) and Gregory (1968). A disadvantage of the 'blue line method' is that not all streams may be represented on the map and contour crenulations may indicate where a valley is present although a blue line may not occur in the base of the valley indicated. Some workers, particularly in the United States (for example, Bowden and Wallis, 1964; Carlston, 1963; Orsborn, 1970) have, therefore constructed drainage networks by first tracing the network of blue lines and then supplementing this network by additional segments according to the pattern of the contour crenulations. With reference to areas in the United States, Morisawa (1957) compared the results obtained by using blue lines on available maps, with those results obtained by including the valleys indicated by contour crenulations, with those on a map obtained from field survey. This comparison showed that there was no significant difference between the contour crenulation and field survey methods but that both did differ significantly from the blue line method. She therefore concluded that the use of blue lines on topographic maps should not be used to give networks for basins in the U.S.A. less than 7 km² of drainage area. Elsewhere, including Great Britain for

Figure 2.3   **Problems of stream network definition**
The problem is indicated by the need to quantify different densities of drainage, classified as coarse, medium and fine by K. G. Smith (1950). The scale of the network is illustrated by four drainage maps (A–D, all same scale) of a small area in south Devon: A is based upon 1 : 25,000 Regular Edition Map and dry valleys are shown by dotted lines (drainage density for streams = 1·72 km/km² drainage density for streams and valleys =2·43), B is based upon 1 : 25,000 Provisional Edition map (D_d = 1·32), C is based upon 1 : 63,360 map (Dd = 1·09), and D is based upon 1 : 250,000 map (D_d = 0·82). Further methods of definition of the network are illustrated by use of radar imagery (E) compared with 1 : 24,000 maps (McCoy, 1971) where X are indicated on both, Y are apparent only on maps and Z apparent only from imagery; and also by comparing networks from 1 : 15,840 maps of New Zealand (Fl) with those obtained from air photos (F2) (Selby, 1963). Ordering of the network necessarily depends upon the precision of network definition (see text).

example, it has been demonstrated (Gregory, 1966) that dry valleys occur not only in areas of limestone outcrop but on other rock outcrops and in deglacierised areas as well. The use of contour crenulations to provide part of the stream network can, therefore, result in a value of valley density which is substantially greater than the value of stream density at the present time. In south-east Devon, for example, the basin of the Otter has a blue line density on maps of 1·04, a field-mapped density of 1·25 and a valley density of 2·16. Clearly the field-mapped density is more closely approximated by the derivation of the stream network according to blue lines on maps. In five basins of Devon the ratio of the blue line density (1·63) to contour crenulation density (3·03) is 1·86, which compares with the ratio of 2·5 quoted for south-west Germany by German (1963).

The method used to obtain the drainage network must depend upon the purpose of the analysis envisaged. The method based upon contour crenulations is particularly appropriate where the study aims to describe the form of the basin and the character of the dissection—this is, therefore, an index of valley or dissection density. Where the density is to be used in relation to the contemporary processes operating in the basin, the blue line method is most appropriate in Britain but elsewhere depends upon map convention. The choice is also influenced by additional qualifications of map convention and map scale because these determine the nature of the network published. Some topographic map series show ephemeral and intermittent streams as well as perennial ones but other series may show only perennial water courses or an intermediate version. The network shown also varies according to map scale and in some cases there can be variations, at the same scale, from one edition to another (Figure 2.3A, B, C, D). Giusti and Schneider (1962) compared maps of different dates and scales for the Piedmont province from Virginia to Alabama and demonstrated that the number of first order streams per square kilometre varied from 1·4 to 3·2 and from 0·43 to 1·9 at two different map scales, and that it varied from 0·14 to 3·2 on map scales ranging from 1:250,000 to 1:24,000. In studies of Malaysia, Eyles (1966) indicated substantial differences between 1:62,500 map values and values obtained from aerial photographs, and he concluded that the use of map evidence alone could lead to erroneous conclusions because topographic maps indicate the highest drainage density values on granite outcrops, whereas air photographs showed that densities on these rocks were the lowest in the area. Similarly Selby (1968) showed in New Zealand (Figure 2.3F) that densities of 5·4 measured from air photographs compared with densities of 2·8 from 1:15,840 maps.

Map convention, map scale and method of determination of the drainage net are necessary qualifications to drainage density values but these values are time-consuming to calculate. More rapid methods of calculation have, therefore, been devised. Carlston and Langbein (1960) proposed a rapid line intersection method of approximating drainage density by drawing a line of known length (L) on a contour map and counting the number of streams (n) which intersect this line. A minimum of fifty contour crossings was advocated to provide an adequate sample. The expression sin 45° . L/n affords an index of the mean normal distance between the channels, and the reciprocal of this gives an approximation to drainage density where $D_d = 1·41n/L$. A second rapid method has been employed by Gardiner (1971), who utilised the number of junctions and the number of sources on 1:25,000 maps of south-west England to produce a map of estimated drainage density for Dartmoor and adjacent areas. A third rapid method entails the use of electronic scanning of air photographs, infra-red or radar imagery (McCoy, 1971). On radar imagery, changes in film density reflect changes in terrain slope and these changes can be detected by a flying spot scanner which shows, on a television tube, the edges or boundaries along which there is an abrupt change in film intensity. These edges can be measured automatically or traced and measured manually. McCoy (1969, 1971) found that, compared with 1:24,000 maps, reliable estimates of stream lengths and drainage density could be made from edge-enhanced images, although this may not be so effective where agricultural patterns exist. Drainage density has increasingly been appreciated to be a fundamental indicator of drainage basin form and a valuable index of drainage basin process. Its future use in the context of more rapid methods to ease its calculation, and in the context of greater understanding of its nature (pp. 294–5), will facilitate its application.

2.1c Basin and channel length

Calculations of drainage density rely upon measurements of total stream length within a drainage basin. An alternative which has sometimes been utilised is the length

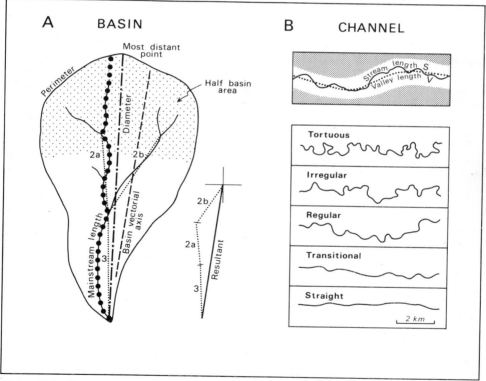

Figure 2.4 Measurement of basin and channel length
Channel patterns are based upon Schumm (1963).

of the main channel which can be related to the basin area (Figure 2.1 C1) and also to total stream length in the basin. However, the length of the main channel in a drainage basin introduces two further considerations in that, firstly, mainstream length is dependent upon the shape of the drainage basin and, secondly, there is not always exact agreement between valley length and river channel length.

Basin length has, therefore, been measured in a variety of ways (Figure 2.4A). For some purposes measurement of the mainstream length has been made and this may be extended to the watershed (Figure 2.4A). Because the stream channel may be very sinuous the length parallel to the main drainage line has been used to give a measure of basin length. This requires a subjective decision where the stream is tortuous or irregular and where the drainage basin possesses an unusual shape and so a further alternative is to measure the longest basin diameter between the mouth of the basin and the most distant point on the perimeter. A rather more time-consuming method, used by some hydrologists (for example, Potter, 1961), has been to identify the point where the line dividing the basin into two equal halves crosses the mainstream and then to join this point (the centre of gravity of the basin) with the mouth to give the centroid direction (Figure 2.4A). This method requires areal measurement but it can be very precise. Ongley (1968) devised a simple property called the basin vectorial axis. This involves, firstly, selecting the highest (n) and all the next highest orders (n−1) in a particular drainage basin. For each stream the vectorial equivalent is drawn in (3, 2a and 2b in Figure 2.4A) and the direction and magnitude of each vector is then used to calculate the resultant by trigonometric or graphical methods as shown in Figure 2.4A. The resultant then gives the basin vectorial axis, which can be drawn on the basin diagram passing through the mouth of the basin.

Similar problems arise when considering part, or the whole, of the channel network of a particular stream. In many cases the stream length (s) may not correspond with the shorter valley length (v) (Figure 2.4B). Accordingly, indices of stream channel sinuosity have been proposed, mainly in the form P = sinuosity = s/v (Schumm, 1963). Schumm (1963) proposed the five categories of channel sinuosity illustrated in Figure 2.4B. This method is commended by its simplicity but is sometimes difficult to apply precisely because of problems of delimiting a channel reach and of determining the valley length where valley margins are not clearly defined. The result may be difficult to interpret because it includes both hydraulic and topographic sinuosity. This arises because the valley length may also deviate from a straight path and so one sinuosity index could be derived for the channel and one for the valley course. Mueller (1968) defined:

CL = length of stream channel
VL = valley length (mid-valley line)
Air = shortest distance between mouth and source of stream,

from which he derived:

Index of total sinuosity = Channel Index = CI = CL/Air
Index of topographic sinuosity = Valley Index = VI = VL/Air
Percentage departure from a straight line course due to hydraulic sinuosity in
   a valley = Hydraulic sinuosity index = HSI = CI−VI/CI−1%

Percentage of a stream's departure from a straight line course due to topographic interference = Topographic sinuosity index = TSI = VI−1/CI
−1%.

The difference between stream and valley networks is, therefore, reminiscent of the
problem encountered in dealing spatially with the drainage network (Section 2.1c)
and it, similarly, includes some consideration of shape.

2.1d Basin, network and channel shape

Shape has proved to be one of the most elusive topographic properties to measure
unambiguously with accuracy, significance and precision (e.g., Boyce and Clark,
1964). Shape of the basin, of the drainage network contained, and of portions of the
channel system can all be influenced significantly by other drainage basin characteristics such as rock type; and they can affect the processes operating, particularly in
that they may determine the potential efficiency of the basin, the network or the
channel. Horton (1932) used a form factor (F) and its reciprocal was used as a shape
factor by the American Corps of Engineers. Horton also referred to a compactness
factor (C) devised by Gravelius, which expresses the ratio of the perimeter of the
drainage basin (p) to the perimeter of a circle of area equal to the area (M) of the
drainage basin. Thus, $C = p/2\sqrt{\pi M}$. Horton did not recommend this compactness
ratio because it is the same for two basins of identical form but having stream outlets
at different positions. The same criticism applies to later proposals for characterising
drainage basin shape (Table 2.3). These have included comparison of the basin area

Table 2.3 Some methods of expressing drainage basin shape

Method	Derived by	Source
Form Factor (F)	$F = \dfrac{A}{L^2}$ where A = drainage area L = basin length	Horton (1932)
Basin circularity (C)	$R_c = \dfrac{\text{Area of basin}}{\text{Area of circle with same perimeter}}$ $= \dfrac{4\pi A}{p^2}$ where p = basin perimeter	Miller (1953)
Basin elongation	$E = \dfrac{\text{Diameter of circle with same area as basin}}{\text{basin length}}$ $= \dfrac{2\sqrt{A/\pi}}{L}$	Schumm (1956)
Lemniscate (K)	Based upon comparison of basin with lemniscate curve $K = \dfrac{L^2}{4A}$	Chorley, Malm and Pogorzelski (1957)

with the area of a circle with the same perimeter, proposed by Miller (1953) as a
circularity ratio, and an elongation ratio (Schumm, 1956). Comparison with a circle

is not realistic because the average basin is probably pear-shaped rather than circular and so, accordingly, a lemniscate ratio (Table 2.3) has been proposed (Chorley, Malm and Pogorzelski, 1957). Other shape measures proposed (for example, Bunge, 1966) may be applied to the drainage basin.

Difficulties confronting the description of drainage basin shape are equalled in the case of network shape and it is perhaps for this reason that the pattern of stream networks has only recently been attempted in quantitative terms. Until 1950, patterns were usually expressed in qualitative terms and classifications (for example, Zernitz, 1932; Howard, 1967) produced categories of pattern such as those illustrated in Figure 2.5. Application of such classifications to contrasting areas by workers with differing interpretations encouraged the need to express pattern numerically so that it could be compared with other basin characteristics and related to processes. Quantitative description of the drainage pattern has been approached in three main ways; by the analysis of component directions and orders, by the analysis of junction angles or by using bifurcation ratios, and by the derivation of generalised patterns. These three approaches have been complemented by simulation studies of drainage pattern development (pp. 79–81) and by observations of actual development over time (pp. 367–8).

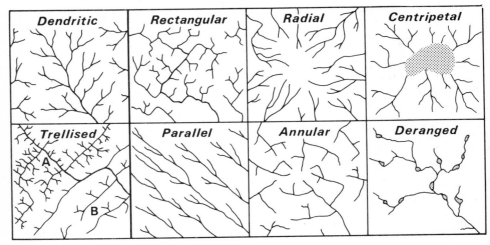

Figure 2.5   Morphological classifications of drainage pattern

Perhaps the simplest way of characterising network shape is to divide each component of the network into arbitrary measured lengths selected according to map scale, measuring the azimuth of each, and then plotting the result as a circular frequency distribution for a basin, which may be of a particular order (Figure 2.6A). In this manner, the overall orientation of the network may be established (for example, Woodruff and Parizek, 1956; Judson and Andrews, 1955; Brown, 1969). The diagram obtained by plotting the total length of channel segments in a given vectorial class has been referred to as a vectorial rosette (Milton, 1965). The examples given in Figure 2.6A are for two basins on the granite outcrop of Dartmoor, where stream directions are related to joint directions. Results can be approximated by counting the number

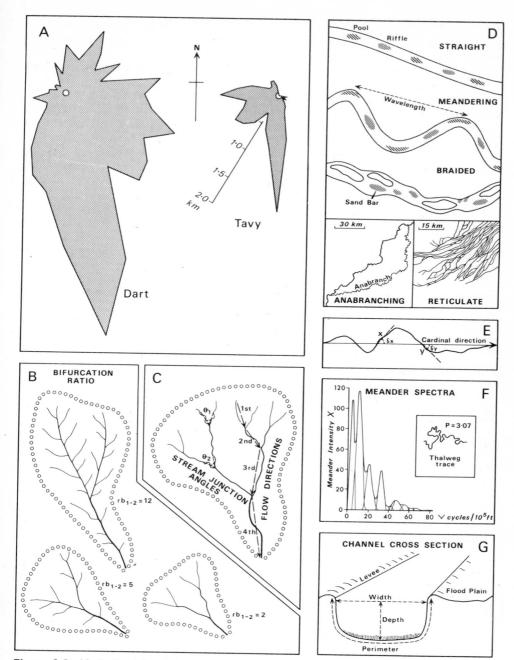

Figure 2.6 Methods of describing network, pattern and channel cross-section shape
The anabranching and reticulate patterns are based upon Dury (1969) and the instance of meander spectra is from Speight (1965).

of streams flowing towards the four quadrants (NE, SE, SW, NW) (Morisawa, 1963), but this method, like the vectorial one, suffers from the limitations that only one diagram is produced per basin, that the result depends considerably upon the arbitrary measured length selected, that it does not clearly distinguish between the stream and valley orientation, and that it does not provide a single value for one basin. This can be remedied by using the mean value, or the coefficient of variation (Bowden and Wallis, 1966), of flow lengths (Figure 2.6A). It is possible to measure the deviation of a particular stream course from its cardinal direction (Schick, 1965) by examining the deviations at equally spaced points, or at points selected by random numbers, to give a frequency distribution of deviations from the cardinal direction (Figure 2.6E). This method can be applied to an entire basin or to part of a network in which case the aim of the index approximates that of the sinuosity index (p. 50).

Bifurcation ratio (R_b) was recognised as an important characteristic of the drainage basin by Horton (1932) and was defined as the ratio of the number of streams of order n to the number of streams of the next highest order (n+1) (Figure 2.6B). It is therefore dependent upon the ordering methods of either Horton or Strahler (Figure 2.2A, B) but it is characteristically between 3·0 and 5·0 in watersheds where geologic structure does not exercise a dominant influence on the drainage pattern (Strahler, 1964). Giusti and Schneider (1965) showed that bifurcation ratio is not independent of stream order and that in a particular drainage basin the ratio should be calculated between the streams of one order lag (major tributaries) with those of two order lag to reduce the difficulties introduced by order and by the size of the area analysed. In a single drainage basin several bifurcation ratios may be available according to the order of the whole basin; thus in a fifth order basin four bifurcation ratios could be calculated. To obtain one weighted mean bifurcation ratio WR_b Schumm (1956) recommended the use of

$$WR_b = \frac{\sum[R_{b_{n:n+1}} \times (N_n + N_{n+1})]}{\sum N}$$

A method considering the orientation of the drainage network has been proposed based upon the measurement of stream junction angles (Lubowe, 1964). Generalisation of the drainage pattern is the basis for measuring the junction angles between streams of different orders and the population of angles may be compared for different orders. Entrance or junction angles were defined as angles projected in the horizontal between the average flow direction, defined by the ends of stream segments at their junctions to an upstream point 0·2 times the average length of second-order streams (Figure 2.6C). Lubowe (1964) showed that the mean junction angle increases as the order of the receiving stream increases.

Consideration of the overall character of the drainage network is possible using flow direction frequency distribution parameters initially devised by Shykind (1956) for analysis of submarine canyons but later applied to subaerial drainage patterns (Figure 2.6C). This entails inserting the flow direction for each order segment by joining the head and the mouth of the segment, and measuring the aximuth of the flow direction of the highest order stream in the basin (Figure 2.6C). This method does not give results substantially different from the vectorial method and in both cases a dendritic pattern will produce a unimodal frequency distribution whereas a structur-

Plate 5 Braided river in basin in Aleutian Range, Alaska
Dimensions of river channel and river pattern may be adjusted to events which occur during a small portion of the year. See p. 257 and Figure 1.7.

ally-controlled pattern could produce a bimodal distribution. Difficulties confronting the precise quantitative description of basin shape derive partly from the three-dimensional nature of the basin and partly from the fact that the measures proposed hitherto (Figures 2.6A, B, C; Table 2.4) have focused either upon the network or upon the basin outline form. Moment measures have been proposed by Ongley (1970) as superior to the basin vectorial axis (Ongley, 1968; Abrahams, 1970) and involve linking each source to the mouth of the basin and then evaluating the frequency distribution of the vectorial quantities. The *basin moment axis* can then be obtained by plotting the mean direction through the basin mouth.

The pattern of channel reaches has similarly been classified into qualitative groups and later supplemented by measurements of sinuosity (Figure 2.4). Channel patterns are qualitatively referred to as straight, where length is seldom more than ten times the channel width (Leopold and Wolman, 1957); braided where the stream reaches bifurcate into two or more anastomosing channels separated by islands or bars; and meandering where the channel is in the form of a number of sweeping curves (Figure 2.6D). In addition to these three patterns Dury (1969) has distinguished deltaic distributary patterns, which may often have been ignored compared with confluent patterns (Allen, 1965), anabranching where distributary streams flow parallel with the trunk stream and often unite after several kilometres, reticulate channels which are similar to braided channels in pattern but are markedly ephemeral in flow (Figure 2.6D), irregular and straight-simulating channels. Parameters of meander dimension (Figure 2.6D; Figure 5.3) have frequently been described and Leopold and Wolman (1966) used the term arc distance to refer to the length measured along the centre line of the stream channel from one point of inflection to the next, whereas sinuosity is the ratio of arc distance to half the meander wavelength.

Quantitative description of braided channels has proved more elusive although Brice (1960, 1964) devised a braiding index (B.I.) which was based upon the sum of the lengths of islands or bars in a particular reach (l) and the length of the reach measured midway between the banks (m) and expressed as B.I. = 2l/m. A value greater than 1.5 was shown, by comparison with field situations, to indicate a braided channel. A difficulty confronting the use of such indices is that the index must represent one section of the stream course or be an average of several meanders or braids. Carlston (1965), for example, first measured only one meander which approximated most closely to the ideal meander form in each area, but later he adopted the mean value of measurements of six or eight meanders. Measurements can be made from large scale topographic maps or from air photographs but the recurrent problems of date, scale, definition and convention are qualifications that often necessitate measurements of channel pattern in the field (Plate 5).

Such indices do not easily express the variation in channel shape along the length of the stream, they require subjective decisions for their application, wavelength is measured along straight lines, and the possibility of a wavelength of more than one amplitude is ignored. Speight (1965) therefore recommended the application of spectral analysis to the plan of a river. This involves measurement of a number (n) of changes of angle of direction of the river, measured at standard distances (l) along the river course to indicate variations in meander intensity X (k) where:

$$X(k) = \frac{1}{n}\left[C_x(0) + \sum_{l=1}^{m=1} \left\{ C_x(l) \left(1 + \cos \frac{\pi l}{m} \right) \cos \frac{\pi k l}{m} \right\} \right]$$

where   l = distance of sampled points apart
              C_x = autocorrelation between successive angles x
              m = number of frequency bands
              n = number of direction angles between regularly spaced points
                  on thalweg trace.

This method has the advantages that it does not merely recognise a single dominant wavelength of meandering in a particular stream pattern, it is not subject to operator variation, and it can be employed easily with computer techniques. Several spectra may thus be detected and in Figure 2.6F two intense peaks are indicated. The efficacy of spectral analysis depends upon the number of points analysed, their spacing and the number of frequency bands used in producing spectrum m. However the technique does afford a means of characterising the pattern of the river in a succinct manner capable of contrasting different rivers or the same river at different periods.

Difficulties confronting the measurement of channel pattern arise partly from the fact that the channel pattern changes with discharge and whereas bars may be apparent in a braided channel at low flow they can be covered at high flows and so the braiding index is time-dependent. On a longer time scale changes in meander position indicate that measurements of meanders are equally time-dependent (Chapter 5) and this is also pertinent in the case of the cross section of the stream channel. The shape of the cross section involves measurements of width and depth of the channel (Figure 2.6G) which are closely related to measurements of size or area of the channel cross section. For any particular cross section it is possible to measure the shape of the entire channel or merely the wetted perimeter at a specific time. The width of the river channel at the bankfull stage is not always easy to identify because the bankfull level may be altered by man-induced modifications of the channel, levées may necessitate a decision as to whether to use the height of the levée top or that of the flood plain, and the level of the flood plain may itself also vary in height. Vegetation lines, indicated by moss or lichens, the upper limit of sand mixed with river boulders, the lower limit of herbs and forbs, and flood debris of sticks and trash may all assist to indicate a flood level (Leopold and Skibitzke, 1967). The bankfull channel width or the water surface width can easily be measured with a tape stretched taut across the channel, but the cross sectional area can most easily be levelled and such measurements may often be required during the course of discharge measurements at a rated section (p. 143). Mean depth is most easily obtained by dividing cross sectional area by the appropriate width measurement, or, alternatively, as the mean of a number of depth measurements. The most likely channel cross sectional form is parabolic but in practice many river channels are trapezoidal in straight stretches, more asymmetric at river bends, and composed of several channels in braided segments. Perhaps for this reason there have been few attempts to express the shape of the cross section of the river channel in the way that shape indices have been developed for the basin, the network and the channel pattern. One index, sometimes employed in relationships with stream channel slope, is the width/depth ratio which compares the channel width with the mean depth as an index of channel form. The detail of the channel cross section may be described as the roughness. This often has to be defined because the estimation of discharge of major floods, evaluated from formulae for water flow in open channels (pp. 127–9) often necessitates the inclusion of a value for channel roughness. The roughness coefficient of a particular channel may be determined by inspection and Barnes (1967) has provided fifty examples of photographs, channel cross sections and roughness coefficients as a standard.

2.1e Channel, network and basin relief

Studies of the hydraulic geometry of stream channel cross sections (p. 239) in addition to employing measurements of width, depth and area, often include measurements of channel slope at a particular station and these parameters together have been employed in at-a-station analyses (Leopold, Wolman and Miller, 1964). To consider the relief aspects of the basin it is, therefore, possible to proceed from the cross section, to the reach, to the network and the basin.

Along the course of a non-meandering river channel there is usually an alternation of deeps or pools with shallows or riffles which are often spaced fairly regularly at distances approximately five to seven times the channel width. In meandering channels the same characteristic obtains, more expectedly, because deep pools occur in asymmetric sections at river bends. This variation must be borne in mind when deriving the channel slope at a particular cross section and in addition it is necessary to distinguish between the slope of the bed of the channel and that of the water surface. The slope of either may be measured by level or by telescopic alidade and expressed in units of metres fall per metre distance. According to Leopold and Skibitzke (1967) the length of the reach over which slope is measured should be not less than twenty and usually between thirty and forty times the channel width.

Measurements of slope along the course of a river can be made by an extension of this method whereby the slope of uniform reaches is levelled and the long profile of the river is plotted, often with a vertical exaggeration. In some studies the long profiles may be derived from the contour information on topographical maps but in this case the long profile of the valley can be drawn by taking the shortest distance along the river, or, alternatively, the long profile of the channel can be computed by following the bends of the river course. Where field survey is employed it is necessary to decide between the long profile of the bankfull, or some other water stage, and the profile of the river bed, which may be required for studies of bedload movement. Since the nineteenth century it has been appreciated that the long profile of a river course tends to be concave upwards and theoretically tends to approach a gradient of zero at the level, often denoted as base-level, towards which the river profile is developed. The tendency for distribution of elevation (H) as a dependent variable to vary with distance (D) as an independent variable has warranted attempts to describe the river profile by equations. Broscoe (1959) identified four equations:

the simple linear form, $H = a - bD$, which is a poor fit because of the upward concavity;

the exponential form, $\log H = a - bD$;

the logarithmic form, $H = a - b \log D$;

the power form, $\log H = a - b \log D$.

Different authorities have found the latter three to give good approximations although difficulty may arise in selecting a meaningful reference point from which the equation constants are derived. This can be accommodated by replacing D by $D+c$ (where c is a constant) so that when $D = o$ the equation can still be solved for H. These equations have been reviewed by Tanner (1971) who considered other forms including a Devdariani heat flow analogy based upon the application of Fourier analysis to river profiles, as attempted earlier by Culling (1960).

These methods are complex for some purposes and because certain profiles may be composite and consist of several concave portions to which an equation may be fitted, simpler indices have been proposed. Horton (1932) used the average fall or slope of streams, obtained by dividing the difference in elevation between the source and the mouth by the length of the stream, and this was applied to the mainstream by Taylor and Schwarz (1952) as their slope factor or equivalent mainstream slope. This method provides a useful index but it can give the same value for a variety of long profiles.

Modification of this average main channel slope of the trunk stream of the basin can be used as an index of drainage basin slope, and this may be effected by extending the line of the trunk stream to the watershed of the basin and calculating average slope along the trunk stream and its extension. Expression of the slope and relief aspects of the basin is complicated further by the fact that one is often attempting to express three dimensional variation in a very simple index. Numerous solutions have been proposed (Table 2.5) and they may be visualised as simple indices of relief, as complex indices, as specific measurements of slope, and as mapping methods. Simple indices have the merit of ease of calculation, they have been used with considerable success and Relief Ratio is most frequently employed to express the relief aspect of a particular basin. More complex indices have been developed such as ruggedness number (Table 2.5) but the utility of these is somewhat restricted by the fact that a single value can also represent a variety of conditions. Selected slope measurements, often derived from the field, have been made (Table 2.5) and measurements of maximum valley side slope angle can give a population of data useful in the analysis of small drainage basins.

Few of these methods include the facility to represent the way in which slope or relief is distributed throughout a particular basin and this has been accommodated by slope and relief maps developed expressly for the purpose. The simplest and most frequently used method of producing a slope map is from a topographical map by dividing the map into areas of uniform slope according to contour spacing and then applying a system of slope categories to the map (Raisz and Henry, 1934) but other methods are available (Zakrzewska, 1967). In cases where a detailed knowledge is required, as in the case of small basins, field mapping is required and morphological mapping may be used in the field (Savigear, 1965; Gregory and Brown, 1966). Field-based and map-based methods are time-consuming and have the disadvantage that no single quantitative index can easily be derived from the maps. In the case of small basins with a fairly simple contour pattern it may be possible to fit landform equations (for example, Troeh, 1965) or more quantitative mapping methods may be used as employed to describe the Gladefield catchment land systems (Speight, 1969).

2.2 Rock and sediment characteristics

Complete description of the characteristics of the drainage basin must include reference to the rocks and sediments beneath the basin. This is necessary because the type of rock underlying the basin will determine the nature and extent of groundwater storage and also the type of material which is available for erosion and transport within the drainage basin. Geology has provided numerous techniques for describing the character of rocks and similarly pedology has devised methods of soil description and both sciences have produced maps at a variety of scales. Such maps are, necessarily, governed in their production by the needs of the disciplines from which they arise and so the bases of classification of rock type, soil types and superficial

Table 2.5 Some methods available for expressing relief and slope characteristics of the drainage basin

Type of Expression	Name	Derived by	Source
Simple index	Mean slope $Sg = 1 \cdot 571 \dfrac{DN}{\Sigma I}$	where D is contour interval, N is number of contours crossed by grid of squares and ΣI total length of subdividing lines	Horton (1932)
	Maximum basin relief = H	Difference between highest and lowest points in basin. Alternatives afforded by (i) basin relief along longest dimension of basin parallel to principal drainage line; (ii) basin relief along basin diameter	Strahler (1952) Schumm (1956) Maxwell (1960)
	Relief ratio R_h	$R_h = H/L$ where L = horizontal distance along longest dimension of basin parallel to principal drainage line	Schumm (1956)
	Relative relief R_{hp}	$R_{hp} = H/P$ where P = basin perimeter. Alternative afforded by $\dfrac{\text{Maximum relief}}{\text{basin diameter}}$	Melton (1957) Maxwell (1960)
Combined indices	Ruggedness number HD_d Geometry number $\dfrac{HD_d}{Sg}$	where D_d = drainage density where Sg is the tangent of ground slope in degrees	Strahler (1958) Strahler (1958)
Selected slope Measurements	Stream channel slope θ_c Maximum valley side slope θ_{max}	Fall per horizontal distance or degrees Measure at intervals along valley walls, often 50 or 100 measurements in each basin	Horton (1932) Strahler (1950)
	Dihedral angle between valley sides		Melton (1957)

deposits on published maps are frequently based primarily upon genesis which is not necessarily the most relevant and appropiate basis for the fluvial geomorphologist. Ancillary techniques therefore have to be utilised to express the character of rocks and sediments relevant to the form and function of the drainage basin. Such techniques have been incompletely explored by geomorphologists compared with the techniques denoting topographic characteristics of drainage basins or describing the fluvial sediments within the basin. Techniques are here treated to describe the characteristics of rock, of superficial deposits and soils, and of fluvial sediments.

2.2a   Rock characteristics

More relevant than the subdivision of rocks into sedimentary, igneous and meta-morphic is the distinction between unconsolidated and indurated materials (Waltz, 1969). Unconsolidated do not include cementing materials in their pore spaces and these usually include most recent and superficial deposits such as aeolian sediments, whereas indurated materials have fewer pore spaces and are more effectively ce-mented. Expressions of the character of such sediments are available in porosity, permeability, and aquifer properties of deposits. Porosity can be defined as the volume of voids in a particular material divided by the bulk volume of the material, and so provides an index of the amount of water which can be held by a particular deposit. Permeability is a measure of the capacity of a material to transmit water under pres-sure and varies according to grain size, shape, variation and packing. An aquifer is a rock which yields significant amounts of water to wells, and this does not neces-sarily reflect porosity or permeability of the material, because a rock which is itself impermeable and not porous can be a good aquifer due to its well-developed jointing. An aquifer is often defined in relation to particular requirements such as domestic or agricultural water supply. In contrast an aquiclude is an impermeable formation which may contain water but is not capable of transmitting significant quantities and an aquifuge is an impermeable formation which neither contains nor transmits water.

Unconsolidated materials usually have high porosities, often between 25 and 65 per cent, but they can be as low as 20 per cent in coarse, poorly-sorted muds and as high as 90 per cent in soft muds and dry organic material. A considerable range of porosity can be found in indurated rocks although sedimentary rocks usually have the highest, between 5 and 25 per cent. Fresh metamorphic or igneous rocks are often less than 3 and frequently less than 1 per cent, although unfractured rocks can show a range from less than 1 to more than 85 per cent in the case of pumice. The porosity of a sample of rock can be found by a laboratory method which requires measurement of the weight of the oven dry sample in gms (w_s), the total volume or mass of the sample (v) in cc, which can be used to give the dry unit weight ($d = w_s/v$). The porosity (P) in per cent can then be obtained as

$$P = \frac{porespaces}{total\ volume} \times 100 = \frac{specific\ gravity - d}{specific\ gravity} \times 100$$

Permeabilities can vary to a much greater degree and in unconsolidated sediments the highest can be more than 10^9 times greater than the lowest, and indurated rocks, such as sedimentary ones which have relatively high porosities, can have very low permeabilities. Permeability or hydraulic conductivity can be measured in the labora-tory or in the field. In the laboratory it is usual to employ a permeameter on a small sample, which entails observing the rate of percolation of water through a sample of known length and cross-sectional area under a known difference in head (Morris and Johnson, 1967). Constant head permeameters are used for samples of medium to high permeability, and variable head permeameters are used in the case of samples of low permeability. The basic law for the flow of fluids through porous media was expressed by Darcy as:

Q = k i A where Q = quantity of water discharged in a unit of time
A = cross-sectional area of sample through which water percolates
i = hydraulic gradient, i.e. difference in head divided by length of flow L
K = coefficient of permeability for material for water.

Therefore $$K = \frac{Q}{iA} \text{ or } \frac{QL}{hA}$$

Alternatively, field measurements may be employed using either pumping tests or tracers including dyes and salts, which are inserted into ground water and the time taken to travel over a known distance is measured. The field method, although often difficult to employ, overcomes the disadvantages attendant upon the use of disturbed samples in laboratory measurements which may give values of permeability bearing little relation to field values.

Two other parameters which may be used to express the character of the rock underlying a particular drainage basin are provided by specific retention of a rock and the specific yield. The specific retention is defined as the percentage of its total volume occupied by water that cannot be drained by gravity and will not be yielded to wells, and the specific yield of a rock is the percentage of the total volume occupied by water that will drain under gravity to wells. Therefore,

specific yield = porosity — specific retention

and in the case of a fine-grained sandstone Morris and Johnson (1967) illustrate that the porosity may vary between 13·7 and 49·3 per cent, the specific retention can vary from 1·2 to 30·8 per cent, and the specific yield can vary between 2·1 and 39·6 per cent. Values of porosity and permeability for the major rock types are summarised in Table 2.6. A drainage basin is frequently underlain by more than one type of rock

Table 2.6 Some average ranges of porosity and permeability (*compiled from various sources including Todd (1970); Waltz (1969)*)

Rock		Porosity (%)	Permeability m/day
Unconsolidated:	Clay	45–60	10^{-6}–10^{-4}
	Silt	20–50	10^{-3}–10
	Sand	30–40	10–10^4
	Gravel	25–40	10^2–10^6
Indurated:	Shale	5–15	10^{-7}–10
	Sandstone	5–20	10^{-2}–10^2
	Limestone	1–10	10^{-2}–10
	Conglomerate	5–25	10^{-4}–1
	Granite	10^{-5}–10	10^{-7}–10^{-3}
	Basalt	10^{-4}–50	10^{-5}–10^{-2}
	Slate	10^{-4}–1	10^{-9}–10^{-6}
	Schist	10^{-4}–1	10^{-9}–10^{-5}
	Gneiss	10^{-5}–1	10^{-9}–10^{-6}
	Tuff	10–80	10^{-6}–10^{-2}

and so in these cases it is usual to indicate the percentage of each rock underlying the catchment.

2.2b *Superficial deposits and soils*

The superficial deposits and the solum overlying the solid rock are also often described in terms of the percentage of several types extending over the area of the drainage basin. For example, Comer and Zimmermann (1969) compared two basins, in the larger of which poorly and very poorly drained soils occupied 44 per cent of the basin area, whereas in the other basin they were responsible for only 22 per cent of the area. However, many characteristics of the soil mantle may be required as a basis for more quantitative assessment of the soil character. Such characteristics include soil texture, organic matter content, grain size, density of grains, particle shape, porosity and drainage characteristics, shear strength, and cohesiveness of particles due to surface forces. Soils can be divided into frictional and cohesive soils (Sundborg, 1956) which occur at the two extremes of a spectrum of variation and grade into one another. In coarse material the forces resisting separation are frictional but in fine material, containing a high clay content for example, cohesive forces dominate. The amount of information collected on soils depends upon the scale and scope of the drainage basin investigation but may be taken initially from available topographical maps. Information on published maps is frequently not directly relevant to the drainage basin study, when a classification distinguishing soil types according to texture, coarse and fine-grained, may be more appropriate than a genetic classification. Therefore it may be necessary either to investigate certain soil properties specifically and perhaps to map these in a way relevant to the study of the drainage basin or, alternatively, it is necessary to employ existing information, develop it and depict the results in a way directly relevant to the fluvial geomorphologist.

An almost bewildering array of soil properties is available for field and laboratory measurement but perhaps these can be most easily grouped as those expressing the character of soil or superficial deposit at a particular time and those which attempt to indicate the way in which a particular soil may change or be changed as conditions affecting it alter. Soil moisture content is illustrative of the first category and indices of soil erodibility are representative of the second. In describing the nature of soils, determinations have to be made at a number of individual sites and this introduces a sampling problem laterally and vertically. It is usual to select a number of sites for detailed investigation, perhaps by random selection, or by stratified random sampling if several terrain types or soil units are expected. At each site a sample is usually taken of the soil profile by a corer which can give an undisturbed core of the soil which can be placed in a sealed container and conveyed to the laboratory for analysis.

Soil is composed of mineral particles, organic matter, air and moisture and it may be necessary to know the size distribution of the first, and the percentage occurrence of the other two. These two parameters can be analysed for a soil sample under particular conditions. The particle size distribution of a sediment can be ascertained by first passing a dried sample of known weight through a series of sieves each of known graduation so that the percentage by weight retained in each sieve can be calculated. Sieves can be used for the material coarser than 0·06 mm diameter but for finer grades other methods involving hydrometers, pipette or sedimentation balance are available. The percentage in each size category (Table 2.7) can be used to plot a

Table 2.7 Particle size distribution

Description	Range of particle diameter (mm)		
Boulders		> 256 mm	
Large cobbles	256	−128	
Small cobbles	128	− 64	
Very coarse gravel	32·0	− 64·0	
Coarse gravel	16·0	− 32·0	
Medium gravel	8·0	− 16·0	
Fine gravel	4·0	− 8·0	
Very fine gravel	2·0	− 4·0	
Very coarse sand	1·0	− 2·0	ϕ value (for
Coarse sand	0·5	− 1·0	0 maximum in
Medium sand	0·25	− 0·5	+1 each case)
Fine sand	0·125	− 0·25	+2
Very fine sand	0·0625	− 0·125	+3
Coarse silt	0·031	− 0·062	+4
Medium silt	0·016	− 0·031	+5
Fine silt	0·008	− 0·016	+6
Very fine silt	0·004	− 0·008	+7
Coarse clay	0·0020	− 0·004	+8
Medium clay	0·0010	− 0·0020	+9
Fine clay	0·0005	− 0·0010	+10
Very fine clay	0·00024	− 0·0005	+11

cumulative frequency curve which illustrates the size frequency of the particular sample. Such cumulative size frequency distributions will illustrate the relative proportion of different size fractions in a soil and so often reflect the texture of the soil. An illustration of the diagram is provided for coarse material in Figure 2.8 (see pp. 68–70). The organic matter content can be determined in various ways which include boiling a weighed, oven-dried soil sample in a 6 per cent solution of hydrogen peroxide (Baver, 1948, 49) and then reweighing to ascertain how much organic matter oxidised. Field moisture content is an important parameter and can be derived in a number of ways, both directly and indirectly (p. 107).

Moisture content exerts a significant influence upon the properties and function of soils. The moisture content per oven-dry weight or per unit bulk volume is designated field capacity (FC) when moisture is largely retained by capillary, rather than by gravitational, forces (see p. 283). When plants extract soil water to such an extent that the plants begin to wilt, the moisture content has reached wilting point and the difference between this value and field capacity is the plant-available water. Soils may also be characterised by Atterberg limits including their liquid, plastic, and shrinkage limits and their plasticity index. The liquid limit is the moisture content, as a percentage of the oven dry weight, at which a soil-water mixture will just flow, whereas the plastic limit refers to the moisture content at which a soil–water mixture may be moulded. A range of techniques is now available (British Standards Institution, 1968; Akroyd, 1958; Capper and Cassie, 1968) which could be employed in future drainage basin studies. Complete description of the soil may require assessment of soil

Table 2.8 Examples of indices of soil erodibility (*based upon R. B. Bryan, 1968*)

Indices based upon properties affecting soil dispersion:

Dispersion ratio	Comparing silt + clay content in undispersed sample with that in sample treated with dispersing agent	Above 15% erodible soils Below 15% nonerodible soils	H. E. Middleton (1930)
Surface aggregation ratio	Ratio between total surface area of particles larger than 0·05 mm diameter and quantity of aggregated silt-clay	Soils on acid igneous rocks 164 Soils on basalt 59	H. W. Anderson (1954)

Indices based on soil properties affecting both soil dispersion and water transmission:

Erosion ratio	$\dfrac{\text{Dispersion ratio}}{\text{colloid content/moisture equivalent ratio}}$	Erodible soils above 10 Nonerodible soils below 10	H. E. Middleton (1930)
Soil erosion	$\dfrac{\text{Constant of proportion} \times \text{ease of dispersion}}{\text{Absorption} \times \text{permeability} \times \text{size of particles}}$		L. D. Baver (1933)
Index of erodibility	$\dfrac{1}{\text{mean shearing resistance} \times \text{permeability}}$		R. J. Chorley (1959)

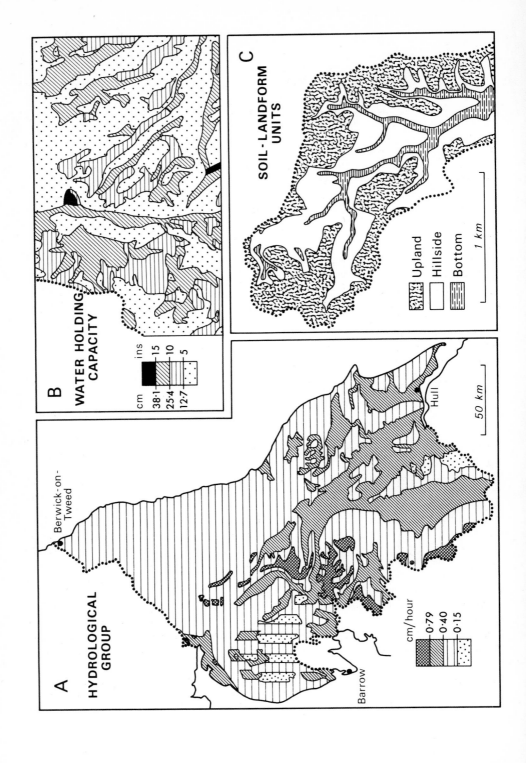

A HYDROLOGICAL GROUP

Berwick-on-Tweed

Hull

Barrow

cm/hour
0·79
0·40
0·15

50 km

B WATER HOLDING CAPACITY

cm   ins
38·1   15
25·4   10
12·7   5

C SOIL-LANDFORM UNITS

Upland
Hillside
Bottom

1 km

strength and Melton (1957) employed a 7·26 kg (12 pound) iron shot, dropped a standardised distance to impart 27·2 kg (60 pounds) of energy normal to the ground surface. The soil strength at a particular site was derived from the measurements of the diameter of the imprint which could be used to give the volume of the imprint. A further method to indicate the resistance of some soils to shearing stress is afforded by a penetrometer which estimates the force required to cause penetration by some type of pointed rod or small foot to a given depth at a specified rate. The cone penetro-meter employs a 90 cm steel shaft tipped by a 30° cone. At the upper end of the shaft is a proving ring in which a strain gauge is mounted. As the cone is forced into the soil the ring is deformed to an ellipse and the degree of strain is registered on the strain gauge (Strahler and Koons, 1960).

Further methods of describing soil in a manner relevant to drainage basin charac-teristics include various indices of soil erodibility which have been grouped by Bryan (1968), firstly into those indices which relate to dispersion of soils and, secondly, into those indices which involve consideration of dispersion of soils and water transmission properties. The percentage amount of silt and clay present in a dispersed state can, as an example of the first type, be evaluated by comparing the amount of silt and clay in an undispersed sample with that in a sample previously treated with a dispersing agent. These two results combined into a dispersion ratio expressed as a percentage were found (Middleton, 1930) to be generally above 15 per cent for erodible soils and below 15 per cent for non-erodible soils. Indices of the second type include water transmission properties as well. Chorley (1959), for example, used the physical pro-perties of moisture content (m), grain size, range of grain size (r), permeability (p) and shearing resistance (s) to give an index of resistance (I_r) in $I_r = \dfrac{\text{soil density} \cdot r}{m}$

and also to give an index of erodibility (I_e) in $I_e = \dfrac{1}{s \cdot p}$. Bryan (1968) concluded that of the various indices proposed (Table 2.8) no single index hitherto combined the merits of simplicity of measurement with reliability in operation and universal application.

In the moorland region of the Peak District, England, Bryan (1969) collected samples from a variety of soils and tested them under simulated rainfall conditions to assess their relative erodibility. The results of these tests using an artificial rainfall simulator permitted estimates to be made of the annual soil loss under specified condi-tions and indicated variations over the area ranging from values near zero to values as much as 395 tonnes/hectare/year. Using 27·2 tonnes/hectare/year as the threshold value for soils liable to serious erosion it was estimated that 58 per cent of the soils were vulnerable and so the percentage area of the catchment susceptible to serious erosion could be quantified.

An outstanding problem is that soil maps are often required for a particular catch-ment in a form appropriate to drainage basin processes. Attempts have, therefore, been made to overcome the limitations of published maps and three maps are illus-trated in Figure 2.7. Edmonds, Painter and Ashley (1970) devised a semi-quantitative

Figure 2.7 Three approaches to the characterisation of soil types relevant to the drainage basin Northern England (A) is based upon Edmunds, Painter and Ashley (1970), a watershed in Nebraska (C) follows England and Holtan (1969), and B is based upon Stephenson and England (1970).

hydrological classification of soils in north-east England (Figure 2.7A). They employed two larger scale maps and classified the soil associations into four main groups according to relative minimum infiltration rates. Although each group was subdivided into two minor groups the four-fold classification is illustrated in Figure 2.7A. Once this hydrological classification of soils had been obtained an average minimum infiltration index was determined for 26 particular catchments according to the area of each soil group included in the catchment. At a different scale England and Holtan (1969) distinguished three groups of soils in an experimental watershed near Hastings, Nebraska. They distinguished (Figure 2.7C) relatively uneroded upland soils, more severely eroded hillside soils including all soils of the steeper slopes, and depositional bottomland soils occurring on footslopes or along stream channels. Knowing the proportion of the catchment occupied by each soil-landform group and applying the exhaustion concept of infiltration it was possible to utilise this grouping as a measure of hydrological performance of the soils of the watershed (Table 2.10). The production of such maps, although very desirable and worthwhile, can be very time-consuming. Accordingly, Stephenson and England (1970) extracted data from existing geological, soils, topographic, range-site and isohyetal maps for a 232 km² Reynolds Creek experimental watershed and at each of 2,565 sample points digitised values of locating co-ordinate, soils mapping unit symbol, elevation, slope tangent, slope aspect, geologic formation, precipitation, and range-site designations. For each sampling point an index of total profile porosity or of water-holding capacity was computed according to soil texture and solum depth given in soil profile descriptions and this index, ranging from 1·25 to 45 cm of potential water-storage capacity was mapped as illustrated in Figure 2.7B. The opportunities presented by such mechanical automated methods can be of considerable future significance, particularly in relation to further integrated attempts of describing basin characteristics (pp. 76–7).

2.2c Characteristics of fluvial sediments

Fluvial sediments can be classified according to their mode of deposition, into alluvium, crevasse splays, point bars, for example, but it is necessary also to express the character of those deposits in the bed, and in the banks of the stream channel. Parameters selected for this purpose should be capable of expressing variation along a particular river course. Techniques appropriate to these needs have usually been directed to the composition, the components and the shape of the components in a particular deposit, and some of the techniques of soil description (pp. 63–4) may be employed.

Samples of the sediment in the bed and banks of a stream channel may be collected and their particle size composition analysed (pp. 63–4) according to the established categories shown in Table 2.7. The result for each sample needs to be plotted or expressed in a convenient form and the cumulative frequency curves (Figure 2.8) cannot easily be compared from one sample to another. Schumm (1961) found that median grain size and absolute grain size values were not as useful as the silt-clay ratio

Plate 6   River bed and channel forms
Illustrating potholes, channel bedload, waterfall over Carboniferous limestone at Aysgarth Falls, Yorkshire, and the gorge of the Kamienna River on the margin of the Krknose Mountains, Poland.

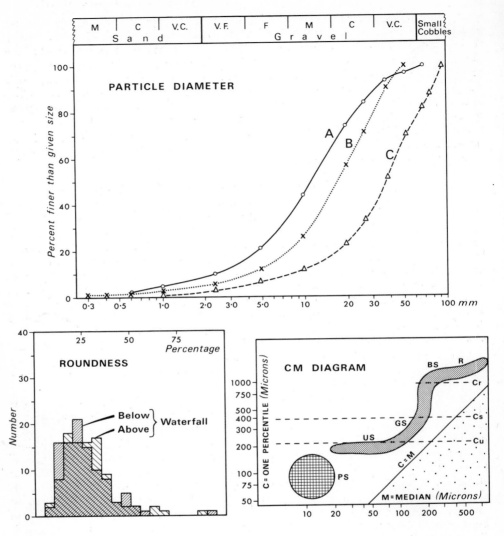

Figure 2.8 Methods describing fluvial sediments

The particle-size curves are plotted from three samples of flint bedload collected in a bedload trap in a small catchment in east Devon following discharges of 0·072 m³/s (A), 0·026 m³/s (B), 0·032 m³/s (C). The weight of bedload in each case was 11·21 kg (A), 3·51 kg (B), and 5·38 kg (C). The roundness (2r/L) of two samples of Yoredale sandstone is compared above and below Aysgarth Falls, Wensleydale, Yorkshire. The CM Passega diagram is derived from Royse (1968) and explained in text.

which appears to be more indicative of some physical properties of the sediment. An advantage of the silt-clay ratio, defined as the percentage of the total sample which passes through a 200-mesh sieve (smaller than 0·074 mm; see Table 2.7), is that the boundary between the coarser fraction and the silt-clay fraction generally corresponds to the lowest practical limit of sieving in particle size analysis. In some cases it is necessary to derive a value of the silt-clay ratio for a complete channel cross section and so several samples can be taken from the channel bed and banks and, according to channel width, some ten to twenty point samples may be selected from the cross section of the channel. Schumm (1960) used the weighted mean percentage silt clay for each cross section (M) as a parameter to express the character of the sediment forming the perimeter of the channel where:

$$M = \frac{\begin{array}{c}(\% \text{ silt-clay in composite channel bed sample} \times \text{channel width}) + \\ (\% \text{ silt-clay in bank samples} \times 2 \text{ channel depth})\end{array}}{\text{channel width} \times 2 \text{ channel depth}}$$

An alternative method employed by Royse (1968) utilised CM relationships devised by Passega (1957, 1964) in which a plot is made of median grain size (M) against the smallest particle size in the coarsest one percentile (C) of the size distribution (Figure 2.8). In this case M is representative of the total range of grain sizes in the sample and C is representative of the minimum competence of the transporting agent. Plotting M against C in a scatter diagram can be employed not only to distinguish types of deposit, between beach and river sands, for example (Friedman, 1967), but they can also be employed to distinguish rolling (R), bottom suspension (BS) and rolling, graded suspension (GS), uniform suspension (US) and pelagic suspension (PS) fractions of the fluvial environment. Passega (1964) illustrated the distinction of these fractions for the Mississippi.

In some cases it is necessary to express the character of the bedload in a particular channel cross section, especially where coarse bed load occurs and where the bed load is affected by contemporary river flows but where the material of the banks is not. A method for sampling coarse bedload has been outlined by Wolman (1954) and applied in his study of Brandywine Creek (Wolman, 1955). A sample may be collected, for example, by using a procedure in which a tape with corks attached is stretched across the stream and stones collected at regular intervals beneath each cork until one hundred have been obtained on a grid-sampling basis. Leopold (1970) has demonstrated that random sampling of surface rocks on a gravel bar may be biased towards the larger sizes and so a weighted sampling procedure can be adopted, based upon conversion of number of stones of a particular type to weight. The fine and coarse particles along the river channel may be arranged in characteristic forms and so deposits may be classified according to these forms (p. 237).

Two outstanding characteristics of fluvial sediments worthy of attention are mineralogical composition of the sediments and the size and shape of the individual constituents. In some situations the composition of the sediments expressed in terms of particular minerals can provide valuable information on the way in which particular rocks or minerals react to transport by fluvial processes. Ljunggren and Sundborg (1968) studied black sand deposits in the valley of the River Lule Alv and employed heavy mineral analysis.

Table 2.9 Some indices for expressing particle and pebble shape

L = major axis (length) of pebble; l = minor axis (width); E = breadth of pebble; r = smallest radius of curvature

COARSE MATERIAL : PEBBLES

Index	Derived as	
Flattening F	$F = \dfrac{L+l}{2E}$	Cailleux (1947)
Roundness R	$R = \dfrac{2r}{L} \times 1000$ where r = smallest radius of curvature measured in main plane of pebble which contains L and l	Tricart and Shaeffer (1950)
Index of rounding P	P = % of smoothed convex circumference of the pebble obtained by careful visual estimation 0–10% angular 41–60% rounded 10–20% subangular over 60% well-rounded 27–40% subrounded Coefficient of variation used to indicate homogeneity of sample	Luttig (1962)
Index of flattening I_F	$I_F = E/L \times 100$	

Total angularity A	Angularity of single corner $= (180° - a)\dfrac{x}{r}$
	
	Where a = measured angle
	  x = distance of tip of corner from centre of maximum inscribed circle
	  r = radius of maximum inscribed circle
	Total angularity A = sum of values for all corners
		Lees (1964)

Morphological index M	Where $a = L/l$, $b = \dfrac{2r}{l}$, $c = E/l$ Then $3K = a+b+c$ and K indicates extent to which pebble departs from spherical form because in a sphere $a+b+c = 3$	Rivière and Ville (1967)

FINE MATERIAL

Index	Derived as	
Roundness	Visual estimation chart. Assign each to appropriate category	Krumbein (1941)
Roundness Ro	$Ro = \dfrac{\text{Average radius of corners and edges}}{\text{Radius of maximum inscribed circle}}$	Wadell (1932)
Zingg shape ratios	Assign to 4 classes according to values of l/L and E/l	Zingg (1935)
Index of roundness X	$X = r/R$ where $R = 1/3\,(L+l+E)$	Ouma (1967)

Size and shape of individual sediment particles are an important way of expressing the character of material at a particular cross section and of facilitating comparison between cross sections and between rivers. Size is frequently expressed by the three main axes of the particle of the pebble, namely, length, width and depth, and samples can be used to indicate the range of dimensions in a particular sample. Indices of particle shape have been devised to allow the quantitative distinction to be made between contemporary and fossil deposits representative of different environments but in the context of the stream channel they can also provide information useful to the study of the mechanics of bed transport, and of transport capacity for different types of material. The problems of describing a three-dimensional particle by a simple index are reminiscent of those confronting the measurement of drainage basin shape (Section 2.1d). Fleming (1964) has suggested that a particle is described by thirteen partially independent parameters but simpler methods are available and some are summarised in Table 2.9. In each method a sample, usually of one hundred pebbles or particles, is taken at each site, each individual is measured, and the resulting frequency distribution can be used to express the character of the deposit (Figure 2.8). The various indices proposed vary from the simple ones such as those of Cailleux (1947), Tricart and Shaeffer (1950) to the more complex advocated by other workers (Table 2.9). Adequate results have been obtained from the simplest indices and from visual estimation as advocated by Krumbein (1941) and Luttig (1962).

2.3 Vegetation characteristics

As in the case of soil types the vegetation character of the drainage basin needs to be specified precisely and this can often be achieved by distinguishing major land use types such as deciduous, coniferous woodland, heathland, grassland, agricultural land and residential areas, and calculating the percentage of each type within a particular watershed. This method can be used to characterise a single watershed or to compare several drainage basins on the basis of their percentage composition according to several land use types or to compare one watershed at different times as it is affected by the progress of land use change. In some cases it is necessary to know the precise nature of each element in the land use pattern and for this quantitative procedures of vegetation description are available (for example, Greig-Smith, 1964). Such methods can be expedited by using subjective, systematic or random sampling of small areas for investigation and then employing a quadrat at each sample point, of a size depending upon the scale and purpose of the study, in which the species are identified and counted. The point quadrat method allows the measurement of the foliage cover of herbs and low shrubs by a point quadrat frame with pins which are lowered to give a basis for the calculation of percentage cover for each species (Brown, 1954). Association analysis may then be utilised to assign a particular quadrat to a specific vegetation group according to the presence or absence of certain species. Such methods depend upon the presence of species rather than upon their relative abundance but they are more objective than field inspection. These methods can be easily applied to small areas although, even when coupled with air photo interpretation, they are not easily extended to the vegetation mapping of larger areas.

In some studies more detailed information is required in addition to the incidence of particular species and their frequency of occurrence. This can include estimations of density, of cover, and of weight or volume. In the case of tree cover number of stems may indicate density and a stand density index may be calculated, and the area of cover may be indexed by the basal area of trunks. Particularly important in relation to interception studies and to the water balance is the amount of leaf cover and this may be indicated by estimating a leaf area index equivalent to the area of leaves carried above a unit area of ground surface. Measurements of the dimensions of sampled plants may be employed to calculate volumes and techniques are also available to relate volume to dry weight to provide an indication of biomass in specific plots (Newbould, 1967). In addition to records of occurrence and to assessments of the parameters of density, cover and volume, the amount and character of plant litter on the surface and the character of roots beneath the surface may be relevant to drainage basin dynamics. Root depth may be examined and root density calculated by washing an exposed pit face.

In some cases one particular form of land use may be of paramount significance and in this case the percentage of this single type within a basin may be used as an index of land use character. The bare area, which may have an influence upon the amount of soil exposed and prone to erosion, is one such index and Melton (1957) used per cent bare area to compare several watersheds in the south-west U.S.A. (Figure 2.10B). This measure was derived by measuring the exposed bare mineral soil with a measuring tape extended for approximately 15 m over the ground in a randomly oriented direction. The ground was inspected at fifty intervals along the length and if it was devoid of all vegetation cover it was given a count of 1. The number of marks out of 50 was multiplied by 2 to give a percentage value and two such line samples were made at each station surveyed in the basin.

In settled areas the presence and magnitude of cultural modification must be recorded. This can be represented by drainage schemes in forested areas or in farmed lands, and may involve systems of subsurface drains or surface drainage channels and ditches whose density may be estimated. Types of farming may be a significant influence where earth terraces have been constructed, or where practices such as mulching, contour ploughing or strip cropping are employed. In urbanised areas the density of buildings, or more pertinently the density of built-over land or impervious area, and the extent, nature and density of stormwater drains and sewers may be calculated. One study has shown that the proportion of impervious area resulting from different degrees of urban and suburban development can be estimated from population density (Stankowski, 1972). Thus vegetation characteristics and land use may require assessment of types of species, their frequency, density, volume and weight and the identification of diagnostic features of the land use. Records are frequently required of the recent history of land cover. The incidence of logging, of afforestation, of fires, of different degrees of grazing or of various types of farming methods have frequently afforded opportunities to compare drainage basins and therefore to investigate spatial and temporal variations (see p. 345).

2.4   Landscape evaluation and response units

It is apparent from the preceding sections that description of topographic form, rock type and superficial deposits, soil and land use, collectively termed basin characteristics, all need to be integrated in some way. Such integration is necessary, firstly,

Table 2.10 Properties of soil-landform groups distinguished by England and Holtan (1969) for a small experimental catchment (see *Figure 2.7C*)

Soil-landform group	Total Area (ha)	Watershed area (%)	Mean depth to B2 horizon (cm)	Weighted average slope	Average length of overland flow	Storage potential (cm)	Computed Infiltration Initial rate (cm/hour)	Final constant rate (cm/hour)
Slightly eroded upland soils on 0–4% slopes	82·23	44·4	35·6	2·1	80·5	8·6	14·66	0·58
Severely eroded hillside soils on 5–12 +% slopes	86·81	46·7	12·7	8·9	67·7	3·8	5·03	0·25
Alluvial and colluvial deposits	16·57	8·9	61·0	1·0	15·9	14·2	28·88	0·58

because there is a measure of correspondence between many characteristics of the basin and, for example, specific types of soil are often associated with a particular position in the pattern of topography, and are covered by a particular form of land use. Thus the soil-landform groups distinguished by England and Holtan (1969) and illustrated in Figure 2.7C and Table 2.10 are one illustration of this interrelationship and the method devised by Stephenson and England (1970) and included in Figure 2.7B provides one device whereby the interrelationship may be expressed. Secondly, integrated methods are required to attempt to simplify the complex interrelationships and, thirdly, methods capable of showing spatial variation of physical characteristics in map form are a necessary prelude to an attempted understanding of drainage basin behaviour. The significance of such attempts is established increasingly with the advent of new concepts of runoff formation (p. 28) and Amerman (1965), for example, shows that the unit source area refers to a subdivision of a complex watershed and that ideally the subdivision has a single cover, a single soil type and is otherwise physically homogeneous.

Realisation of the need for integrated descriptions of drainage basin characteristics in fluvial geomorphology and in hydrology has been paralleled by similar approaches in other fields of geography and in other disciplines. Such approaches have been grouped into landscape ecology and landscape evaluation by Marosi and Szilard (1964). Landscape ecology involves the physical characterisation of the earth's surface and therefore of its drainage basins and is, accordingly, of direct relevance to the description of drainage basin characteristics. Landscape evaluation involves the description and evaluation of the physical environment with reference to potential use. Although landscape evaluation is not obviously directly relevant it is apparent that many of the methods devised for the evaluation of physical landscape are capable of application to the description of drainage basin characteristics. Progress made towards methods of landscape evaluation has been reviewed in a volume published in Australia (Stewart, 1969) where much work has been done in this field, particularly by the Commonwealth Scientific Industrial Research Organisation (C.S.I.R.O.).

Distinction of units within the physical landscape or in large drainage basins may be achieved in three main ways; by description of the several elements of the physical environment, directly by classification of the physical environment according to process, or by evaluation of the drainage basin or its parts with respect to a particular purpose. Description of the physical environment can be illustrated most easily by mapping those features relevant to understanding of the nature and function of the drainage basin. An impressive scheme of mapping was initiated in Poland in 1950 (Klimaszewski, 1956; Galon, 1964; Wit-Józwik, 1968) based upon field mapping at the 1:25,000 scale followed by publication at 1:50,000. The map aims to show the distribution and cause of hydrographical phenomena and, therefore, to show the interrelationships between hydrographical features and geological structure, relief, climate and vegetation (Galon, 1964). Initially using three, and later seven, colours together with 150 symbols the maps provide information on rock type, including a classification according to permeability; on relief by contours; on the detail of the drainage pattern including distribution of ephemeral, intermittent and perennial streams; together with data on measurement stations, and on groundwater levels.

The spatial variation of drainage basin characteristics is well-evidenced on such hydrographical maps but an alternative method of describing the interrelationships between physical character of the environment has sometimes emerged from studies of the interrelationships between characteristics. Such studies (Section 2.5) utilise techniques of factor analysis, multiple discriminant and cluster analysis to indicate how groups of drainage basin characteristics are interrelated and then apply knowledge of these relationships to distinguish different portions of major drainage basins (for example, Mather and Doornkamp, 1970).

The sum total of drainage basin characteristics can also be characterised according to their combined influence upon basin response. In such studies, measures of drainage basin process discussed more fully in Chapter 5, may be employed to distinguish between different drainage basins in contrasted areas and the differences in response may be attributed partly to differences in drainage basin characteristics. Therefore, Ledger (1964) was able to distinguish hydrological regions in West Africa (Figure 6.12) and Sopper and Lull (1965) distinguished the units which affected streamflow in the north-east of the United States.

The third method whereby the overall character of the drainage basin is described is when it is evaluated with respect to a particular purpose. A method for providing a quantitative inventory of riverscape has been proposed by Leopold and Marchand (1968) which endeavours to characterise river landscape not as a basis for relationship to river or to drainage basin processes but as a possible basis for evaluating choices in river basin development. The method was applied to 24 minor valleys in the vicinity of Berkeley, California and some 26 parameters in three categories pertaining to physical and chemical character, biological character and to human use and interest were derived for each site. A uniqueness value on a scale from 0 to 1·0 was calculated according to the frequency of occurrence of a particular value. Consideration of the total uniqueness scores for each site enabled comparison to be made between the 24 valleys. If a particular characteristic had the same value at all 24 sites the uniqueness was expressed as 1/24 or 0·04 but if one site was the only one to have a particular value its uniqueness would be 1/1 or 1·0. This approach, although as yet preliminary in nature, is of considerable significance in describing the overall character and appearance of a drainage basin and could be developed and employed much more in the future as a basis for decisions about developments taking place within watersheds during the course of river basin development.

Methods of integrating topographic characteristics of drainage basins are necessarily difficult to achieve because the extent and nature of the interrelationships is incompletely understood, the time required for the completion of integrated surveys is appreciable despite the advances in the fields of automated data processing, and the basic information is seldom readily available in an appropriate form. Sufficient progress has been made, however, to demonstrate that further advances will be made in the light of an increased understanding of the relationships which exist between the drainage basin characteristics and also between these characteristics and the processes which they influence and determine.

2.5   Association between drainage basin characteristics

Techniques of landscape ecology and landscape evaluation have been developed partly in response to a need to simplify and to represent the amount of interrelationship which exists between the basin characteristics described above. Interrelationships are present at several levels. Firstly, because certain attributes of the drainage basin may be expressed in different ways there will be relationships between these alternative parameters. Thus a broad correlation should exist between the various indices proposed to express basin shape or basin relief. Secondly, because of the size of different drainage basin units certain associations will exist, for example, between basin area and total length of stream channels. Thirdly, because it is unrealistic to separate the drainage basin system arbitrarily into components there will be association between different basin measures. Therefore on a single rock type, a particular type of network or basin shape may be produced according to the local relief and so a correlation may exist between network shape and basin relief or relative relief. Fourthly, certain relationships may occur between drainage basin characteristics because of their interaction with basin process. Thus particular rock types may be associated with certain types of stream network pattern due to weaknesses in the rock and, equally, certain rock types may be associated with particular densities of streams. This may arise because of the effect of rock type upon soil character and hence upon infiltration and runoff, which in turn is related to stream density. This introduces, fifthly, the possibility that strong associations apparent between characteristics may merely reflect dependence upon a third or fourth intermediate characteristic. For example, a measure of relief could be related to drainage density but it may be rainfall, which is related to relief and also to drainage density, which accounts for the association. Numerous relationships exist between drainage basin variables and the nature and extent of the influence of several independent, upon one dependent, variables pose a problem analysed by a variety of statistical techniques ranging from simple regression to multiple regression, factor analysis and cluster analysis (Doornkamp and King, 1971).

Understanding of the relationships between basin characteristics is necessary because relationships can be compared from one area to another and this knowledge is necessary before extensions can be made spatially or before relations between particular drainage basin characteristics and drainage basin process can be sought. The multi-variate relationships amongst drainage basin characteristics are further complicated when the morphological system is related to process in process-response systems when feedback relationships occur. Process-response systems must be visualised in two ways because process can influence form and form can influence process, at different time scales. The time scales of Schumm and Lichty (1965) are pertinent in this context (Table 1.4), especially in view of the fact that as a result of climatic changes or the influence of man, contemporary processes may not be in harmony with present landforms. The relationships between drainage basin characteristics are, therefore, illustrated briefly in this section and referred to more specifically in subsequent chapters. At each of the three levels, of the basin, the channel and the channel cross section, interrelationships may be discerned and in addition there are relationships between all three system levels. Each of the three systems has been studied in three ways; statically investigating statistical relationships between numbers of variables, stochastically based upon information from statistical sources, and dynamically by appreciating that the characteristics are subject to constant change.

2.5a   Drainage basin relations

The drainage network offers an excellent illustration of this three-fold approach. Within the context of the drainage basin the network was first studied by the laws of morphometry, later these were complemented by stochastic analysis based upon simulated channel networks and more recently awareness of the dynamic character of the drainage net is a necessary pre-requisite for appreciating the effect which the network has upon stream channel process (Section 5.4a).

Horton (1945) formulated two laws of drainage composition. The law of stream numbers was based upon the inverse geometric series which exists between number of streams of each order and order, and the law of stream lengths involved the geometric series between mean length of streams of each order and order. These two laws were supplemented by the law of stream slopes expressed as an inverse geometric series between average slopes of streams of a particular order and order (Horton, 1945), by the law of drainage basin areas (Schumm, 1956) relating mean area of a particular order to order, and by the law of contributing areas (Schumm, 1956), which logarithmically related the drainage areas of each order and the total stream lengths which they contained and supported. The five laws were originally conceived in terms of the Horton method of ordering (Figure 2.2) but after 1952 the Strahler ordering system necessitated a modified form of the 'laws' of drainage composition (Bowden and Wallis, 1964). This modification could be achieved in the case of the law of stream lengths by relating cumulative mean stream length of streams of each order with order (Broscoe, 1959). Subsequently, consideration of allometric growth led Woldenberg (1966) to suggest that the river open system grows allometrically according to the general equation $y = ax^b$, and Milton (1966) argued that some drainage net laws were geomorphically irrelevant, that the law of stream numbers is a statistical probability function that automatically derives from the definition of stream order, and that other networks such as a plum tree accord with the laws of drainage composition. Shreve (1966) simulated stream networks and compared these with 172 published sets of stream numbers and concluded that the law of stream numbers is largely a result of random development of the topology of channel networks according to the laws of chance. He speculated that the law arises from the statistics of a large number of randomly merging stream channels in a fashion somewhat similar to the law of perfect gases arising from the statistics of a large number of randomly colliding gas molecules.

Although attempts have been made to explain drainage networks in a deterministic way it can be argued that the actual junction of drainage channels in a network occurs in a stochastic fashion (Scheidegger, 1966). Interest in the topological properties of channel networks has led partly to a shift of interest from the properties which could be explained by geological and other controls to those which can be accounted for by topological randomness (Smart, 1969). Understanding has been greatly facilitated by statistical generation of stream networks on a probability basis. This was first attempted by generating random walk game networks and Leopold and Langbein (1962) generated the stream pattern in Figure 2.9A employing a gambler's ruin model. Such a network can be generated on a square grid from a number of sources selected by random numbers along a base line and then by employing simple rules such as no backward moves allowed, equal probability of moving in one of three directions. This was employed by Leopold and Langbein to produce the network in Figure 2.9A. Subsequently, more sophisticated approaches have been employed and modifications have been introduced, firstly, by using digital computer-derived networks (Schenck, 1963; Smart, Surkan and Considine, 1967) and, secondly, by utilising constraints which allow generation of simulated networks according to the type of control which may be envisaged in actual stream systems. It is thus possible to incorporate a greater probability of a move in one direction (forwards) than that to right or left, which introduces a slope variable into the simulation model and, equally, it is possible to include geological controls.

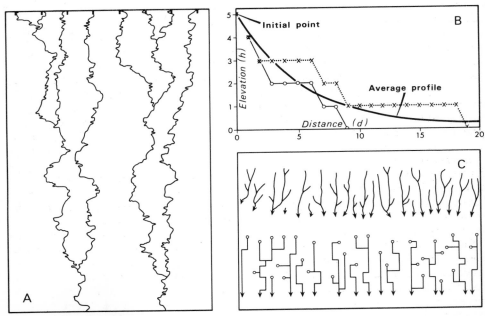

Figure 2.9 Simulated drainage basin properties
A and B based upon Leopold and Langbein (1962) and C derived from Smart and Moruzzi (1971).

An example of this type is afforded by a study of drainage patterns which occur on a homoclinal ridge (Smart and Moruzzi, 1971). The random walk headward growth model was tested by comparison with an actual system, Clinch Mountain, Virginia, which possesses a uniform lithology and structural parameters which vary along the strike (Figure 2.9C). This method like the one developed by Howard (1971) has employed a headward-growth model which generates the network from the outlet rather than from the sources. Although water flows from the sources it is more appropriate to develop a network from the outlet of the basin to simulate the extension of a drainage network. A more recent refinement has been developed by Smart and Moruzzi (1971), who have proposed a headward-growth model in a region of uniform lithology and uniform slope in which probability of growth is made dependent upon the area contributing to runoff at the stream tip. Progress in the simulation of stream networks has thus reached the stage at which it is possible to model networks under varied but controlled conditions, and the technique has recently been applied to deltaic networks (Smart, 1971).

Emerging from simulated networks, sometimes when compared with actual stream networks, has come a greater understanding of the properties of stream nets and in some cases new relations to supplement or modify those proposed by R. E. Horton. The law of stream numbers has been shown to be a statistical relation (Shreve, 1966) and a concave relation has been shown to be expectable where the Strahler method of ordering is adopted (Smart, 1967). In examining the variation of basin area (A) and mainstream length (L'), Smart and Surkan (1967) demonstrated that the deviation of n' from $\frac{1}{2}$ in the usually accepted relationship of $L' = CA^{n'}$ is explicable in terms of variation in mainstream sinuosity because this increases downstream and so leads to n' values which are often closer to 0·6 than to 0·5. This feature has also been invoked by Gosh and Scheidegger (1970) to account for the fact that edge length increases with Strahler order. When values for link length of streams are plotted against order, the increase has been shown to be in the form of a geometric progression (Gosh and Scheidegger, 1970). The link length ratio provided by the slope of the regression equation is a constant in one basin and found to vary between 1·04 and 2·34 for twenty different basins. A further illustration of the development of new relations is provided by Yu-Si Fok (1971) who proposed a law of stream relief in which the average stream relief was found to be a semi-logarithmic function of stream order.

A variety of relations have been demonstrated at the level of the basin (Figure 2.10A, B), and these are paralleled by relations for the channel network (Figure 2.10C, D, E) and for the channel cross section (Figure 2.10F, G). These samples of various relations illustrated in Figure 2.10 are representative of the several categories outlined above (p. 78). In the basin system the topographic characteristics are all intricately related (see Figures 2.1, 2.10) and, therefore, relief ratio exercises an influence over drainage density as shown by the relation between these two parameters for thirty-one small basins on a uniform rock type on the margin of the east Devon plateau, England (Figure 2.10A). In other cases a more direct association may occur, such as that between stream direction (Figure 2.6A) and joint direction, or that demonstrated by Melton (1957) between drainage density and an index of extent of vegetation cover, per cent bare area, for twenty-one basins in south-west U.S.A. (Figure 2.10B). Bivariate relationships such as these are limited by the fact that other significant variables are not included and so Woodruff (1964) examined the values of several morphometric properties in 25 randomly selected drainage basins in four distinct physiographic areas of the U.S.A., namely, the Basin and Range, the Sand Hills, Lower Michigan and the Ridge and Valley provinces. Although no single one of the parameters employed was statistically different between all regions examined, it was found that all areas could be distinguished by at least one index. In a more extensive study of Uganda (Mather and Doornkamp, 1970; Doornkamp and King, 1971) it was possible to examine a large number of variables which were then combined into groups to distinguish between areas having different combinations of morphometric properties. Such methods can indicate the ways in which drainage basin variables are grouped or clustered and in their Uganda study Mather and Doornkamp (1970) found that 18 morphometric properties of 130 third order basins could be reduced to 6 factors which account for 95 per cent of the original variance. The first factor summarised the basin size variables and explained 48·1 per cent of the

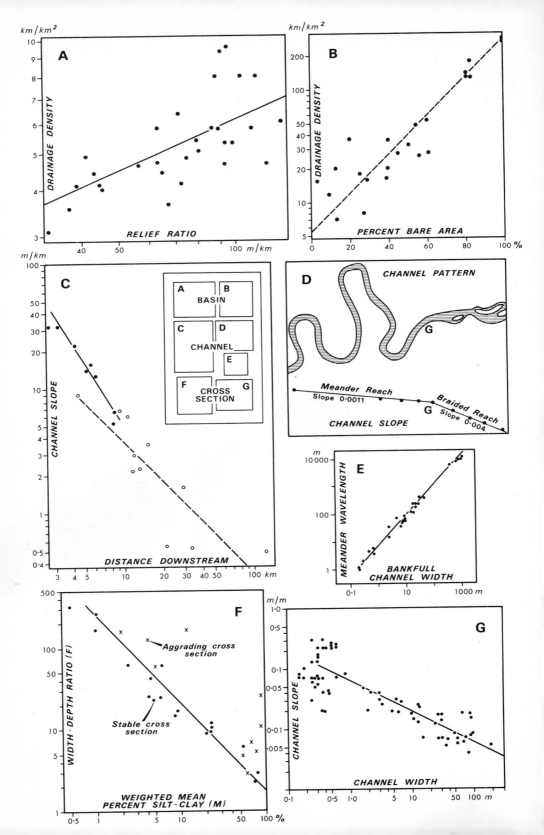

total variance, and the second factor represented stream number and explained 21·4 per cent. Because morphometric properties are so closely inter-related it is not always easy to decide what property the clusters or groups represent and care should be taken in the selection of the original variables. A sample of 410 fourth order drainage basins from west Malaysia were classified by cluster analysis by R. J. Eyles (1971) into six groups. All pairs of basins in the samples are inter-correlated over five morphometric indices, namely, basin area, drainage density, basin relief, average slope and hypsometric integral. This analysis was the basis for a map showing drainage basin types which reflects the main features of the macro-geomorphology of west Malaysia and Singapore.

These studies show that it is possible to produce maps based upon quantitative evaluation of morphometric parameters. Such maps are a useful topographic complement to methods of landscape evaluation and they should be of great value in distinguishing regions that reflect the sum total of environmental characteristics and which could be expected to behave as response units from the point of view of fluvial processes.

2.5b   Channel reach relations

Comparable relations exist in the case of lengths of river channels and at the first level the way in which channel slope varies with downstream distance may be plotted and is illustrated in Figure 2.10C, based upon the work of Hack (1965) comparing two rivers in the Eastern Highland rim of Tennessee. Slope may also exert an influence upon channel pattern and a classic diagram of Leopold and Wolman (1957) demonstrated in a particular area the difference in stream slope for a meandering reach and a braided reach (Figure 2.10D). The character of the deposits present in the stream channel will also vary along the river channel. Several studies have examined the variation of fine and coarser fractions of bed material as they occur along the stream length (see p. 281). In the case of coarse fractions some authorities have shown no distinguishable change of shape (for example, Sneed and Folk, 1958; Brush, 1961) while others have noted that shape does change downstream. In Figure 2.11 the roundness of 16 samples of 100 pebbles of sandstone lithology is compared at sample points along the Ure valley, Yorkshire, England. This demonstrates that in this case the roundness of bed material does increase downstream as indicated by the trend line but that superimposed upon this are variations which appear to be related to the energy environment at particular locations. For instance, the roundness of material decreases from site 10 to site 11 when material is transported over a waterfall at Aysgarth Falls (Figure 2.8). In analyses of this kind distance transported is one factor but others including availability of additional material and size of material present are other complicating factors. At site 14 in Figure 2.11 the high roundness may be attributed to

Figure 2.10   Relationships between drainage basin properties
Drainage density is related firstly (r = 0·69) to relief ratio of 31 basins on the margin of the east Devon plateau (A), and secondly to vegetation cover (B) indexed by per cent bare area (Melton, 1957). The relationship between distance downstream and channel slope (C) for Doran Cove (upper) and Elk River (lower) in Tennessee is based upon Hack (1965). Association of channel pattern and channel slope (D) is from Leopold and Wolman (1957), and F is from Schumm (1960).

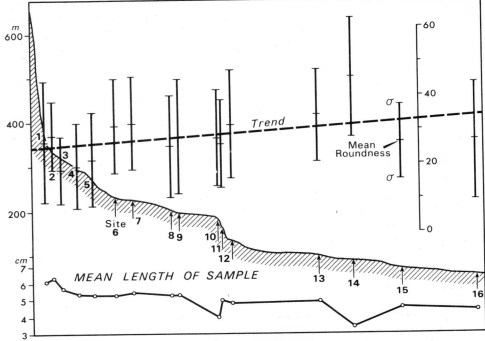

Figure 2.11   Bedload roundness variation downstream
The example is from the Ure river, Yorkshire and the mean value and standard deviation of the roundness of each sample is indicated. Mean length is shown in the lower diagram.

the presence of smaller particles. This illustration demonstrates the way in which relations between drainage basin characteristics may arise consequent upon the process conditioned by one variable (distance downstream) reflected on the other (in this case, particle shape).

Some relationships, therefore, reflect process directly and the relation demonstrated between meander wavelength and bankfull channel width (Figure 2.10E) probably foreshadows the relation of bankfull discharge to meander geometry (see pp. 247–57). This relationship also indicates that relations can occur between parameters of a length of channel, in the form of meander wavelength, and the value of an attribute of the cross section. At a particular cross section a relationship is often found between channel slope and channel width reflecting the fact that slope is related to water velocity and, therefore, that cross sectional area decreases as slope increases (Figure 2.10G). In alluvial channels, relationships will obtain between the hydraulic geometry of the cross section and the material in the bed and banks. Schumm (1963) demonstrated relationships between the channel shape (width-depth ratio) and the composition of the channel bed and banks (weighted mean per cent silt-clay) from basins on the Great Plains (Figure 2.10F).

The multiform relations which exist between drainage basin characteristics are

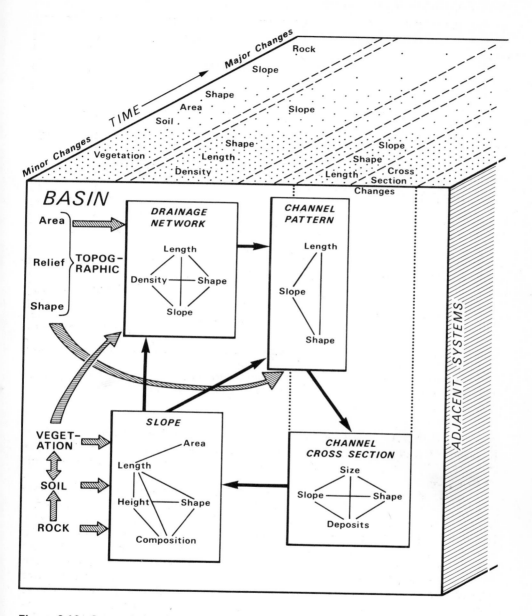

Figure 2.12   Some relations in the drainage basin morphological system
Major interrelationships are indicated.

portrayed in Figure 2.12 by representing the drainage basin as a square. The drainage basin characteristics are all inter-related and these are all further inter-related with the four subsystems which are shown. In each subsystem there are internal relationships between parameters such as length, density, shape and slope and in addition there can be external relationships between the subsystems. It is unrealistic to represent the basin system as a square because variations occur in space and also in time and for this reason the second and third faces of the cube (Figure 2.12) can be employed to introduce two dimensions more fully explored in Chapters 6 and 7.

2.5c   Dynamic drainage basin characteristics

Many of the inter-relations of the essentially morphologic drainage basin characteristic system underline the fact that relationships may exist between many variables because of the relationships of these variables with process. For this reason it is necessary to know what processes occur, how they can be measured (Chapter 3), how they can be evaluated (Chapter 4) and how and why they are related to drainage basin characteristics (Chapter 5). However, the dynamic character of many characteristics is increasingly appreciated and must be considered. The notion that the drainage network is composed of perennial, intermittent and ephemeral streams (depicted in Figure 1.3) is long-established but only since 1960 have the full implications of this notion been realised. Realisation has accompanied the development of the interflow and contributing area models of drainage basin mechanics (pp. 28–30). Three illustrations are provided in Figure 2.13. The first illustrates the presence of three types of component in the drainage network of a small experimental catchment in south-east Devon. The second illustrates a small forested watershed, 0·25 km² in area, in the south-eastern Piedmont, U.S.A. (Carson and Sutton, 1971), which experienced a threefold increase in drainage density during a 104 mm rainstorm and yet the channel network still occupied less than 1 per cent of the total basin area. The third illustrates the implication of the first two as it shows how the drainage network is associated with contributing areas of varying extent during different conditions (Hanwell and Newson, 1970).

The dynamic drainage network places drainage basin characteristics in perspective in two senses. Firstly, the drainage basin is a dynamic unit and the basin characteristics 'stored' within the overall system (Figure 2.12) will only operate when called and a characteristic will operate to an extent determined by the extent of the calling signal. Secondly, and arising from this, the extensive range of drainage basin characteristics which have been proposed (pp. 38–74) will not all be relevant at a particular time to influence a specific process. The student of the drainage basin can, therefore, take heart that the basin does not need to be specified completely in all available ways and he can, therefore, visualise the reason why some characteristics have proved to be more successful and more easily applied to field situations than others!

2.6   Example

It is impossible within the scope of a single example to illustrate the numerous methods of, the diverse difficulties attendant in, and the varied applications of, measurements of drainage basin characteristics. However, the drainage basin of the

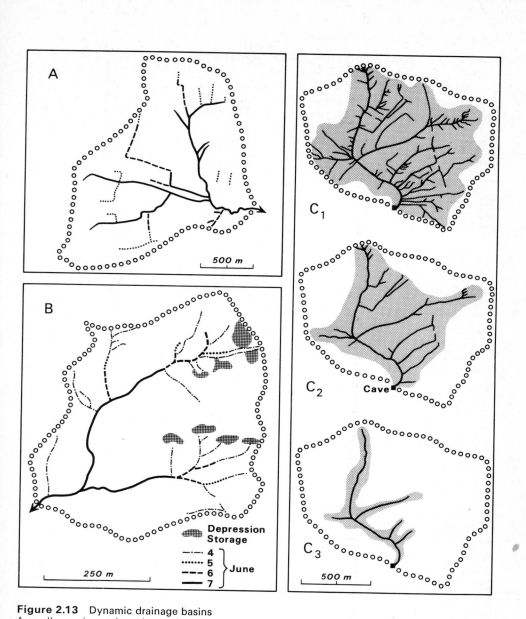

Figure 2.13 Dynamic drainage basins

A small experimental catchment in south-east Devon illustrates perennial (solid line), intermittent (dashed) and ephemeral elements (dots) in A. A basin 0·25 km² in area in the south-eastern Piedmont is illustrated in B after a 104 mm rainstorm which produced a 3-fold increase in drainage density (Carson and Sutton, 1971). Swildon's Hole, a basin 1·18 km² in area in the Mendips, represented during flood situation (C1), average situation (C2) and drought situation (C3) after Hanwell and Newson (1970).

Dove above Rocester weir (Figure 2.14) serves to exemplify some of the objectives and methods. This basin is a useful illustration because it embraces a variety of contrasted rock types which include the permeable and massive Carboniferous limestone immediately beneath the surface of 35 per cent of the catchment area, permeable Triassic sandstone (9 per cent), a variety of grits and shales comprising the Millstone Grit (32 per cent) and the similarly impermeable Keuper marl and lower limestone shales (24 per cent). The contrast in permeability of these lithologies highlights the problems of basin and network definition and the lithological contrasts are also partly responsible for the contrasts in the distribution of the dry valleys which are frequent on the outcrops of permeable Limestone, Sandstone and Pebble Beds but much less common on the outcrops of shales, grits and marls. A final reason for selecting this basin is that in addition to the lithological variety the basin is largely drift-free and so the rock types are largely responsible for the contrasts, and the basin is tributary to a gauging station for which stream flow records are available.

The watershed of this basin is inserted according to the surface configuration and therefore on the basis of the contour pattern shown on available topographic maps. Although this procedure is effective in the areas underlain by relatively impermeable rocks it is less satisfactory for the permeable ones and particularly in the case of the areas of limestone outcrop significant differences have been reported between the surface divides and the phreatic divide below the surface which defines the basins of the sub-surface limestone streams.

Problems also confront the delimitation of the drainage network within the drainage basin and the network shown in Figure 2.14 was obtained by tracing the watercourses shown as blue lines on the Provisional Edition of the 1:25,000 Ordnance Survey maps. Where valleys were indicated by contour crenulations these are indicated (Figure 2.14), by dashed lines, as dry valleys. This method is not infallible because it necessitates the inclusion of some discontinuous streams (A in Figure 2.14) and several instances where man-induced modifications by diversion, regulation and drainage have complicated the network (B in Figure 2.14), particularly along the course of the Henmore Brook in the south-east. The network of streams and valleys produced is thus a reflection of the natural network modified by human action. In addition, the network shown is time-dependent but the significance of time scale probably varies from one lithology to another. Thus on the impermeable shale outcrops instances of dry valleys are indicated by contour crenulations (C in Figure 2.14) but several of these may reflect the mapping and survey procedure and streams are usually mapped at 'normal winter level'. On these outcrops drainage networks expand and contract, as illustrated generally in Figure 2.13, and the extent indicated at any one time is representative of a particular situation. On the permeable rock outcrops occasional flow may occur in the valleys, shown as dry on the topographic maps, but this may be a much less frequent occurrence than flow in apparently dry valleys on the other rock outcrops.

The character of the drainage network in this basin therefore emphasises the fact

Figure 2.14   Dove drainage basin above Rocester weir
Based upon 1 : 25,000 Provisional Edition maps and illustrating drainage network in relation to rock types (shown in inset) and further analysed in Figure 2.15.

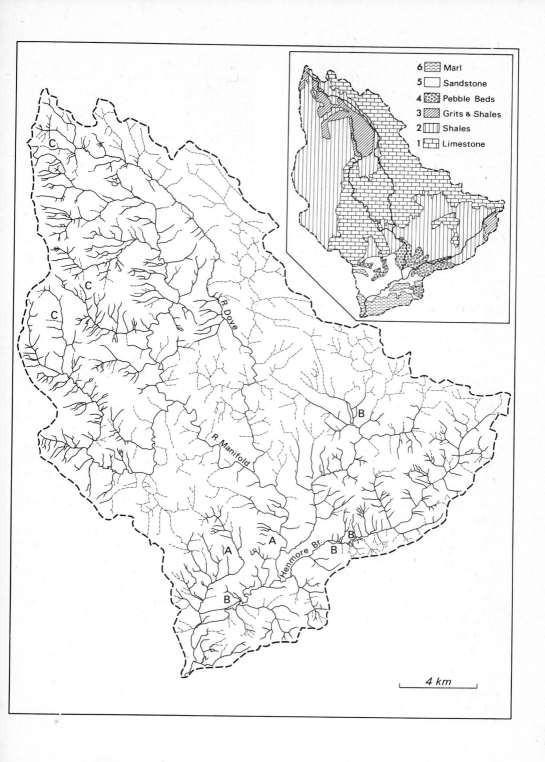

6	Marl
5	Sandstone
4	Pebble Beds
3	Grits & Shales
2	Shales
1	Limestone

C

C

C

R. Dove

B

R. Manifold

A

A

B

B

Henmore Br.

B

B

4 km

that the network must be defined according to criteria determined by the purpose of the analysis anticipated. If analysis is directed towards the contemporary stream network and contemporary process it is necessary to use a method of definition similar to that adopted to produce Figure 2.14, but if the purpose of the analysis is to focus upon the pattern of dissection and the distribution of valleys the network could be produced embracing dry valleys as equivalent to the valleys where streams are shown. This problem is reflected in ordering of the network. If the dry valleys are ignored the main stream is fifth order according to the Strahler method (p. 41) but if the dry valleys are included the basin is sixth order.

Purpose of the analysis must be borne in mind when deciding how to analyse the topographic character of the sample basin and the main purposes which may be identified are concerned with network characteristics, inter-relations of variables in the morphological system and mapping of the distribution of key parameters. Network characteristics may be illustrated by the 'laws' of drainage composition and the relation between number of streams and order, the law of stream numbers, is illustrated in Figure 2.15A. When illustrating these relations the unit of study should be a complete basin of a particular order and so the relation between cumulative mean stream length, the second 'law' of drainage composition, is drawn (Figure 2.15B) for the Dove basin to its confluence with the Manifold. In both cases the relation is illustrated for the stream (dashed line) and the valley network (solid line). Whereas exploration of these laws of morphometry was a subject of major concern in the mid-twentieth century, subsequent attention has been directed to the relations between morphometric properties of drainage basins. This can be achieved by subdividing the large basin (Figure 2.14) into a large number of small component basins and for each measuring the range of topographic, rock, soil and vegetational characteristics as outlined in this chapter. This would then provide a vast amount of data which could be analysed simply by relating any one index with any other. Thus basin area could be related to stream or to valley length and in Figure 2.16C this is illustrated for 15 constituent basins. The difficulty of such bivariate relations is that there are multiple relations between drainage basin characteristics, and multivariate techniques can be employed to document the relations between measures of basin size, network dimensions, basin shape and basin relief and these can be evaluated against the pattern of rock, soil and land use types and possibly also according to variation in climatic parameters.

Multivariate analyses can sometimes yield maps regionalising the combined pattern of drainage basin characteristics and mapping of this kind is obviously important as a means of identifying the distributions of landform character or of runoff-producing characteristics. Such maps can also be produced by employing a single index and drainage density has generally been thought to occupy a central position because it is a sensitive indicator of inputs to the basin system and equally it reflects the sum total of other drainage basin characteristics. Mapping drainage density is not easily achieved based upon small constituent drainage basins because drainage density varies with scale and therefore with order (for example, Figure 2.1 C3) and so a grid-based data-collection method has many advantages (for example, Gardiner, 1971). A map of estimated drainage density can be produced according to number of stream links (Figure 2.15D) and this map demonstrates the regional variation of stream network density which largely mirrors the pattern of rock types (Figure 2.14). To indicate

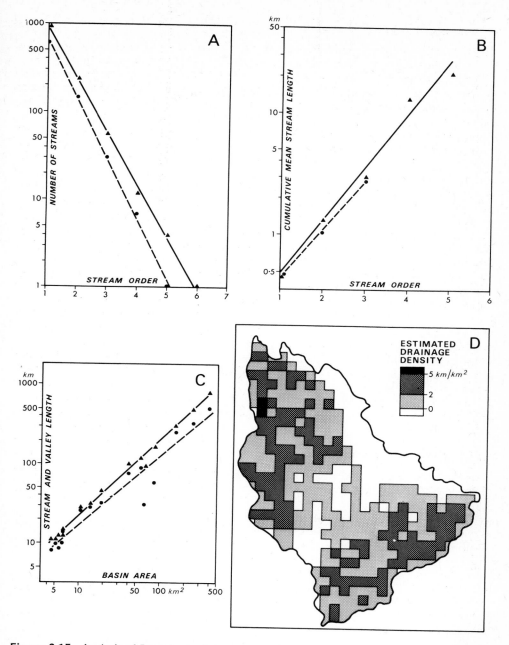

Figure 2.15 Analysis of Dove basin drainage network
Four simple examples of analysis of the drainage network shown in Figure 2.14 include the 'laws' of stream numbers (A), and of cumulative mean stream lengths (B), the relation of network length to basin area (C), and a map of estimated drainage density (D) compiled according to number of links per kilometre grid square.

the general trend contained within this pattern (Figure 2.15D) smoothing techniques could be employed to produce successively generalised patterns of drainage density.

Whereas these measurements focus upon the total basin or upon the variety indicated by its component basins, other analyses could concentrate upon the details contained and, for example, upon the stream channel network to see how the detailed pattern of the stream may correspond to rock jointing (cf. Figure 2.6A) or to see how stream sinuosity varies throughout the basin and particularly to compare the Dove and the Manifold river courses, for example. Especially along the Manifold are large meanders which anomalously appear larger than those downstream of the confluence of the two rivers and this may occur because the present stream pattern in part contains features relict from a past morphogenetic system. This problem is perhaps the most interesting for the fluvial geomorphologist, and is obviously related to the fossil drainage network indicated by the dry valleys. To approach a solution to this problem it is necessary to relate drainage basin characteristics to drainage basin processes because understanding of their interrelations is one basis for interpretation of past situations. It is therefore necessary to be able to measure drainage basin processes and methods whereby this may be achieved are described in the subsequent chapter.

Selected reading

The early advances are included in:

R. E. HORTON 1945: Erosional development of streams and their drainage basins: hydrophysical approach to quantitative morphology. *Bulletin of the Geological Society of America*, **56**, 275–370.

Topographic characteristics are more recently summarised by:

A. N. STRAHLER 1964: Quantitative geomorphology of drainage basins and channel networks, In v. T. CHOW (ed.), *Handbook of Applied Hydrology*, Section 4–11.

Additional reviews are available in:

J. C. DOORNKAMP and C. A. M. KING 1971: *Numerical Analysis in Geomorphology*, 3–96.
P. HAGGETT and R. J. CHORLEY 1969: *Network Analysis in Geography*, 8–31, 57–82, 90–105.
M. G. WOLMAN 1967: Two problems involving river channel changes and background observations. *Quantitative Geography Part II Physical and Cartographic Topics*, Northwestern University Studies in Geography, **14**, 67–107.

3 The Measurement of Drainage Basin Processes

This is the method one has to pursue in the investigation of phenomena of nature. It is true that nature begins by reasoning and ends by experience, but, nevertheless, we must take the opposite route, we must begin with experiment and try through it to discover the reason. *Leonardo Da Vinci*

In any field of process geomorphology, measurement of the processes operating is essential. Without measurements, ideas, theories and estimates cannot be substantiated, and quantitative relationships between form and process cannot be established. Within the scope of fluvial geomorphology, drainage basin instrumentation can provide quantitative data on the dynamics of particular processes and on their magnitude and frequency. Recent years have seen an increased awareness of the need for field measurements of fluvial processes, and the Vigil Network Scheme (Leopold, 1962) and the Representative and Experimental Basins established within the programme of the International Hydrological Decade (International Association of Scientific Hydrology, 1965) were in part inspired by this need. A necessary outcome of this documentation has been the development of new and more sophisticated instruments and techniques (U.S. Geological Survey, 1967; International Atomic Energy Agency, 1968) and a desire to establish certain standard practices (Leopold and Emmett, 1965; Toebes and Ouryvaev, 1970; World Meteorological Organisation, 1965).

3.1   The measurement problem

In many countries, measurements made for assessment of water resources and for general hydrometric schemes may provide useful data on drainage basin dynamics, but, for the most part, the geomorphologist will require additional or more specialised instrumentation. Small drainage basins, the most valuable unit for many studies, are rarely included in national schemes of data collection, although they may be instrumented within a scheme of Representative and Experimental catchments (for example, New Zealand Ministry of Works, 1970; Australian Water Resources Council, 1969). Faced with the task of instrumenting a drainage basin, the geomorphologist must select from a great number of parameters and processes that may be documented by an ever-increasing catalogue of instruments and techniques. Cost may be an important factor in any choice, but clearly defined objectives are also required. Many of the measurements that may be made are largely, if not exclusively, hydrological or hydrometeorological in context and of no great importance to geomorphological studies, and more particularly, a specialised study may not require a full scheme of instrumentation. The important measurements are those concerned with the input of precipitation into, the interactions within, and the output of water, sediment and solutes from, the drainage basin. More specifically, the geomorphologist is interested in both the land and channel phases of the production and movement of runoff, sediment, and solutes. Certain secondary measurements, of such parameters as soil moisture, infiltration and groundwater levels may be required if the former measurements are to be evaluated fully. Measurements may, therefore, be characterised as *primary* and *secondary*. However, for the purposes of this chapter, division will be made into, firstly, the measurement of processes concerned with runoff dynamics, and, secondly, measurements of sediment and solute dynamics, overall yields and rates of erosion.

Techniques of instrumentation vary in complexity and degree of sophistication, and a distinction can be made between *point* and *continuous* measurements. The selection will often depend primarily on cost, because continuous measurement is more expensive in terms of capital expenditure. Recent developments in electrical chart recorders, digital punched tape recorders, magnetic tape loggers, and telemetry, and associated advances in transducer technology, have greatly increased the scope for continuous monitoring of drainage basin processes, and this is in most cases to be preferred, because no interpolation of data is required and a recorder trace is usually more objective than manual observations. However, the vast increase in data associated with continuous monitoring may itself pose problems and create the necessity for specialised translation and computer processing facilities.

Despite technological innovations, certain general difficulties will always remain in the realm of drainage basin instrumentation. One of the greatest problems is that of *sampling* in both space and time. The use of continuous monitoring will overcome many of the problems of sampling in time, but, when point measurements are used, careful thought should be given to the scheme employed. Observations made on a systematic basis, perhaps at weekly intervals, could miss many of the extreme events associated with a parameter that varied markedly through time (for example, suspended sediment concentration). Moreover, it can be easily demonstrated that manual observations of river stage, even if taken at 6 hour intervals, will rarely provide an adequate picture of the precise form of the discharge hydrograph during flood events. On a small stream, observations at 5 minute intervals might be required (for example, Figure 4.12). The details of a particular sampling scheme will depend on the character of the process being investigated, on the aim of the study, and on the nature of the equipment or apparatus being employed. Furthermore, any data collected from a short period of study is essentially a non-random sample of a time series, and it may require many years of record before data is sufficiently reliable to provide valid conclusions about the magnitude and frequency of specific events and processes. For example, estimates of the rates of sediment transport from a watershed (Tonnes/km^2 . yr) may be invalidated by a rare flood event that transports more material in one day than in the previous decade. Statistical analysis may be required to assess the reliability of data in time and to synthesise and generate long-term records from a short period of measurement.

The problems of sampling in space are equally relevant. Apart from measurements of the output of channel flow and the associated sediment and solute yields from a drainage basin, most data collected is sample data and only pertains to small areas or points within the watershed. For example, 10 Standard Meteorological Office pattern raingauges (rim diameter 13·25 cm), installed within a catchment of 1 km^2, provide data from only $1·4 \times 10^{-7}$ of the total area. Similarly, data from one thousand erosion pins scattered over the slopes of a catchment represent an infinitesimally small proportion of the total number of possible locations. Most measurements, therefore, embody techniques of random sampling to provide data which is to some degree representative of the total area, and the planning of an instrumentation network is an important aspect of any study (International Association of Scientific Hydrology, 1965). Measurement is often not an end in itself but merely the prelude to detailed statistical analysis as a means of assessing the reliability of the sample data and of extrapolating the results. Other difficulties involve considerations of availability of suitable locations for instruments in terms of accessibility, long-term freedom from disturbance, and security from interference. Availability of 'wilderness' sites representative of an environment unchanged by man, if indeed such places exist, must progressively diminish, and most of the data collected will not pertain to natural processes in the true sense of the term.

MEASUREMENT OF RUNOFF DYNAMICS

In this context, the geomorphologist is concerned with measurements of precipitation input, evapotranspiration losses, interactions within the catchment, and output of runoff.

3.2   Precipitation input

Because water is essential for the operation of fluvial processes, quantitative assessment of the input into a drainage basin should be viewed as a primary measurement. Data is required on the nature and, more particularly, on the amount and timing of the precipitation and the variation of these values over the study area. The greatest difficulty involved is that any measurement will be an areal point sample and the data will require extrapolation. Recent developments in radar analysis do, however, indicate the possibility of studying areal variations more directly. Amounts are expressed as the depth to which the precipitation, or its liquid equivalent, would cover a horizontal projection of the land surface, and are reported in millimetres (mm). Intensity values are reported in millimetres per hour (mm/hr). Description of the various methods employed for measurement is best achieved by considering the different forms of precipitation, whilst concentrating on rainfall since this is the most common. In addition, it should be appreciated that standard methods generally provide estimates of *gross precipitation* and that techniques are also available for estimating *net precipitation*. In all measurements, the geomorphologist relies heavily upon the methods and instruments developed by meteorologists and hydrometeorologists (for example, World Meteorological Organisation, 1965).

3.2a   Measurement of rainfall

The basic instrument for rainfall measurement is the rain gauge, which samples the incidence of rainfall at a specific point, through an orifice of known area. In its simplest form it consists of a cylindrical container set in the ground and topped with an orifice of accurately known dimensions, and usually formed by a sharp-edged rim bevelled to the outside. In detail the gauge will be specially designed to meet such requirements as prevention of outsplash, and minimisation of evaporative loss from the collection vessel, and there is considerable variation in precise dimensions and designs between different countries (Rodda, 1969). Rain gauges may also be classified according to degree of sophistication into *non-recording* and *recording gauges*. The non-recording gauge is simply a storage device that collects the quantity of rainfall caught between visits by the observer, perhaps daily, weekly or monthly. In view of its low cost it is the most common, and the standard British Meteorological Office gauge and the United States Weather Bureau standard gauge are illustrated in Figure 3.1. In a recording

Figure 3.1   Instruments for measuring precipitation
This diagram illustrates two commonly used non-recording instruments, the standard British Meteorological Office gauge (A) and the U.S. Weather Bureau standard gauge (B); a Nipher-type gauge shield (C) and a wire gauze cylinder which can be used to assess horizontal interception (D). The construction and principle of operation of three recording instruments is also shown along with an example of the chart produced by a tilting-siphon recording gauge.

NON RECORDING INSTRUMENTS

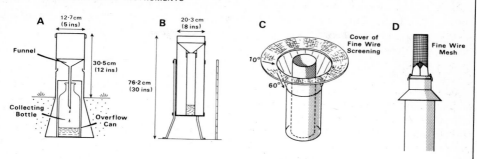

A
12·7cm (5 ins)
Funnel
30·5cm (12 ins)
Collecting Bottle
Overflow Can

B
20·3 cm (8 ins)
76·2 cm (30 ins)

C
Cover of Fine Wire Screening
10°
60°

D
Fine Wire Mesh

RECORDING INSTRUMENTS

TILTING-SIPHON

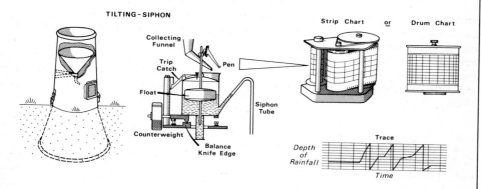

Collecting Funnel
Trip Catch
Pen
Float
Siphon Tube
Counterweight
Balance Knife Edge

Strip Chart   or   Drum Chart

Depth of Rainfall
Trace
Time

TIPPING-BUCKET

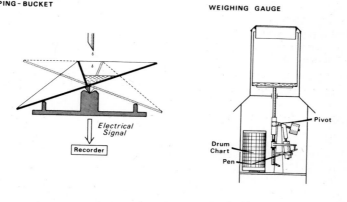

Electrical Signal
Recorder

WEIGHING GAUGE

Pivot
Drum Chart
Pen

gauge there is a mechanism to measure the volume or weight of rainfall reaching the gauge and its timing. This mechanism is usually based on one of several principles, namely, the float and siphon, the tipping bucket, and the weighing collector (Figure 3.1), each of which has its own particular advantages and disadvantages. The tilting siphon autographic gauge is probably the most widely used, and the record is provided by a drum chart or an open-ended strip chart (Figure 3.1). In addition to the recording mechanisms described above, instruments have also been developed to provide direct records of rainfall intensity (for example, Road Research Laboratory, 1968).

The major problems involved in the use of a rain gauge are not those of obtaining records, because the instruments are generally quite reliable, but of assessing the possible errors inherent in the values obtained (Green, 1970; Rodda, 1967; Struzer et al., 1965). Gauges of different design will give different results, and although gauge sites are usually specially chosen to minimise sources of inaccuracy, the very presence of the gauge can disturb the pattern of airflow and affect the results obtained. Shields (Figure 3.1), fences and turf walls have been used to reduce the effects of turbulence around the gauge orifice. Recent studies using ground level gauges (Rodda, 1969), which overcome many of the exposure problems, indicate that standard rain gauges may underestimate the rainfall by as much as 20 per cent, although in lowland areas the value may be nearer two to 5 per cent. No absolute standard for rainfall measurement exists, but the ground level gauge, consisting of a normal gauge set with the rim at ground level and surrounded by an antisplash surface or grid, may provide the most representative results.

Further problems arise when attempting to extrapolate point results to the whole drainage basin. Several techniques exist for calculation of mean rainfall values, using the data from several gauges, and these include the use of Thiessen Polygons (for example, Linsley, Kohler and Paulhus, 1949), subjectively drawn isohyets, and, perhaps more promising, trend surface analysis (Mandeville and Rodda, 1970). Areal extrapolation of rainfall intensity values is more difficult, although a computer routine has been described by Kelway and Herbert (1969). In the future, monitoring of areal rainfall may be made possible by the use of radar (Ryan, 1966), but at present problems exist over calibration of the signals and their range. Meanwhile, the conventional rain gauge must suffice, and in order to use the point data to the best advantage, a carefully designed network of measuring points is required (Corbett, 1967). In many cases a random pattern will be used, and this may be stratified according to certain physiographic variables. Hamilton (1954) has detailed the facet method of network design whereby the catchment or area under consideration is subdivided into a series of facets of uniform physiographic characteristics, and gauges are sited at points representative of these facets.

3.2b   Measurement of snowfall

In many areas snowfall occurs only occasionally and measurements may be made with a non-recording rain gauge by melting the trapped snow with a known volume of warm water. Where snowfall is the dominant form of precipitation, special instruments are required. Non-recording storage gauges are similar to the equivalent rain gauge, although in detail they generally possess a larger orifice, the collecting funnel is omitted, the storage capacity is greater, and the storage container may contain anti-freeze solution. Care must be taken to place the orifice above the maximum snow level, and because the necessary height introduces problems of turbulence and eddy effects, gauge shields are often used. Recording snow gauges are mostly based on the weighing principle (Figure 3.1), because the other devices normally used for rainfall would be subject to problems of freezing. It is also possible to estimate amounts of snowfall by measuring the depth of accumulation of fresh snow with a rule and assuming an approximate water equivalent conversion of 1 cm snow to 1 mm of water.

In certain studies, where sudden thaw can release vast quantities of water from temporary storage as snowpack, the water potential of the snowpack covering a watershed may be an important measurement. This may be assessed by using snow courses or permanently marked sampling transects, and taking depth measurements against permanent snow stakes or with probes. More accurate data can be provided by core samplers which provide a core that can be weighed in the field to determine its water equivalent. The Mount Rose Snow Sampler (Church, 1935) is widely used in the United States. The greatest problem associated with snow course measurements is that of access and several developments have been made in the field of continuous monitoring of snowpack at remote sites. Radioisotope techniques based on the attenuation of gamma rays by the snow, and pressure pillow devices which record the pressure exerted by snow accumulating on a liquid-filled rubber bladder would seem to be the most promising approaches (Warnick and Penton, 1971). Telemetry and radio transmitters can be used in conjunction with these devices.

3.2c   Measurement of other forms of precipitation

In general, the contribution to the drainage basin input of forms of precipitation other than rainfall and snowfall and their associated variants will be very small and not worthy of measurement. When required, attempts may be made to measure the contribution of dew, mist and fog, although considerable difficulties are often encountered. Methods for measuring dewfall include blotting techniques, drosometers or hygroscopic blocks, the Duvdevani Dew Gauge (Duvdevani, 1947) and more sophisticated recording-weighing apparatus (Jennings and Monteith, 1954). Wire gauze cylinders, mounted above the rim of a rain gauge (Figure 3.1), have been used by Nagel (1956) to assess the importance of horizontal interception of fog and mist droplets by vegetation as a source of moisture.

3.2d   Measurements of net precipitation

The methods of precipitation measurement outlined above are generally made in open spaces and provide values of gross precipitation above the vegetation canopy. If there is a thick vegetation cover, and data is required on rainfall reaching the soil surface for study of runoff or sediment production, gross precipitation data may be misleading. Methods have been developed for local measurements of net rainfall, or rainfall beneath the vegetation cover, which is composed of both throughfall and stemflow. In the case of tree cover, randomly placed rain gauges or troughs can be used to measure throughfall, although, because the incidence of throughfall may vary considerably in space, a large number of gauges will be required to provide a representative result. Stemflow may be measured by attaching a collar around a tree-trunk and collecting the volume of intercepted water (Figure 3.2). Summation of the estimates of throughfall and stemflow provides the estimate of net rainfall.

Assessment of net rainfall beneath shrub and herbaceous vegetation is more difficult. Rain gauges can be sited beneath tall species (Gregory and Walling, 1971), but with shorter vegetation this will often be impossible, and, furthermore, the gauge orifice must be positioned far enough above the ground to avoid insplash. Measurements of stemflow using collars would be largely impractical. One approach to overcome these problems (for example, Crouse et al., 1966) has been to completely seal the surface beneath the plant canopy with plastic or rubber, and to collect the surface runoff as representing net rainfall (Figure 3.2). With larger species such as bracken (Pteridium aquilinum) it is possible to install drainage collars around the stem bases to isolate the throughfall component (Leyton Reynolds and Thompson, 1965). The size of the plot used for this type of measurement is obviously restricted, although artificial sprinkling could be employed to extend the scope of data collection. Investigation of interception by surface litter is probably best carried out by collecting samples of litter in wire baskets and weighing, wetting and reweighing, in

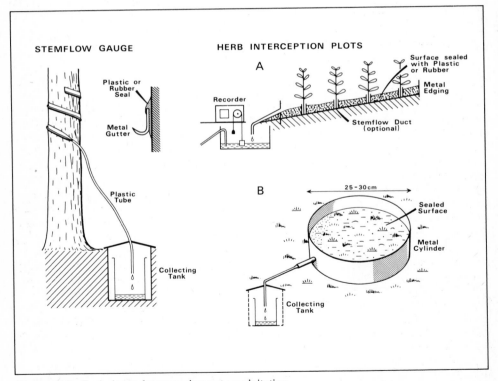

Figure 3.2   Techniques for assessing net precipitation
The construction of a simple stemflow gauge and the apparatus which could be used to monitor interception by large herb species (A) and smaller plants and grasses (B).

order to calculate the amount of moisture retained. In many studies, field measurement of net precipitation will be impractical, and reliance must be placed on the extrapolation of published interception data.

3.3 Losses

Measurement of evapotranspiration loss and its components of evaporation and transpiration is very much the concern of the hydrologist and hydrometeorologist who seek to determine the water balance or moisture budget of a catchment. To the geomorphologist these are secondary measurements, although, nevertheless, of interest in assessing the amount of water available for runoff and in evaluating changes in moisture storage. In many cases, published data may be available, but if this is employed it is important to know the manner in which it was derived and, therefore, its limitations. Detailed assessments of catchment losses are amongst the most difficult of measurements, and the various methods available may be classified into direct and indirect approaches. In the latter group, estimates are derived from mathematical equations incorporating data on certain controlling meteorological variables. All values are expressed as water equivalents, in the same manner as precipitation, and are reported in millimetres (mm). A distinction is often made between *actual* and *potential* values of loss where the latter is the theoretical amount that would occur without limitations of moisture availability.

3.3a *Direct measurements*

Estimates of rates of evaporation from open water surfaces can be obtained using evaporation pans. These are small tanks of water in which measurements are made of the surface lowering or the amount of water required to maintain a constant level, taking account of replenishment by rainfall. There are many different types of pan, and they vary in the precise details of shape, dimensions and installation. Two commonly used designs, the United States Weather Bureau Class A Pan and the British Meteorological Office Pan, are illustrated in Figure 3.3. A small isolated body of water will usually exhibit greater evaporation rates than a larger natural body, and correction coefficients must be applied to pan data to provide representative results. Coefficients for the U.S. Class A Pan are reported as lying between 0·69 and 0·74, whilst those for the Meteorological Office Pan range between 0·93 and 1·07.

Evaporation from soil surfaces may be assessed by using percolation gauges (Lapworth *et al.*, 1948). These consist of a cylindrical soil section encased within a buried metal container (Figure 3.3). A closely sited rain gauge is used to measure rainfall input into the cylinder, and percolation is collected from the base. The difference between rainfall input and percolation can be attributed to evaporation, although this should be a long-term value to avoid the possibility of changes in storage within the container influencing the result.

Measurements of plant transpiration are extremely difficult, although attempts may be made using laboratory-based potometers and phytometers, and in the field by placing a closed container or plastic tent over a plant and recording the resultant increases in humidity or the water vapour produced. Recently, studies have been

EVAPORATION PANS

A

121·9 cm (4 ft)

25·4 cm (10 ins)

Stilling Chamber

Wooden Platform

B

Hook Gauge

183·0 cm (6 ft)

183·0 cm

61·0 cm (2 ft)

Stilling Chamber

PERCOLATION GAUGE

Vegetation

Bare Soil

Soil

Rock

Gravel

c. 1 m

Collecting Pit

c. 3 m

LYSIMETERS

A

Oil

Pump

Recorder

Soil

Air

Reservoir Tank

Stilling Well

Water

Gravel

Ballast

c. 2 m

B

Lifting Hooks

25 cm

Soil

50 cm

Perforated Base

Collector Can

conducted on individual trees using measurements of heat pulse velocity within the water conducting xylem as indicative of total transpiration (Swanson, 1970).

Assessment of water loss from vegetated areas is usually achieved by considering the total evapotranspiration values, without any attempt to measure the relative contribution of the component parts. Evapotranspiration may be measured directly by using percolation gauges and lysimeters. Percolation gauges could be the same as those used for soil evaporation measurements, except that the surface of the soil would be vegetated (Figure 3.3). Values of both actual and potential evapotranspiration can be obtained if two containers are used and if one of these is irrigated to maintain soil moisture storage at field capacity. Lysimeters are similar in principle to percolation gauges, but they provide more detailed information by making provision for the weighing of the soil and vegetation column. Small changes in storage can therefore be detected. Small lysimeter containers can be lifted and weighed (Figure 3.3) and changes in weight equivalent to losses of 0·1–0·2 mm can be detected. With larger installations mobile lifting cranes are sometimes used, but more often a hydraulic system is used to monitor changes in weight (Figure 3.3). If a suitable watertight catchment is available, this can be used as a vast percolation gauge with estimates of evapotranspiration derived from the simple water balance equation (Edwards and Rodda, 1970). The extension of this principle to a runoff plot is illustrated by the Stocks Lysimeter described by Law (1957).

3.3b   Indirect estimation of losses

Many formulae have been developed for estimation of losses (Ward, 1971), although they are usually restricted to evaporation and evapotranspiration, because transpiration is difficult to treat in isolation. Furthermore, the estimates are generally potential values since derivation of actual values requires knowledge concerning soil moisture status. These formulae vary considerably in degree of sophistication, from simple empirical relationships to complex mathematical approximations of the physical processes involved. The Thornthwaite formula (Thornthwaite, 1948) and the Penman formula (Penman, 1963) have been widely used, and the basis of the two equations is illustrated in Table 3.1. The Thornthwaite equation is largely empirical and based upon observed relationships between evapotranspiration and air temperature and day length. Its principal advantage is that sophisticated instrumentation is not required to provide the necessary meteorological data. The Penman equation provides a more theoretical approach, and combines both the turbulent transfer and energy budget components of the physical processes of evapotranspiration. Because of its more complex nature, computer routines are available for calculating the Penman Formula (Chidley and Pike, 1970), and there are many possible modifications to the basic equation to allow for such factors as vegetation type and soil moisture status.

One important characteristic of many of the formulae available is that they require detailed meteorological data for their evaluation. Therefore, although instrumenta-

Figure 3.3   Installations for measuring evapotranspiration losses
This figure shows two types of evaporation pan, the U.S. Weather Bureau Class A Pan (A) and the British Meteorological Office Pan (B); a percolation gauge which could be used to assess evaporation from bare soil and evapotranspiration from vegetated surfaces; and a sophisticated hydraulic lysimeter (A) and a simple weighing lysimeter (B).

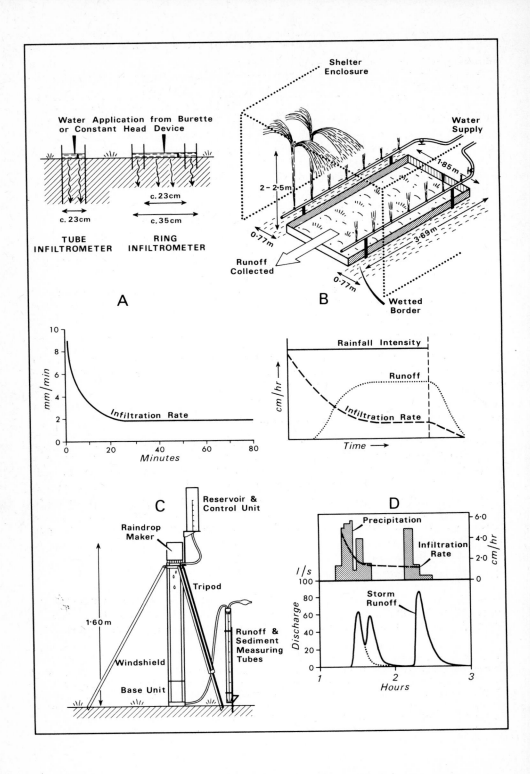

Water Application from Burette or Constant Head Device

c. 23cm

c. 23cm

c. 35cm

TUBE INFILTROMETER

RING INFILTROMETER

A

Shelter Enclosure

Water Supply

2 – 2·5m

1·85 m

0·77m

3·69 m

Runoff Collected

0·77m

Wetted Border

B

10

8

6

4

2

0

mm/min

Infiltration Rate

0   20   40   60   80

Minutes

Rainfall Intensity

cm/hr

Runoff

Infiltration Rate

Time

C

Reservoir & Control Unit

Raindrop Maker

Tripod

1·60 m

Runoff & Sediment Measuring Tubes

Windshield

Base Unit

D

Precipitation

Infiltration Rate

6·0

4·0

2·0

0

cm/hr

l/s

100

80

60

40

20

0

Discharge

Storm Runoff

1   2   3

Hours

Table 3.1 Evapotranspiration equations

Description	Time basis	Units	Equation
Penman	Daily	mm	$PE = \left(\dfrac{\Delta}{\gamma}H + E_a\right) \Big/ \left(\dfrac{\Delta}{\gamma}+1\right)$
			Where $E_a = 0.35(e_a - e_d)\left(1 + \dfrac{U_2}{100}\right)$
Thornthwaite	Monthly	mm	$PE = 16b\left(\dfrac{10T}{I}\right)^a$

where PE = Potential evapotranspiration in mm
Δ = the slope of the saturation vapour pressure curve
γ = the constant of the wet- and dry-bulb hydrometer equation
H = the net radiation at the earth's surface
e_a = the saturation vapour pressure of water at mean air temperature
e_d = the saturation vapour pressure of water at dew-point or the actual vapour pressure at mean air temperature
U_2 = the mean daily wind speed at a height of 2 metres
b = correction factor to account for unequal day length between months
T = the mean monthly temperature
I = the annual heat index
a = a function of I

tion for direct measurement of losses is avoided, other measurements may still be necessary, and if radiation is to be estimated as accurately as possible, sophisticated radiometer instruments will be required.

3.4 Precipitation/watershed interactions

This sphere of measurement is concerned with those aspects of catchment dynamics interposed between precipitation input and output of channel flow. In some geomorphological studies, data on input and output from a drainage basin will be sufficient, but when processes are to be analysed in detail many of these secondary measurements may be necessary. For convenience, division can be made into measurements of, firstly, water in the soil or aeration zone, secondly, water in the rock or saturated zone, and, thirdly, watershed runoff production as involving the pre-channel phases of runoff.

3.4a Water in the soil

Assessment of the major processes occurring in this zone involves measurement of the uptake of water by the soil or infiltration, measurement of soil moisture status at different points within the soil, and measurement of movement and downward

Figure 3.4 The measurement of infiltration
Simple flooding infiltrometers provide a direct measurement of infiltration rates over a period of time (A) while careful analysis of sprinkler application rates, plot runoff and surface detention is necessary with sprinkling infiltrometers such as the Type F infiltrometer illustrated here (B). Hand-portable sprinkling infiltrometers (C) are useful in remote areas and in some small catchments estimates of infiltration rates can be obtained from analysis of rainfall and runoff data (D).

percolation of soil water. Infiltration data is useful both in terms of the physical processes operating within a drainage basin and as a quantitative index for description of soil characteristics.

Measurements of infiltration are usually expressed as depths of water entering the soil during a given time (cm/hr), and a special derivative is the term *infiltration capacity* defined as the maximum rate at which rainfall can be absorbed by a soil in a given condition. In general, measurements of infiltration are difficult to make, because, unless data is collected for a large flooded area, results will reflect lateral boundary seepage as well as downward vertical movement of water. Approaches to measurement are twofold, firstly, the use of flooding infiltrometers and, secondly, the use of sprinkler or rainfall simulator infiltrometers. As an alternative to direct measurement, estimates may be obtained from hydrograph analysis.

The basic principle underlying the flooding infiltrometer is the flooding of a small area enclosed by a metal or plastic cylinder, and the measurement of the rate of water uptake, by observing either the lowering of the head, or the amount of water required to maintain a constant head. Distinction may be made between Ring Infiltrometers, which consist of cylinders inserted a few centimetres into the soil to the minimum depth required to prevent leakage, and Tube Infiltrometers, where the cylinder is jacked into the soil to depths of 35–50 cm to provide an enclosed soil core (Figure 3.4A). Ring infiltrometers often incorporate two concentric cylinders, with the outer one used as a flooded buffer zone to overcome some of the problems of lateral flow, although Hills (1971) has described a procedure for correcting the results from a single cylinder. Infiltration rates decrease rapidly during the first few minutes of water application, and the measurement run is usually continued until a constant rate is achieved. The results may be presented as a graph of infiltration rates versus time (Figure 3.4A), whilst values of infiltration capacity are usually based on the near-constant infiltration rates at the end of a test run.

The problems of small sampling area, soil disturbance during emplacement, boundary effects and the lack of similitude between waterflow from a burette and natural rainfall, associated with flooding infiltrometers, are overcome to some extent by the use of sprinkling infiltrometers. With this apparatus, infiltration is measured indirectly as the difference between the amount of water sprinkled on to a plot of known area and the surface runoff collected from that plot (Figure 3.4B). Raindrops are simulated by sprinkler nozzles (Sharp and Holtan, 1940) or drip screens (Barnes and Costell, 1957), and the rates and pattern of the simulated rainfall can be easily controlled. A fall distance of 5 to 8 m is required if the raindrops are to attain natural fall velocities, although this will depend on the size of the drops. Plot sizes range from small rings 5–10 cm in diameter to areas as large as 0·1 hectares which necessitate complex arrangements of nozzle supply pipes (Hermsmeier *et al.*, 1963) or rotating booms (Swanson *et al.*, 1965). Rates of water application are usually measured before and after the test run by using an impervious cover to the plot, or small troughs and rain gauges. Runoff is collected in a tank or, in the case of a large installation, measured by a small flume. The Type F sprinkler infiltrometer developed by the U.S. Department of Agriculture is a widely used semi-portable apparatus (Figure 3.4B) and the United States Geological Survey have developed a hand-portable apparatus or micro-rainulator (Figure 3.4C) for use in remote areas where water supply is restricted (McQueen, 1963; Selby, 1970).

Infiltrometers do not provide absolute data, and the results obtained from both flooding and sprinkling infiltrometers are of prime value in comparative analysis. Different results may be obtained from different instruments used at the same site (Table 3.2), and in general a sprinkling infiltrometer would seem to give the more representative results.

Table 3.2 Comparison of the results provided by two types of infiltrometer (*after Musgrave and Holtan, 1964*)

Infiltrometer type	Infiltration rate (cm/hr)	
	Overgrazed poor cover	Grazed good cover
Type F rainfall simulator	0·36	2·87
Concentric ring flooding infiltrometer	2·82	5·97

Information on the general infiltration behaviour of a small catchment can also be obtained from analysis of storm runoff hydrographs and the associated storm rainfall records (Sharp and Holtan, 1940; Horner and Lloyd, 1940) (Figure 3.4D). However, it should be recognised that this technique is closely linked to the traditional infiltration theory of runoff formation (Horton, 1933) and may be questionable in view of recent theories on the importance of variable source areas and throughflow in the production of storm hydrographs (pp. 28–31).

Measurement of soil moisture status is a field of study which primarily concerns the agronomist (Holmes, Taylor and Richards, 1967) and the hydrologist (King, 1967), but it can provide useful background data for the geomorphologist. The collection of data on soil moisture status involves measurements of both soil moisture content and suction or tension. Moisture contents are usually expressed as weight per cent values—a percentage of the oven dry weight of the soil. Values of soil moisture tension reflect the suction force with which the moisture is held within the soil, and, therefore, decrease with increasing moisture content. This force is generally expressed in terms of the head of water or mercury required to produce a suction corresponding to that at a point within the soil (cm Hg, cm H_2O, or atmospheres). Alternatively, and to overcome the problem of large units, the logarithm of the head of water is reported as a pF value.

The basic technique for the determination of soil moisture content is the gravimetric method (Reynolds, 1970), and this is often used to calibrate other instruments. In this procedure a sample is transported to the laboratory, weighed, dried at 105 °C until a constant weight is achieved, and reweighed. The loss of weight is used to calculate the moisture content. This technique possesses many problems, notably those of the destructive and time-consuming nature of the sampling process, the transportation of samples without changing the moisture content, and the difficulty of obtaining a continuous record.

Several types of instrument have been developed to overcome the above difficulties, and these provide for continuous records or frequent field readings without disturbance or removal of the soil after the initial installation. Electrical resistance methods are based upon the principle that the electrical resistance of a block of porous material in moisture equilibrium with the soil can be used as an index of soil moisture content.

The porous blocks are made of gypsum, nylon or fibreglass and contain two electrodes between which the resistance is measured. A quantitative calibration may be made against the gravimetric technique, although this may not be required if a simple record of relative variations is sufficient. In general, porous blocks are best suited to low moisture contents below field capacity and because of instability, hysteretic and temperature effects the results may not be of high quantitative accuracy.

Plate 7 A neutron soil moisture probe This 'Wallingford' soil moisture probe was specially developed by the Institute of Hydrology to provide a reliable, portable instrument suitable for use in difficult terrain.

Neutron scattering techniques have also been applied to the measurement of soil moisture content (Bell and McCulloch, 1966; Van Bavel, 1965) and the neutron probe provides a very useful though expensive instrument (Figure 3.5B). The method is based on measurement of the slowing or moderation of neutrons emitted from a fast neutron source into the soil. This deceleration is due, principally, to collision with elements of low atomic weight, notably hydrogen, which is contained in the soil water molecules. The count of slow neutrons can be directly related to the moisture content of the soil, and, unlike the methods described previously, the values obtained apply not to a small point in the soil profile but to a bulb-shaped zone with dimensions varying from 14–34 cm in the vertical and 42–53 cm in the horizontal at moisture contents ranging from 43 to 5 per cent respectively (de Vries and Kring, 1961). The instrument, therefore, consists of three main components, the radioactive source, a rate counter, and a slow neutron detector. An essential accessory is the access tube, which is installed permanently in the ground and down which the probe is lowered. Calibration may be carried out using specially constructed standards (Bowman and King, 1965) or the gravimetric technique. Many of the instruments produced have been heavy and, therefore, provide problems if the probe needs to be carried across rough terrain, but the Institute of Hydrology (1968) have developed a lightweight apparatus for this purpose (Plate 7). Despite several technical problems, the neutron probe is a very valuable tool, which will provide quantitative data, which is especially suited to repeated field measurements and which provides considerable labour-saving over gravimetric methods. Cohen and Tadmore (1966) have suggested that the work accomplished in 36 days with a neutron probe might take 244 days using gravimetric techniques.

The soil moisture suction or tension at a point within the soil may be measured directly using a tensiometer. This consists of a water-filled porous ceramic or semipermeable plastic cup connected by a water column to a manometer, or vacuum gauge (Figure 3.5A). The cup is buried in close contact with the soil so that water in the cup attains equilibrium with the soil water. Water flows out of the cup as the soil dries out and develops an increased suction, and back into the cup as the soil becomes wetter and the tension decreases. Changes in soil suction, as reflected in movements of water between the cup and the soil, are indicated on the manometer or vacuum gauge. Tensiometers function best in the moist conditions between field capacity and saturation; with drier conditions and tensions greater than 0·9 atmospheres, air enters the system and the instrument no longer functions. Tensiometers may also be used to determine soil moisture content if a calibration is established between tension and moisture content.

To overcome the difficulties of using tensiometers in dry soils, developments have been made recently in the use of permeable membranes. The C.S.I.R.O. (Peck and Rabbidge, 1966) have produced an instrument that will measure suctions of up to twenty atmospheres and consisting of an aqueous solution placed in a container sealed with a permeable membrane. Water can move through the membrane into the container, but the solute (polyethylene glycol) cannot, and a pressure transducer is used to measure the solution pressure as indicative of soil moisture tension.

Measurement of soil moisture movement and percolation can also be undertaken. Theoretical estimates of soil moisture movement may be derived from detailed analysis of values of soil moisture tension. For example, if Darcy's law (p. 61) is applied to the movement of soil moisture in unsaturated conditions it can be demonstrated that movement of soil water is controlled by the capillary conductivity of the soil and the resultant gradient of the combined suction and gravitational forces within the soil. In general, there will be a movement of soil moisture from moist soil into dry soil or, more specifically, from areas of low suction to areas of high suction. Because of the relationship of capillary conductivity to soil moisture content, this movement will tend to be more rapid in moist than in dry soils. Although a detailed theoretical approach is useful (for example, Brooks and Corey, 1966), results are uncertain and measurements are still required to provide the basic data. Direct measurement may be more convenient and meaningful.

Movement of soil moisture may be inferred from measurements of changes in soil moisture content at different locations and horizons within the soil, and the neutron probe provides a useful means of obtaining the necessary frequent readings. Direct

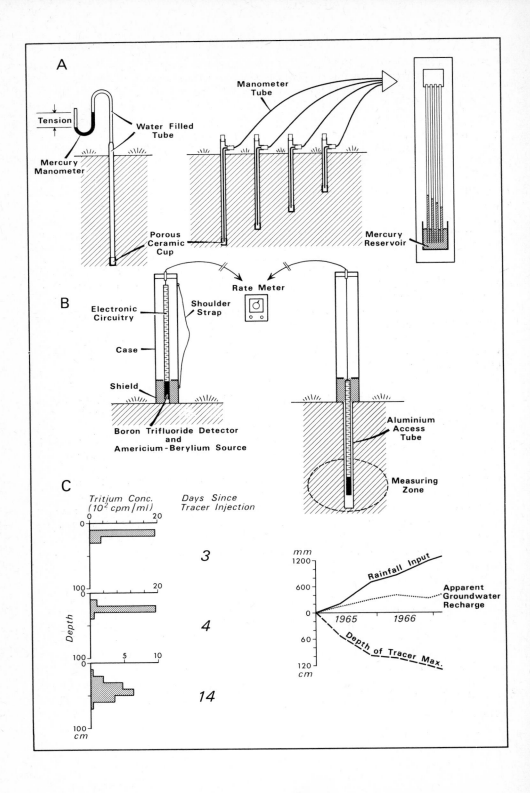

A

Tension

Mercury Manometer

Water Filled Tube

Manometer Tube

Porous Ceramic Cup

Mercury Reservoir

B

Electronic Circuitry

Shoulder Strap

Case

Rate Meter

Shield

Boron Trifluoride Detector and Americium-Beryllium Source

Aluminium Access Tube

Measuring Zone

C

Tritium Conc. (10² cpm/ml)

Days Since Tracer Injection

0   20

3

100   20

Depth

4

0   5   10

14

100
cm

mm
1200

Rainfall Input

600

Apparent Groundwater Recharge

0

1965   1966

60

Depth of Tracer Max.

120
cm

Table 3.3 Properties of the ideal tracer

---

1 The tracer must be capable of quantitative measurement at very low concentration

2 It must be transported in the environment to which it is added in the same way as the naturally occurring water (no density and viscosity differences)

3 Its introduction to the environment and its withdrawal must not modify the naturally occurring transport phenomena

4 It must not be sorbed by the materials with which it comes into contact

5 It must not react to form a precipitate

6 It must not be modified by the action of organisms

7 It should be cheap and readily available

8 It should not be present in appreciable concentration in the natural water

9 It should not create a hazard or interfere with later investigations

---

Source: Elrick and Lawson (1969)

measurements of percolating water may be obtained from lysimeters and percolation gauges, but the results may be influenced by degree of disturbance of the soil column. The most valuable technique for the direct monitoring of movements of soil water is essentially non-quantitative and consists of the use of tracers. The extent of movement can be deduced from the pattern of tracer diffusion. The tracer used should conform to several requirements and in Table 3.3 the properties of an ideal tracer are listed. Tracers may also be classified according to the method of detection into mechanical, chemical, dye, and radioactive tracers.

Reynolds (1966) has described the use of Pyranine Conc., a fluorescent dye, to trace the percolation of rainwater below a woodland floor. The dye was applied to the surface and the depth of penetration determined by excavation of trial trenches—the dye fluoresces under wet conditions and ultra-violet photography can be used to provide a permanent record. Radioactive tracers have also been used, namely, radioactive Iodine (^{131}I) (Benecke, 1967) and Tritium (^{3}H) (Blume *et al.*, 1967). A syringe can be used to inject the tracer below the surface in order to reduce wastage, and augured soil cores are generally used to detect tracer movements. Blume *et al.* (1967) have also attempted to obtain quantitative data on volumes or depths of groundwater recharge by observing the downward passage of the radioactive tracer maximum through the soil and calculating the amount of water displaced downwards by this movement (Figure 3.5C). In addition, throughflow is a particular form of downslope soil moisture movement occurring above a percolation limiting horizon under saturated or near-saturated conditions (see section 3.4c).

3.4b   Water in the rocks

Precise distinction between soil and rock is often difficult and, for the purposes of this account, an alternative definition might be 'water in the saturated zone'. The most

Figure 3.5 The measurement of soil moisture status and percolation
This diagram shows the principle of operation of single and multiple tensiometers (A) and a portable neutron probe (B), and illustrates the way in which estimates of percolation and ground-water recharge can be obtained by monitoring the downward movement of a tracer maximum (C, based on Blume *et al.*, 1967).

important measurements required are those of the level of the water table and of movement within the groundwater body.

Measurements of groundwater levels are usually expressed as heights above mean sea level or ordnance datum or as depths below the surface (m). The water table can be located by the geophysical techniques of resistivity and seismic surveying (Galfi and Palos, 1970), but most data is collected more directly as measurements of standing water level in an observation well or borehole. Boreholes are usually specially constructed for this purpose, with perforated strainer sections and gravel packing to ensure a close representation of water levels in the surrounding rock (Figure 3.6A). In general, small diameter wells and boreholes provide the best data since they are more sensitive. When pre-existing wells are used, care should be taken to evaluate the representativeness of the results, and even with specially constructed wells the data must be interpreted carefully. Hewlett (1961) has cited the interesting case of observation wells in the Coweeta Hydrological Laboratory in the Southern Appalachians, U.S.A., which demonstrated the expected characteristics of accretion and depletion, but which were subsequently found to be no more than storage cisterns charged by excess water from heavy rain, and with leakage providing the appearance of water table decline.

Methods for monitoring the standing water level may be classified into manual and recording techniques. Manual methods make use of a graduated tape or line equipped with a device for sensing the water level. Electrical contact devices (Figure 3.6) are the most widely used and these are coupled to a meter, buzzer or bulb circuit. Most recording gauges are based upon the float and counterweight principle (Figure 3.6) and closely resemble river stage recorders (Figure 3.11). In boreholes less than 10 cm in diameter the small size of the float required presents problems because it will not provide enough force to operate the recorder pulley, and several devices have been developed to overcome this limitation. The Koopman Ferret Gauge and the Drescher Float attachment (Shuter and Johnson, 1961) both utilise a small electric servomotor to take up the slack in the float cable and therefore to actuate the recorder.

Groundwater movement is a three-dimensional phenomena, but for convenience it is usual to consider the two-dimensional flow fields, namely, movement in a horizontal plane and movement in a vertical plane. Patterns of horizontal movement in the zone of the water table can be derived from water table contour maps, with flow lines constructed downslope and perpendicular to the contours. In order to study horizontal movements at depth and flow patterns in the vertical plane, measurements of piezometric pressure or fluid potential at points within the groundwater body are required. This potential consists of two components, a pressure-head component and an elevation-head component, and may be measured by means of a piezometer. A piezometer is a small diameter tube or pipe with an open end and within which the water level reflects the total head or potential at the point in the groundwater body where the tube is open (Figure 3.6B). Sometimes the piezometer tube will have a porous ceramic or plastic tip to overcome problems of blocking. Often a piezometer

Figure 3.6   Apparatus and techniques for studying groundwater dynamics
The constructions of a groundwater observation borehole (A) and a piezometer (B) are illustrated, and the manner in which Meyboom *et al.* (1966) utilised piezometer data to assess the direction of groundwater movement beneath a river is shown in (C).

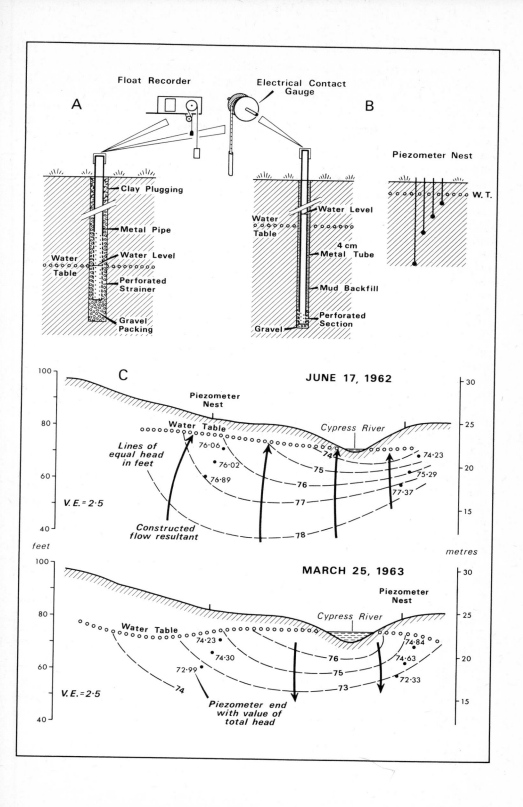

Float Recorder

Electrical Contact Gauge

A

B

Clay Plugging

Metal Pipe

Water Level

Water Table

Water Level

Perforated Strainer

Gravel Packing

Piezometer Nest

W. T.

Water Table

4 cm Metal Tube

Mud Backfill

Perforated Section

Gravel

C

JUNE 17, 1962

100

30

Piezometer Nest

Water Table

80

Cypress River

25

Lines of equal head in feet

76·06

74·...

74·23

76·02

75

20

60

76·89

76

75·29

V. E. = 2·5

77

77·37

40

15

Constructed flow resultant

78

feet

metres

100

30

MARCH 25, 1963

Piezometer Nest

80

Cypress River

25

Water Table

74·23

74·84

74·30

76

74·63

60

20

72·99

75

72·33

V. E. = 2·5

74

73

15

Piezometer end with value of total head

40

nest will be installed, consisting of several tubes terminating at different levels (Mey-boom *et al.*, 1966). This data is used to plot lines of equal piezometric head or equi-potentials on a vertical cross section of the aquifer, and flow lines are drawn normal to these lines in the direction of minimum head, because water will move from areas of high pressure to areas of low pressure (Figure 3.6C). Similarly, equipotentials or piezometric surfaces could be plotted for several horizontal planes at different depths within the groundwater body, and flow lines constructed.

Further quantitative data on groundwater movement may be obtained by combin-ing the above data with theories of saturated flow through porous media such as the Darcy equation (p. 61).

In addition to theoretical studies, groundwater movement may be assessed more directly using tracers. The tracer used should conform to the requirements listed in Table 3.3, and radioactive tracers (e.g., ^{3}H, ^{131}I and ^{82}Br) are particularly useful because of the high sensitivity of detection. In the simple point to point technique, the tracer is introduced into a well and its progress down gradient is monitored with observation boreholes. The borehole dilution technique (Halevy *et al.*, 1967) is particularly suited to radioactive tracers and involves measuring the dilution rate of a tracer solution homogeneously distributed within a borehole. The decrease in activity or dilution is proportional to the flow velocity. If the measuring probe can be slowly moved along the borehole then it is possible to determine a vertical profile of flow velocities. Vertical flow velocities can be assessed by timing the passage of a tracer up or down the borehole. Radioactive tracers may also be used to determine the direc-tion of the horizontal flow component within a borehole because the maximum con-centration of the tracer cloud or maximum adsorption of tracer by the walls of the borehole occurs in the direction of flow. However, a problem of all flow measurements utilising boreholes is the relationship of the results to conditions within the surround-ing aquifer, because the presence of the borehole may distort the natural flow pattern.

Limestone areas provide a specialised field for measurement of groundwater move-ment because flow may be concentrated in caves and passages, and it is frequently necessary to determine the flow route from, and the rising associated with, a parti-cular sink (Figure 3.7). Tracers can be employed to reveal flow routes and to measure the associated flow velocities. Both fluorescent dyes (Fluorescein and Pyranine Conc.) and Lycopodium spores, a mechanical tracer which can be dyed for recognition pur-poses, have been used (Drew and Smith, 1969).

3.4c   *Measurement of watershed runoff production*

This field of measurement involves the pre-channel flow phases of runoff. Documenta-tion of channel flow is comparatively easy because the runoff is concentrated and confined within the stream channel, but any attempt to measure the pre-channel phases faces considerable practical and theoretical difficulties. Sampling constitutes a serious problem because the whole of a watershed may contribute runoff to channel flow, but the precise contributions may vary considerably in both space and time. Furthermore, the areas contributing to individual flow components may constitute only a part of the total catchment, and these contributing areas may vary in extent through time. Values of watershed runoff are expressed either as flow rates at the

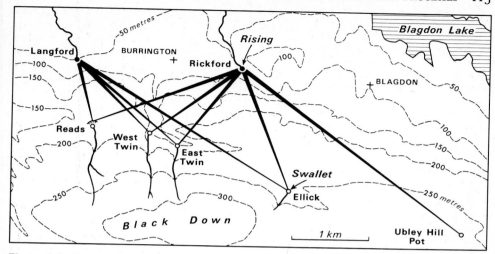

Figure 3.7   Tracing of underground drainage in limestone areas
Dyed Lycopodium spores were used to trace the major underground flow paths between swallets and risings in this area of the Mendip Hills, England (based on Drew, Newson and Smith, 1968).

collecting or measuring point (l/s or cc/min) or as equivalent depths over the assumed contributing area (mm or mm/hr).

Division may be made into techniques for measuring, firstly, surface runoff or overland flow; secondly, subsurface runoff from the soil and including both through-flow and interflow; and, thirdly, outflow from the groundwater reservoir. These divisions are arbitrary because, for example, throughflow from the upper portion of a slope might emerge as surface runoff at lower levels, and it may be difficult to separate the flow components, but they nevertheless provide a useful classification appropriate to one point on a hillslope.

Measurements of surface runoff production may be made by installing a trough on the watershed slope and collecting the intercepted water in a container or recording apparatus (Figure 3.8A). Care must be taken to ensure a good seal between the upper lip of the trough and the soil surface, and the trough is usually roofed to exclude direct precipitation. The Gerlach trough (Gerlach, 1967) has been specially designed for this purpose. The volume of water collected can be related to the area of the slope segment uphill from the trough for computation of runoff depth, although problems may arise due to the areal non-uniformity of runoff production and to the difficulty of delimiting the precise contributing area in zones of contour concavity or convexity. Some of these problems may be surmounted by the use of a runoff plot, which is surrounded by a cutoff wall. The downslope boundary is shaped to convey the surface runoff to an outlet point where it can be measured. Runoff plots are often used to document soil erosion as well as runoff production (see Section 3.6). Artificial sprinklers may be employed with both plots and troughs to extend the range of measurements.

A different approach involves the use of tracers to indicate the movement of water across the surface, and Hills (1971) has described the use of Pyranine Conc., a

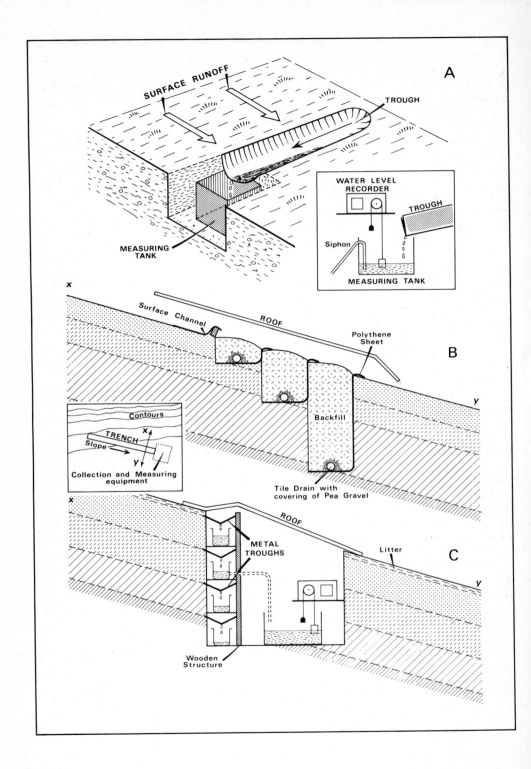

A

SURFACE RUNOFF

TROUGH

MEASURING
TANK

WATER LEVEL
RECORDER

TROUGH

Siphon

MEASURING TANK

B

x

Surface Channel

ROOF

Polythene
Sheet

Backfill

y

Contours

TRENCH

x

Slope

y

Collection and Measuring
equipment

Tile Drain with
covering of Pea Gravel

C

x

ROOF

METAL
TROUGHS

Litter

y

Wooden
Structure

fluorescent dye, for this purpose. However, this technique is essentially non-quantitative and only provides an indication of the presence or absence of surfaceflow. Pilgrim (1966) used Gold-198 (^{198}Au) and Chromium-51 (^{51}Cr), two radioactive tracers with low energy radiation and low biological toxicity, to study the time of travel of surface runoff within a catchment.

Direct measurements of subsurface runoff within the soil can be made to complement data obtained from overland flow troughs. For this purpose, a trench is dug across a section of a hillslope, and the water flowing out of the different soil horizons, usually distinguished by steps in the side of the trench, is collected. This technique is generally attributed to Whipkey (1965), although later workers including Ragan (1967) and Dunne and Black (1970) have further developed the method. Distinction may be made between installations where the trench is left open (Figure 3.8C) and troughs are constructed to collect the flow from the different horizons (Whipkey, 1965; Ragan, 1967), and those where the trench is backfilled and the runoff intercepted by tile drains (Dunne and Black, 1970) (Figure 3.8B). The runoff may be collected in a simple storage container or recorded by some more sophisticated apparatus. The length of trench used has varied considerably in different studies from the 1-metre section employed by Weyman (1970) to the 84-metre installation described by Dunne and Black (1970). Both natural and artificial rainfall may be used with interceptor trench studies, and Takeda (1967) has described an interesting attempt to induce subsurface flow by placing a water reservoir in the soil upslope of the trench. Trench measurements could be adapted to permit the assessment of water movement in subsurface seepage lines, such as those described by Bunting (1964). Attempts might also be made to monitor flow of water in soil pipes (Jones, 1971). The precise method used would depend on the scale of the feature and, once excavated, small weirs, flow meters, siphoning-collectors and tipping bucket mechanisms could be employed (see plate 8).

In order to study the mechanics of subsurface runoff in more detail, it is usual to combine the interceptor trench results with data from tensiometers, piezometers, neutron probes and other apparatus measuring conditions within the soil. In addition, trench and pipe measurements may be related to the pattern of channel flow by gauging the upper and lower points of a channel reach, calculating the inflow into that reach, and comparing the inflow with the estimates of slope runoff (Weyman, 1970). Evaluation of results in terms of channel flow can be very valuable if false conclusions regarding the contribution of subsurface runoff to the storm hydrograph are to be avoided. Because of the problems associated with defining the slope zone contributing to a trench, the use of clearly defined plots, with boundary walls extending down to impervious strata, may be a useful although costly alternative. Installations of this type have been constructed at the Valdai Hydrological Laboratory in the U.S.S.R. (Toebes and Ouryvaev, 1970), and Hewlett (1961) has also described the construction of an artificial sloping soil profile.

Groundwater outflow or runoff is more difficult to measure than surface runoff and subsurface throughflow. In most cases the movement occurs at considerable depths below

Figure 3.8 Installations for measuring runoff production
Surface troughs (A) can be used to intercept and measure surface runoff and backfilled (B) and open (C) trenches can provide complementary data on subsurface flow.

Plate 8   Measurement of subsurface runoff in soil pipes
These field installations are part of an intensive study of subsurface pipeflow being carried out by
the Institute of Hydrology at Plynlimon (*cf.* Figure 3.27). The flow through several pipes has been
diverted into flexible tubes for monitoring. At low flow levels siphoning reservoirs (*left*) are
used to record discharge rates, and at high flows orifice mounted current meters (*right*) are
employed. Signals from the various monitoring sites are fed to a central data collection station.

the surface, and although theoretical estimates of outflow may be made from borehole and piezometer measurements, direct measurements are restricted to areas of
surface outflow, namely, springs, seepage zones and effluent river channels. Springs
and seepages may sometimes be gauged using streamflow measurement techniques or
collection devices, and the extent of effluent seepage into a channel reach may be
assessed by measuring the discharge increment within that section, and estimating the
groundwater contribution. Data on the location of sites of groundwater influx along a
stream bed can be obtained by analysis of water table levels and piezometer data
(Pluhowski and Kantrowitz, 1962; Lang and Rhodehamel, 1962). Another promising
technique might be the careful monitoring of changes in water quality along a channel
reach, because groundwater will generally possess different physical and chemical
properties from the channel flow. Tentative estimates of influx volumes could be
based upon rates of change of the water quality parameters.

Hydrograph separation studies provide an alternative approach to the determination of
the relative importance of different flow origins in runoff production, and involve
measurement of the hydrograph of channel flow and apportionment of the total flow
volume to different sources. This apportionment may be largely arbitrary, as in the
case of traditional hydrograph separation, or more physically significant when the

hydrograph trace is combined with data on water quality variations within the stream (cf. section 4.3b).

Traditional hydrograph separation procedures (Linsley, Kohler and Paulhus, 1949) are essentially empirical, and are generally based upon the assumption that the beginning of a hydrograph rise is the result of the arrival of surface runoff in the stream channel and that one or more lines may be drawn through the hydrograph to distinguish the surface runoff from the interflow and baseflow contributions (Figure 3.9). The precise method of drawing the separation lines varies considerably. In some cases a straight line is constructed from the point of hydrograph rise or a point beneath the hydrograph peak, to a point on the recession limb. However, selection of the latter point is somewhat arbitrary and may merely be based on inspection of the recession limb for changes in curvature. Barnes (1939) has used a more complicated procedure of semilogarithmic hydrograph plotting, based upon backward extension of straight line recession curve segments, to distinguish the individual flow components (Figure 3.9). The scheme of hydrograph separation proposed by Hibbert and Cunningham (1967) avoids much of the recent controversy over the relative importance of overland flow and throughflow in the formation of storm hydrographs (Kirkby, 1969; Weyman, 1970) by abandoning genetic terms and using a time-based separation. A storm hydrograph is divided into Quickflow and Delayed Flow components by a line drawn upwards from the point of hydrograph rise at a gradient of 0·55 litres per second per square kilometre per hour (Figure 3.9). The precise origins of the two components will vary between different catchments.

Hydrograph separation based upon geochemical or water quality parameters is more objective, although its use is relatively recent. The underlying principle is that water from different sources will possess different chemical characteristics, and that the relative contributions of the different sources can be evaluated by measuring both the stream discharge and the chemical quality of the mixed water flowing in the stream. Assuming that there is no chemical reaction when solutes contained in runoff from different sources combine, a mass balance equation may be applied i.e.

$$C_t Q_t = C_1 Q_1 + C_2 Q_2 + \ldots C_n Q_n$$

where C_t = solute concentration of mixed water
$\quad\quad\quad Q_t$ = discharge of mixed water
$\quad\quad\quad C_n$ = solute concentration of a particular runoff component
$\quad\quad\quad Q_n$ = discharge of a particular runoff component.

In the simple case where total runoff (Q_t) with a measured solute concentration (C_t) is composed of surface runoff (Q_s) and groundwater flow (Q_{gw}), the groundwater contribution (Q_{gw}) can be calculated by solving an equation derived from the mass balance equation cited above:

$$Q_{gw} = Q_t \left(\frac{C_t - C_s}{C_{gw} - C_s} \right)$$

where C_s = solute concentration of surface runoff
$\quad\quad\quad C_{gw}$ = solute concentration of groundwater flow.

The values of C_s and C_{gw} are often assumed to approximate to the solute concentrations of the stream at maximum and minimum flow respectively, although values could be obtained from runoff collected in overland flow troughs and similar apparatus (p. 115).

This approach can be applied using data on the specific conductance of streamflow and the runoff components (for example, Kunkle, 1965) and an example of the resultant hydrograph separation is presented in Figure 3.9A. More detailed analysis is possible using the concentrations of individual ions. Newbury, Cherry and Cox (1969) used specific conductance and sulphate ion concentrations to distinguish a long-term groundwater storage component and an interflow and transient groundwater component (Figure 3.9B); and analysis of the variations in silica, bicarbonate, chloride, nitrate, sulphate, calcium, iron, magnesium, potassium and sodium ions was used by Pinder and Jones (1969) to determine the groundwater component of storm hydrographs from small watersheds in Nova Scotia. Russian workers (Skakalskiy, 1966; Voronkov, 1963) have made extensive use of chemical-based hydrograph separation on short-term and annual river discharge records (Figure 3.10), and Skakalskiy (1966) has produced maps of European Russia showing the annual contribution of different runoff origins (Figure 6.14).

3.5   Water in channels

Measurement of channel flow is of particular relevance to the geomorphologist because streamflow comprises a major component of output from the drainage basin system, and because it is linked closely to processes of channel erosion and development and transport of sediment and solutes. In principle, the measurements are easier to make than those of most other aspects of drainage basin dynamics because the flow is concentrated within the channel and it is possible to isolate the total volume of water discharging from a catchment of known area. In detail, however, the precise methods of measurement are complex and the associated science of Hydrometry (Troskelanski, 1960) has developed into a specialised discipline. British Standard Practices have been developed as detailed specifications for several types of measurement (British Standards Institution, 1965) and in several other countries stream gauging manuals have appeared (Corbett, 1943; Hiranandani and Chitale, 1960; U.S. Geological Survey, 1967; Church and Kellerhals, 1970). The major streamflow parameters that are measured are those of stage or water level, velocity, and discharge, and their variations through time.

3.5a   Measurement of stage

This parameter of streamflow is important both in itself and, more particularly, as an essential element of several methods of discharge measurement. The units used will depend on water depths and the range of variation encountered (m, cm, or mm), and

Figure 3.9 Hydrograph separation
Techniques of hydrograph separation range in complexity from the construction of simple arbitrary separation lines to delimit two or three components, through more complex semilogarithmic plotting to detailed and objective water quality-based analyses as carried out by Kunkle (1965) (A) and Newbury *et al.* (1969) (B).

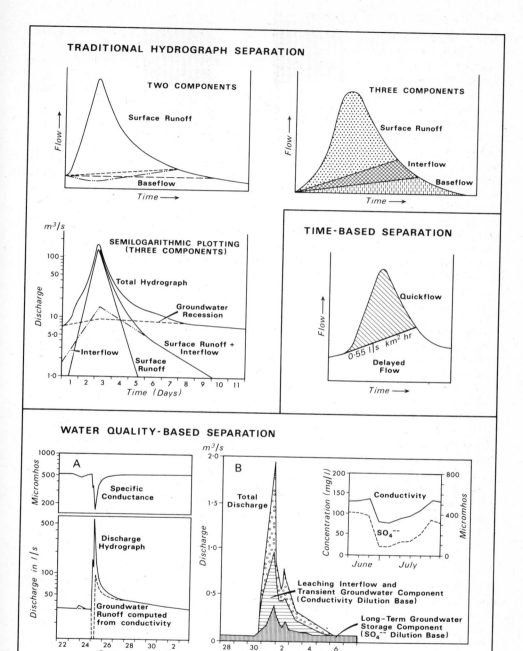

TRADITIONAL HYDROGRAPH SEPARATION

TWO COMPONENTS

Flow →

Surface Runoff

Baseflow

Time →

THREE COMPONENTS

Flow →

Surface Runoff

Interflow

Baseflow

Time →

m^3/s

SEMILOGARITHMIC PLOTTING (THREE COMPONENTS)

Discharge

100
50

10
5.0

1.0

Total Hydrograph

Groundwater Recession

Surface Runoff + Interflow

Interflow

Surface Runoff

1 2 3 4 5 6 7 8 9 10 11
Time (Days)

TIME-BASED SEPARATION

Flow →

Quickflow

0.55 l/s km² hr

Delayed Flow

Time →

WATER QUALITY-BASED SEPARATION

Micromhos

A

1000

500

200

500

100

50

Specific Conductance

Discharge Hydrograph

Groundwater Runoff computed from conductivity

Discharge in l/s

22 24 26 28 30 2
Sept Oct

m^3/s
2.0

B

Discharge

1.5

1.0

0.5

0

Total Discharge

Concentration (mg/l)
200
150
100
50
0

Micromhos
800

400

0

Conductivity

SO₄⁻⁻

June July

Leaching Interflow and Transient Groundwater Component (Conductivity Dilution Base)

Long-Term Groundwater Storage Component (SO₄⁻⁻ Dilution Base)

28 30 2 4 6
June July

express the height of the stream surface above a certain datum level which may be the stream bed, a weir crest, or a more arbitrary point. A great variety of techniques and instruments of varying accuracy are available, and a distinction may be made between manual and automatic methods. Automatic instruments and their associated installations are expensive but they possess the advantage of providing a continuous and objective record.

The simplest procedure for the manual measurement of stage involves the use of an enamel gauge plate or a board with painted divisions (Figure 3.11). These can be easily read by eye to an accuracy of 0·5 cm or better, although turbulence may present problems. To suit particular channel configurations, sloped and stepped gauge boards may be used (Figure 3.11). Where a bridge is available for measurements, a wire weight, float, or electric contact gauge may be used. Crest stage gauges are a more sophisticated type of manual gauge designed to indicate the maximum stage attained during a certain period of time. Many techniques have been used including non-return ratchets or friction catch attachments to float and counterweight devices (Stevens, 1942), staffs painted with water sensitive paint (Doran, 1942), vertical sequences of cups (Guy, 1942) and non-return float indicators (Gregory and Walling, 1971) (Figure 3.11). However, the most widely adopted crest stage gauge, and the one advocated for use in Vigil Network instrumentation (Miller and Leopold, 1963) is the device developed by the United States Geological Survey. This consists of a tube containing an index rod and powdered cork (Figure 3.11), the crest stage being marked by the cork 'tidemark' on the index rod.

Continuous stage recording instruments may be classified according to the principle used for sensing the water level. Float and counterweight chart recorders are the most widely used, and these are installed over a stilling well to overcome problems of water turbulence and associated float oscillation. The float and counterweight are connected by a cable or metal tape passing over a pulley system, and movements of the pulley operate the recorder (Figure 3.11). An example of a stage recorder chart is provided in Figure 3.12. If a vertical recorder is used, this may be equipped with a strip chart mechanism to provide greater time detail and a longer interval between chart changing. Recent developments have also produced analogue-to-digital recorders which punch out the stage values at predetermined intervals on to paper tape (Figure 3.12) and therefore simplify the subsequent data processing. The stilling well is an integral part of any float-operated stage recorder installation, and this consists of a vertical tube or well constructed in or against the stream bank and connected to the channel by one or more small diameter pipes (Figure 3.11). The water level in the tube or well is the same as that in the river, but the turbulence is damped. The detailed design of a stilling well may be quite complex and involve provision for flushing the connecting pipes in order to avoid blockage by silt.

Although the float and counterweight is the most commonly used water sensing mechanism, other devices have been employed, especially in circumstances and condi-

Figure 3.10 Runoff components of the Pyalitsa River, U.S.S.R., 1960
Water quality data on the concentration of total dissolved solids (2) and bicarbonate (3), sulphate (4) and chloride (5) ions has been used by Skakalskiy (1966) to apportion the total annual runoff hydrograph (1) into components of water of slope origin (a), soil-ground origin (b) and ground origin (c). In this river, water of slope origin causes an increase in chloride and sulphate concentrations and a decrease in the bicarbonate concentration.

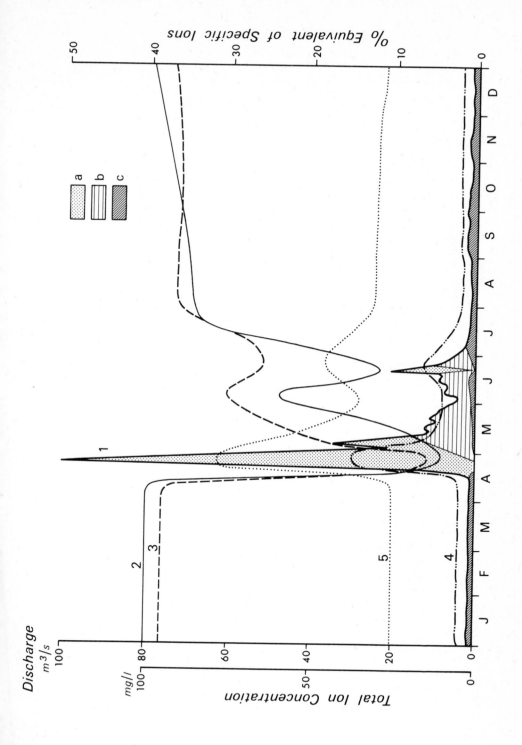

Plate 9 Water stage recorders
This water stage recording instrument developed by R. W. Munro Ltd., incorporates a Munro IH 94 vertical-type chart recorder and a Fischer and Porter analogue-to-digital punched tape recorder (by kind permission of R. W. Munro Ltd. London).

tions where traditional installations would prove difficult. A Pressure Bulb recorder consists of a pressure bulb or diaphragm installed within the channel below minimum water level and connected to a pressure recording device by small bore tubing (Figure 3.11). Because these instruments do not require a stilling well, they are semi-portable and the recording device can be sited away from the stream. However, leakage and poor response characteristics may provide problems. More accurate results are provided by another pressure recording device, the Bubble Gauge. In this case gas is bubbled at a constant rate from a dip tube in the stream, and changes in stage are marked by variations in pressure at the outlet of the dip tube, which can be recorded. One simple form of Bubble Gauge apparatus utilises a compressed air supply and records the pressure at the dip tube orifice directly (Figure 3.11). The more sophisticated gauge used by the United States Geological Survey utilises a nitrogen gas supply and a servo-monometer recording device in which the height of one limb of a mercury manometer is adjusted by an electrical drive to balance the pressure at the orifice (Figure 3.11). The movements of the manometer limb are recorded to provide a record of variations in stage. This particular apparatus is reported as capable of operating for periods of up to one year from one cylinder of nitrogen and 2 ×6 volt dry batteries. Bubble Gauges are especially useful where river bank character and freezing conditions would preclude the operation of a float recorder.

With all the above recording instruments, careful consideration should be given to the size, range and speed of the recording chart. A clear, detailed and accurate record is required, and with small catchments this may require a chart speed in excess of 2·5 cm/hr. Strip charts provide a useful means of obtaining these speeds without excessive chart changing. In certain instances telemetry apparatus may be of value in providing immediate data for remote sites.

3.5b The measurement of velocity

Several methods are available for measurement of the velocity of channel flow, and the choice will depend upon the magnitude and character of the channel and associated flow, cost, and the accuracy required. Values of velocity are usually reported in metres per second (m/s) or centimetres per second (cm/s).

The current meter is the most widely used apparatus. This consists of a rotor or series of cups which rotate at a speed proportional to the flow velocity. If the revolutions can be counted or recorded over a fixed period of time, then velocity can be computed from the calibration data. More specifically, current meters are usually classified into cup meters with a series of cups rotating around a vertical axis, or propeller-types where the axis is horizontal (Figure 3.13). The revolutions are indicated by a system of spring contacts or magnetic reed switches, and may be counted acoustically or by an electrical counter. A period of 60 seconds is usually used for timing the revolutions, in order to provide an integrated value of velocity, although direct reading meter circuits have been developed. The meter body may be mounted on a wading rod (Figure 3.13) or suspended from a cableway. In the latter case a tail fin attachment and a sinker weight are required to orientate and position the meter (Figure 3.13). Small wading rod meters such as the 'Mini Ott' meter are available for use in shallow streams.

MANUAL METHODS

GAUGE BOARDS

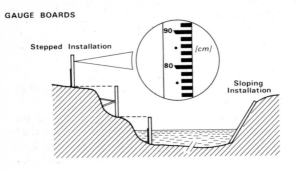

Stepped Installation

Sloping Installation

90
80
[cm]

CREST STAGE GAUGES

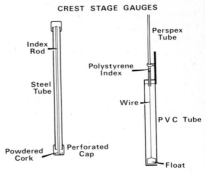

Index Rod

Steel Tube

Powdered Cork

Perforated Cap

Perspex Tube

Polystyrene Index

Wire

PVC Tube

Float

RECORDING INSTRUMENTS

VERTICAL FLOAT RECORDER

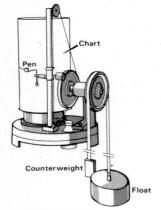

Chart

Pen

Counterweight

Float

STILLING WELL INSTALLATION

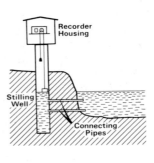

Recorder Housing

Stilling Well

Connecting Pipes

HORIZONTAL FLOAT RECORDER

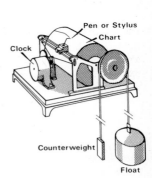

Pen or Stylus

Chart

Clock

Counterweight

Float

PRESSURE BULB RECORDER

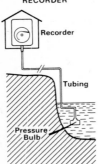

Recorder

Tubing

Pressure Bulb

BUBBLE GAUGE APPARATUS

A

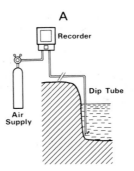

Recorder

Dip Tube

Air Supply

B

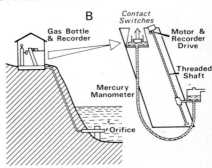

Contact Switches

Gas Bottle & Recorder

Motor & Recorder Drive

Mercury Manometer

Threaded Shaft

Orifice

A current meter can easily be used to measure the velocity at any point in the cross section, and several techniques have been evolved for measuring the mean velocity in the vertical. The Integration Method involves lowering and raising the meter at a constant rate not exceeding 5 per cent of the estimated mean flow velocity or 0·04 m/s, and timing the overall number of revolutions. More usually, however, the mean velocity is calculated from velocity values taken at specific depths. In the Two Point Method, the mean of the velocities at 0·2 and 0·8 depth is used, whilst the Three Point Method utilises the values at 0·15, 0·5 and 0·85 depth, and is employed where the velocity distribution is more irregular. When depths of less than 60 cm occur, the velocity at 0·6 depth is generally used to represent the mean velocity.

Floats provide a very simple means of measuring velocity, involving only data on time and length of travel, but they have the disadvantage over the current meter that they cannot be used to measure velocities at a specific point in the cross section. The value of velocity obtained will be integrated over the measuring distance and will reflect the flow path and the depth of flotation. According to constructional details, floats are usually classified into surface and rod types. Surface floats occupy less than 0·25 of the depth from the surface, whilst rod floats sink deeper. Correction coefficients can be used to obtain estimates of mean velocity in the vertical. The major application for floats is for measurement under flood conditions when other methods would be impractical, and cases have been reported where floats were dropped from aircraft flying over inundated areas.

Several other devices have been developed for velocity measurements in rivers, and these include the Pendulum current meter (Roche, 1963) and the velocity head rod (Wilm and Storey, 1944), a simple instrument for measuring flow velocities in shallow swift streams. A rather different approach to velocity measurement, and one especially suited to shallow turbulent streams where normal instruments are difficult to apply, is the use of tracers and, more particularly, the salt-velocity technique. Velocity data may be computed from the time of travel of a tracer between two points (for example, Calkins and Dunne, 1970). Common Salt (NaCl) is often used for this purpose, because the passage of the salt pulse past the measuring point can easily be detected using conductivity measuring apparatus. Velocity may also be calculated on a theoretical basis using equations of open channel flow, of which the Manning Formula is a simple example, viz.:

$$V = K \frac{R^{2/3}S^{1/2}}{n}$$

where   K = constant depending on units used
         V = mean channel velocity
         R = hydraulic mean radius
             (stream cross sectional area/wetted perimeter)
         S = Slope of the water surface
         n = manning's roughness coefficient

Figure 3.11   The measurement of stage
This figure shows a gauge board and a stilling well installation, and the principle of operation of two types of crest stage gauge and several types of stage recorder.

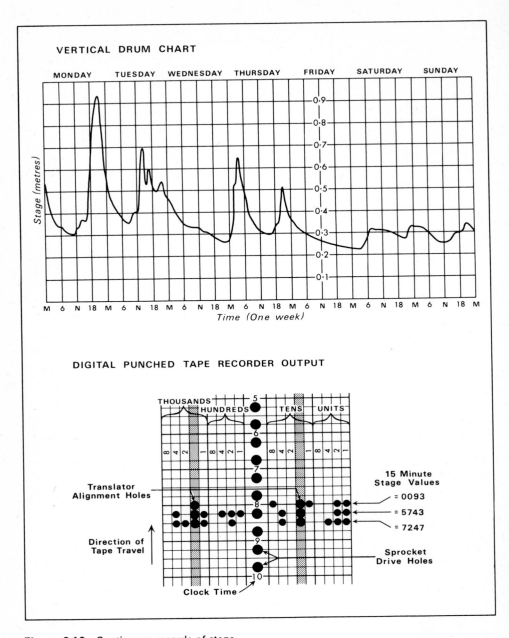

Figure 3.12 Continuous records of stage
A typical chart from a vertical float recorder and a portion of the paper tape output from a Fischer and Porter-type analogue-to-digital stage recorder.

Typical values of Manning 'n' are given in Table 3.4. This theoretical approach is particularly valuable when trying to study flood flow conditions after the event. The 'tidemark' left by the flood water may give sufficient indication of the flooded cross section and long profile to evaluate the Manning formula.

Table 3.4 Some values of Manning 'n' for natural stream channels

Type of channel	Minimum	Normal	Maximum
(a) Streams on plain*			
i Clean, straight, no deep pools	0·025	0·030	0·033
ii Clean, winding, some pools and shoals	0·033	0·040	0·045
iii Sluggish reaches, weedy, deep pools	0·050	0·070	0·080
(b) Mountain streams			
i Bottom: gravels, cobbles and few boulders	0·030	0·040	0·050
ii Bottom: cobbles, large boulders	0·040	0·050	0·070
(c) Flood plains			
i Pasture, short grass	0·025	0·030	0·035
ii Cultivated, no crop	0·020	0·030	0·040
iii Cultivated, mature crops	0·030	0·040	0·050
iv Medium to dense brush in winter	0·045	0·070	0·110
v Medium to dense brush in summer	0·070	0·100	0·160
* Top width at flood stage < 30m			

Source: Chow (1959)

The methods of velocity measurement described above are largely restricted to single points in time or integration over a short period. There has been little development in the field of continuous monitoring, although Finlayson (1969) has described the use of a continuous monitoring velocity transducer based upon a rotor and light chopper coupled to a phototransistor measuring circuit.

3.5c Measurement of stream discharge

Discharge is the most important parameter of channel flow and its measurement usually involves consideration of both stage and velocity. Units used are those of volume/time, and values are generally reported in cubic metres per second (m^3/s) or for small catchments in litres per second (l/s). The most commonly encountered methods of discharge measurement are the velocity area technique, dilution gauging, volumetric gauging, the slope area technique, weirs and flumes, and the rated section.

The velocity area technique is the most widely used method for spot measurements of discharge. Discharge is by definition the product of velocity and cross-sectional area

CURRENT METERS

VERTICAL AXIS
(Price type)
CABLE SUSPENSION

Conical Cup

Streamlined Weight

HORIZONTAL AXIS
(Ott type)
WADING ROD MOUNTED

Wading Rod

of flow, and this procedure evaluates these two terms for a stream section at a parti-cular point in time. It is difficult to determine the mean velocity of the overall cross section, and the channel is therefore divided into a number of segments, for each of which the mean velocity, the cross-sectional area, and the section discharge are determined. Total discharge is computed as the sum of the section discharge values.

A current meter is generally used for the velocity measurements, and the basic procedure is to select a series of verticals of known spacing and to determine the depth and mean velocity in each vertical (Figure 3.14). The use of a cableway facilitates this procedure on a larger river. The verticals are usually uniformly spaced, but the major consideration is to provide an adequate representation of the bed level elevations and the horizontal variations in velocity. As a general rule, the interval between any two verticals should not exceed 5 per cent of the overall width, and the discharge between two verticals should not exceed 10 per cent of the total discharge (World Meteoro-logical Organisation, 1965). In very uniform channels a smaller number of verticals may suffice. Two procedures are available for the calculation of section and total discharge data from the values of spacing, depth and velocity for the series of verticals, and these are the Mean-Section Method and the Mid-Section Method (Figure 3.14). This measurement technique can be time-consuming, and for accurate results the measuring section should conform to certain requirements, some of which are listed in Table 3.5.

Table 3.5 The velocity area technique—some requirements for a good measuring section

1 A regular and stable streambed
2 Velocity lines parallel and normal to the stream cross-section
3 Velocities exceeding 10–15 cm/sec
4 A depth of flow preferably in excess of 30 cm
5 No aquatic growth

Dilution gauging is often used as an alternative to velocity-area measurements at sites where excessive turbulence, high velocities and rocky or shallow sections would make the operation of a current meter difficult. The principle involved is that dis-charge may be calculated from the degree of dilution by the flowing water of an added tracer solution. There are upper limits on the size of a river that may be gauged, because the injected solution must mix uniformly with the flow and the result must permit detection of concentration. The choice of tracer is important, because it should conform to several specific requirements (Table 3.6). Sodium Chloride (NaCl) and Sodium Dichromate ($NaCr_2O_3$) are the two commonly-used chemical tracers which satisfy most of these conditions. Radioactive tracers such as Gold 198 (^{198}Au) possess very good detection properties, but may involve a health hazard.

There are two gauging techniques, namely, constant rate injection and gulp injec-tion. In both cases the tracer is injected at an upstream site and samples are collected at a downstream station where complete mixing has occurred. Dyes such as fluo-rescein and empirical formulae can be used to determine the length of channel reach

Figure 3.13 Current meters
This diagram demonstrates the principle of operation of vertical and horizontal axis current meters, and wading rod and cable suspension mounting of the meter body.

Table 3.6 Dilution gauging: some tracer requirements

The tracer should:
1 Dissolve readily in stream water at normal temperatures
2 Be absent or present only in small quantities in the natural streamflow
3 Remain stable in streamwater and not be absorbed
4 Be easily detectable at low concentrations
5 Be harmless both to handler and while in the river
6 Be of reasonable cost

required to attain complete mixing. In the constant rate injection procedure (Figure 3.15) the tracer solution is injected upstream at a constant rate, using a Mariotte bottle or constant head device, and measurements are made of the rate of injection, the concentration of the injected solution, and the base concentration of the stream.

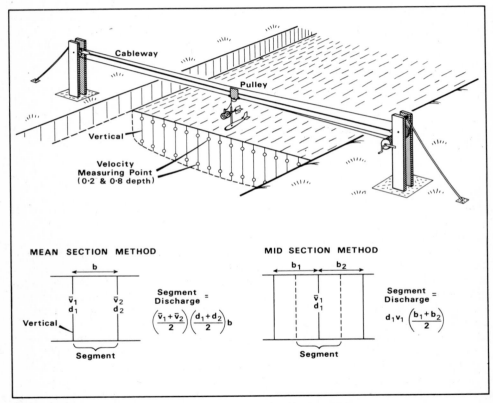

Figure 3.14 The velocity area technique of discharge measurement
A cableway is used on large streams for positioning the current meter in the verticals and a special cable drum can be used to obtain accurate readings of depth and spacing of the verticals. The mean section and mid-section methods are commonly used to compute the discharge of the individual segments.

Samples of the downstream mixture are taken at several points in the cross section and discharge may be calculated by evaluating the formula:

$$Q = q\frac{Ci - Cd}{Cd - Cb}$$

where Q = discharge of the measuring reach
   q = rate of injection of the tracer solution
   Ci = concentration of injected solution
   Cb = base concentration of the stream
   Cd = downstream sampled concentration

The gulp injection procedure is sometimes termed the 'Ionic Wave' technique, and in this method of gauging a slug or gulp of the tracer solution is introduced instantaneously into the stream and the passage of the slug past the downstream site is measured (Figure 3.15). Conductivity recording apparatus may provide a valuable means of recording the downstream passage of the tracer slug. Discharge (Q) is calculated as follows:

$$Q = \frac{(Ci - Cb)V}{\int_0^\infty (Cd - Cb)dt}$$

where V = volume of injected solution
   t = time
   Remaining notation as above

The $\left(\int_0^\infty (Cd - Cb)\ dt\right)$ term represents the area beneath the graphical plot of the 'Ionic Wave' passing the downstream station.

Volumetric gauging is probably the most accurate method of measuring discharge, but because it involves the collection of the total volume of flow over a given period of time (Discharge = Volume/Time), it is limited in application and presents several practical problems. The most obvious application is in measuring the outflow from a small runoff plot where the water can be collected in a large tank, and the accretion to this tank assessed by recording the increases in water level.

The slope area method involves theoretical estimation rather than direct field measurement. The Manning equation can be used to calculate the mean velocity of a channel reach (p. 127), and if this value is combined with the average cross sectional area, discharge values can be computed. Discharge is thereby estimated from data on water surface *slope* and channel cross sectional *area*. The method is particularly valuable for estimating flood flow after the event, when the 'tidemark' provides the necessary basis for deriving the slope and area values. Where floods are associated with inundation of flood plain zones, it may be necessary to treat the overbank portion of this discharge separately from the main channel.

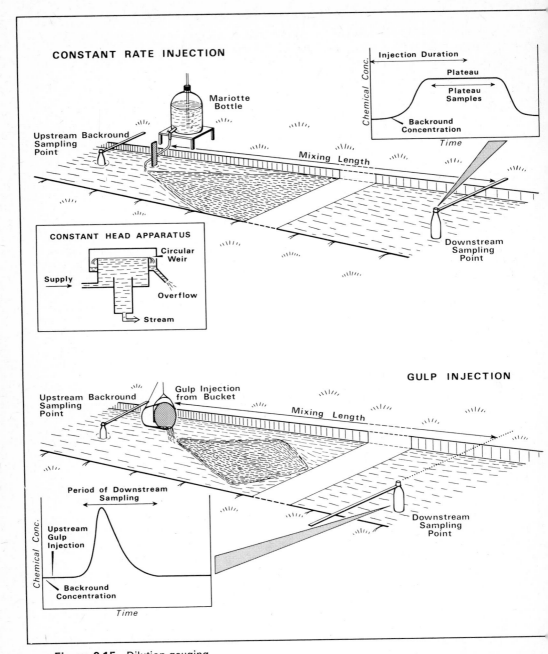

Figure 3.15 Dilution gauging
This figure illustrates the procedure used for dilution gauging by the constant rate injection and gulp injection techniques. A Mariotte Bottle or constant head apparatus is used in the constant rate injection technique in order to obtain a uniform rate of flow of tracer into the stream.

Control structures, or more specifically weirs and flumes, involve a similar theoretical basis to the velocity area method of discharge measurement. However, cross-sectional area is accurately known from the water level and the dimensions of the structure through or over which the water flows, and the flow velocity can be computed from the form of the control and the head of water above it. It is therefore possible to establish a theoretical and stable relationship between stage and discharge for a particular control structure, so that a simple stage measurement will be sufficient to provide a value of discharge. In contrast to the velocity area method, discharge measurements are therefore virtually instantaneous, although the high capital investment means that structures can only be built at a small number of permanent gauging points. The type of structure used, and the details of its design will depend upon the purpose of measurement, the nature of the stream channel, and the resources available (for example, U.S. Department of Agriculture, 1964).

Weirs may be classified into sharp-crested or thin plate weirs (Figure 3.16), and broad-crested weirs (Figure 3.17). In the former case the weir notch and crest is formed by a sharpened metal plate, whilst in the latter case a thicker construction, usually made of concrete, is used. Each type of weir may be further classified according to the form of the crest, and terms such as triangular or vee-notch, rectangular and trapezoidal are well known.

Sharp-crested weirs are most commonly used in small catchments, and the triangular form is especially suited to the accurate measurement of flows as low as 0·1 l/s, and small changes in discharge. The 90° notch is the most common triangular weir form, although 120°, $\frac{1}{2}(90°)$ and $\frac{1}{4}(90°)$ notches are also employed. Rectangular and trapezoidal forms are used where a greater discharge capacity is required, and a distinction is made between suppressed rectangular weirs where the vertical sides of the notch are formed by the channel walls, and contracted rectangular weirs where the width of the crest is less than the width of the approach channel (Figure 3.16). If a considerable range of flows are to be gauged, a compound notch may be used (Figure 3.16).

Many discharge formulae or rating equations are available for the different forms of sharp crested weir, and the final choice must depend upon the accuracy required and the similarity between a weir and the installation, often in a laboratory, for which a formula was developed. Some simple formulae are given in Table 3.7, and more detailed examples can be found in the work of Kindsvater and Carter (1959), King (1954), Troskelanski (1960) and the British Standards Institution (1965). In certain cases, field rating, using other methods of discharge measurement, and model tests may be required to substantiate or modify theoretically derived formulae. Furthermore, particular care must be taken in the siting and installation of a weir, if it is to provide accurate results. The measurement of the head of water above the weir crest (H, Table 3.7) is usually carried out at a distance two to three times the maximum head behind the crest, in order to avoid the effects of drawdown.

Broad crested weirs (Figure 3.17) have the advantage over sharp crested weirs that they are not easily damaged by debris, and they are, therefore, usually to be preferred for larger catchments. However, there is not such a great body of experimental data for use in deriving rating equations, and field rating is often necessary. Notch and crest design will often closely resemble that of the sharp crested weir, although certain special forms also exist. A Crump weir incorporates a crest with a

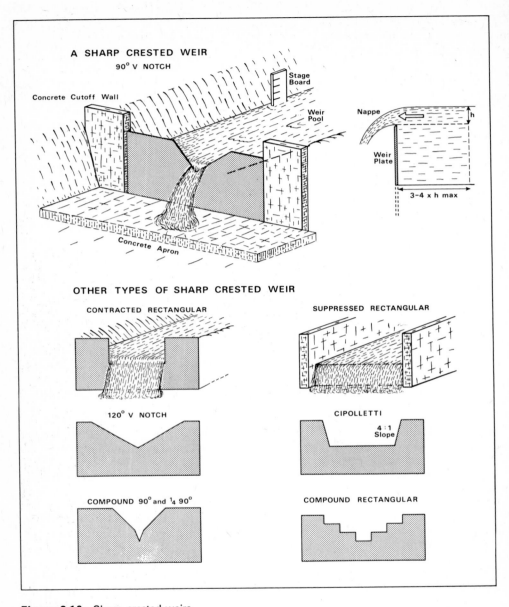

Figure 3.16 Sharp-crested weirs
The form and installation of a 90° V-notch weir and the crest shapes of several other types of sharp-crested weir are illustrated.

Table 3.7 Discharge formulae for sharp-crested weirs

Notch form	Formula
90° V-Notch	$Q = 1.38 \ H^{2.5}$
120° V-Notch	$Q = 2.47 \ H^{2.5}$
$\frac{1}{2}(90°)$ V-Notch	$Q = 0.69 \ H^{2.5}$
Rectangular	$Q = 1.86 \ BH^{1.5}$
Cipolletti	$Q = 1.86 \ BH^{1.5}$

where Q = Discharge in m^3/s
H = Head of water in metres
B = Width of weir crest in metres

(The constants and exponents cited in these formulae are only approximate values)

$1:2$ upstream slope and a $1:5$ downstream slope, and the flat-vee design combines the Crump crest form with a wide angled, triangular cross section in order to permit the accurate gauging of low flows (Figure 3.17 and Plate 11). Compound designs may

Plate 10 Stream gauging structure
A small standing wave trapezoidal flume installed by the Kent River Authority on the River Teise at Bewl Bridge. The stage recorder is housed in the hut on the far bank.

BROAD - CRESTED WEIRS

TRIANGULAR

Stage Board

1:2 Slope

Concrete Structure

h

FLAT - VEE

Stage Board

Concrete Crest

1:5   1:10

1:2

Crest Construction

COMPOUND CRUMP

Crest Dividing Walls

Lower Central Section

1:5   1:2

Crest Section

involve crests at different levels or a combination of crest forms. The general formula for flow over a rectangular broad crested weir is:

$$Q = 1 \cdot 7BH^{1 \cdot 5}$$

See Table 3.7 for notation.

Although concrete is generally used for the weir crest, other materials have been used and Smith and Lavis (1969) have described the use of marine plywood for constructing a flat-vee weir.

The flume differs in concept from a weir in that flow velocity and head are controlled not by the water falling over a crest under the influence of gravity, but by the passage of water through a constricted section where critical velocity is attained. The constricted section or throat is formed by raising the channel floor into a hump, or by contracting the sides, or by a combination of both measures. The cross-sectional shape of the throat may be varied in a similar manner to a weir notch, depending on the range and magnitude of flow, and triangular, rectangular and trapezoidal forms are commonly used (Figure 3.18). The advantage of a flume over a weir lies in the much smaller loss of head or difference in upstream and downstream water levels, and the largely self-cleaning properties. Weir installations are often plagued by the accumulation of silt and debris behind the crest, but this material is carried through a flume.

Several types of flume have been described in the literature. The critical depth or standing wave flume is the simplest form (Figure 3.19 and Plate 10) where measurements of head upstream of the throat may be directly related to discharge. Parshall and Venturi flumes can be used to measure flow under submerged or non-critical flow conditions because head is measured both within and upstream of the throat, and the difference in head related to discharge. The HS, H, and HL type flumes were specially developed by the Soil Conservation Service of the U.S. Department of Agriculture (Harrold and Krimgold, 1943), in order to provide a prefabricated metal structure for installation at the end of a spillway channel (for example, a runoff plot) and requiring no pondage (Figure 3.18). Their major application is for measurement of intermittent flow, and they are not recommended for perennial streams with sustained low flows because of head measuring difficulties. The San Dimas flume (Wilm *et al.*, 1938) was specially designed for use in mountain streams with a high debris load, and the Hydraulics Research Station in Britain has developed a special flume for use in steep-floored streams (Harrison, 1966) (See Plate 12).

Flume construction requires considerable care and expertise, because the dimensions are often very critical and a smooth finish to the throat constriction is required. Concrete is normally used, but marine plywood (Barsby, 1963) and glass fibre (Drost, 1966; Wright, 1967) have been used to build prefabricated sections in order to avoid the need for expensive shuttering. Because of constructional difficulties and high cost, flumes are usually restricted to small catchments where very accurate flow records are required. Calibration may be achieved by theoretical formulae, model tests or field rating. Although weirs and flumes have been considered separately,

Figure 3.17 Broad-crested weirs

This figure demonstrates the form and installation of three types of broad-crested weir.

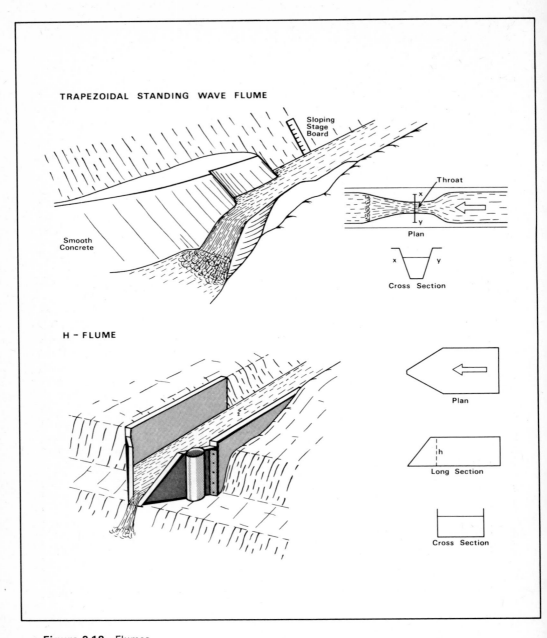

TRAPEZOIDAL STANDING WAVE FLUME

Sloping Stage Board

Throat

Smooth Concrete

Plan

Cross Section

H - FLUME

Plan

Long Section

Cross Section

Figure 3.18 Flumes
Two commonly used types of flume are illustrated. The trapezoidal flume is used in natural stream channels whereas the simpler H-flume is used to gauge intermittent streams and runoff plots where the watercourse can conveniently be contained within a spillway channel.

compound gauging structures may be built combining both, and Plate 11 shows an installation with a flume to measure low flows extending into a broad crested weir to contain the high flows.

The exact choice of control structure involves many considerations, but the magnitude and range of flows to be measured is perhaps the most important. In Table 3.8 the approximate maximum and minimum flow levels are given for several types of structure. All installations require careful planning, site investigation, and construc-

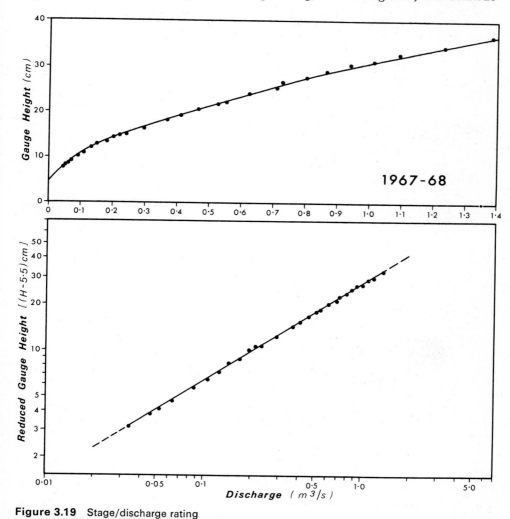

Figure 3.19   Stage/discharge rating
This diagram illustrates the rating data for a small stream section used by the authors as a gauging station. In A the data has been plotted on linear coordinates and in B a straight-line relationship which can be tentatively extended has been obtained by logarithmic plotting and reduction of the gauge height values so that zero stage corresponds to nil discharge.

Table 3.8 Approximate maximum and minimum discharge levels for various control structures

Type	Minimum	Maximum
Sharp-crested weirs		
90° V-notch, 0·5 m high	0·20 l/s	240 l/s
120° V-notch, 0·5 m high	0·45 l/s	430 l/s
½(90°) V-notch, 0·5 m high	0·10 l/s	120 l/s
Rectangular, 1·0 m high, 3·0 m wide	30·0 l/s	5·6 m³/s
Cipolletti, 1·0 m high, 3·0 m wide	30·0 l/s	5·6 m³/s
Broad-crested weirs		
Triangular, 1 : 2 side slopes, 2 m high	0·50 l/s	15·0 m³/s
Rectangular, 1·0 m high, 3·0 m wide	27·0 l/s	5·1 m³/s
Flat-vee, 1·0 m high, 3·0 m wide	2·8 l/s	6·0 m³/s
Flumes		
H-type, 30 cm high	0·10 l/s	56·0 l/s
H-type, 1·37 m high	0·40 l/s	2·4 m³/s
Trapezoidal, 10 cm wide throat, 45° sidewalls, 0·5 m high	0·90 l/s	280 l/s
Trapezoidal, 25 cm wide throat, 30° sidewalls, 1·2 m high	4·0 l/s	10·0 m³/s

tion. Cutoff walls, piling, upstream box structures and grout curtains may be necessary to provide a leakproof gauging station, and in spite of these measures the possibility of deep-seated leakage beneath the structure must not be ignored. Downstream, bank piling, rip rap, concrete aprons, and bed stabilisation with piles and large boulders may be necessary to prevent scour and undercutting.

The rated section technique enables the principles underlying control structures to be extended to situations where use of a structure would be impractical. A rated section utilises a natural control to provide a stable relationship between stage and discharge. The control may be the outcrop of a hard rock band across the stream, or more simply a stable channel reach, and bed and bank stabilisation measures may be employed to improve the stability of the control. The stage/discharge relationship is derived by field rating, using in most cases the velocity area technique, and most large rated sections are equipped with current meter cableways for this purpose (for example, Figure 3.14). A rating graph (Figure 3.19) is constructed from a large number of measurements of discharge and associated stage, taken at all levels of flow. A logarithmic plot is generally employed in order to obtain a straight line rating or a series of straight line segments, and several methods exist for adjusting the data to provide a straight line plot and for tentatively extending the line beyond the limits of field rating. The greatest problems associated with a rated section are those of instability of the control necessitating frequent re-rating, the long period needed to complete the rating plot, and the possible need for extension of the plot to include the maximum levels of flow. The distinction between a control structure and a rated section is not always well

Plate 11 Stream gauging structures
Above, a flat-vee Crump weir used by the authors for measuring discharge from a small catchment in east Devon. The stilling well and recorder box are located on the far bank.
Below, a compound structure comprising a broad-crested weir and central flume notch used by the Kent River Authority for gauging the River Beult at Stile Bridge.

Plate 12   Stream gauging structure
This flume structure installed by the Institute of Hydrology on one of the Wye sub-catchments at
Plynlimon (*cf.* Figure 3.27) was specially designed for use on steep mountain streams. Problems
to be overcome included bed gradients of about 1 in 45, Froude numbers in the range 0·75–1·2,
considerable bedload transport at high discharges and flood/drought discharge ratios of about
1000. Energy dissipating baffles have been placed on the approach ramp.

defined because a broad crested weir may require a complete field rating for calibration purposes. Furthermore, in some cases a control structure may be used to measure
low flows and a rated section employed for the higher flows when the structure is
drowned out.

Continuous measurement of discharge must also be considered because the methods of
measurement referred to above have been described primarily in the context of spot
or point measurements in time. However, the logical and necessary development of
any programme of measurement is towards continuous records. The principle underlying continuous measurement is the use of a stage/discharge relationship so that the
continuous record of river stage provided by a water stage recorder (Figure 3.12)
can be converted to discharge. Control structures and rated sections provide the
necessary stage/discharge relationships, and it is unlikely that either of these installations would be used solely for point measurements. A gauging station can, therefore,
be established at the outlet of a drainage basin or at a specific point on a stream, in
order to provide a continuous record of streamflow. Rated sections are used extensively for gauging large rivers where structures would prove costly, but this may,

nevertheless, sometimes prove to be a false economy in view of the high recurrent expenditure necessary to obtain and maintain the field rating data. In general, a good control structure can provide a discharge record accurate to within $\pm 1-2$ per cent, whilst the errors involved in a good rated section will be about ± 2 per cent. Accurate discharge records depend heavily upon a good stage record as well as the stage/discharge relationship.

MEASUREMENT OF SEDIMENT AND SOLUTE DYNAMICS

The documentation of sediment and solute dynamics within a drainage basin is important to the geomorphologist in providing an index of the effects of water on the landscape and the rates of operation of fluvial processes. Many of the measurements described previously are important to an understanding of these dynamics, but there are also certain techniques directed specifically towards the assessment of the processes of erosion, transportation and deposition within a catchment. A description of these techniques is best achieved by subdivision into studies of, firstly, sediment dynamics and involving the land or slope phases, the channel phases and the overall yields; secondly, solute production and transport; and thirdly, the associated rates of erosion or degradation.

3.6 Sediment on slopes

Slope processes have provided an important focus of study for the geomorphologist and many techniques of documentation have been developed. However, a study of fluvial geomorphology must place particular emphasis on those processes concerned with the erosive action of water on the slopes and the supply of material to the channel at the foot of the slope. Data is usually concerned with depths of material removed (mm), or, where removal is more local, with volumes of material (m³ or tonnes). Rates of removal may also be assessed (m³/yr, mm/yr).

3.6a Rainsplash erosion

Several workers have attempted to measure the detachment or splash of soil particles under the impact of raindrops, but this is a difficult task since it involves assessing the movement of particles in the zone above the soil surface. Any measuring apparatus must not disturb the incident rainfall. High speed photography may be used to demonstrate the explosive effect of raindrop impact (Ellison, 1950), but more quantitative measurements are usually required. A simple gravimetric approach to the direct measurement of soil splash involves the use of trays of soil or small field plots surrounded by metal frames or troughs to catch splashed material. Osborne (1953) has developed this technique further to include the use of a rainfall simulator. The portable rainfall-simulator infiltrometer described by McQueen (1963) (Figure 3.4) also has provision for a splash collecting ring surrounding the circular infiltration plot. However, this approach only assesses the amount of splash across the boundary of the tray or plot. Another, although more indirect, technique is the use of radioactive tracers which are added to a small and well defined area of the soil surface and which

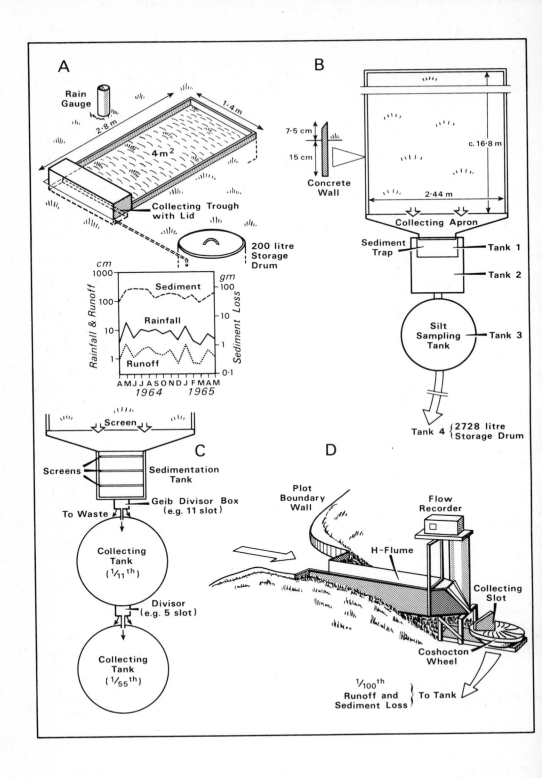

A

Rain Gauge

2·8 m
1·4 m
4 m²

Collecting Trough with Lid

200 litre Storage Drum

cm
1000
Sediment
gm
100
Rainfall
100
10
10
Rainfall & Runoff
1
1
Sediment Loss
Runoff
0·1
A M J J A S O N D J F M A M
1964   1965

B

7·5 cm
15 cm
Concrete Wall

c. 16·8 m
2·44 m

Collecting Apron

Sediment Trap
Tank 1
Tank 2

Silt Sampling Tank
Tank 3

Tank 4 { 2728 litre Storage Drum

C

Screen

Screens
Sedimentation Tank

To Waste
Geib Divisor Box (e.g. 11 slot)

Collecting Tank (1/11 th)

Divisor (e.g. 5 slot)

Collecting Tank (1/55 th)

D

Plot Boundary Wall

Flow Recorder

H-Flume

Collecting Slot

Coshocton Wheel

1/100 th Runoff and Sediment Loss } To Tank

enable the movement of splashed material to be traced using scintillation detectors and radiographic films. Iron 59 (^{59}Fe) is a useful tracer because it is not easily leached into the soil and it possesses a short half-life, and this has been used by Coutts et al. (1968) and Wooldridge (1965).

3.6b Sheet erosion

Soil splash is a component of sheet erosion, and in many instances it will be sufficient to measure the overall effects of sheet erosion. Apparatus for collecting sheetflow and the associated entrained sediment can be essentially the same as that employed for collection of overland flow (Section 3.4c). Gerlach troughs may be used for simple measurements on hillslopes, and runoff or erosion plots and portable rainulator devices (Kazo and Klimes-Szmik, 1962) may be used for more intensive studies. Erosion plots may be as large as 0·1 hectares (Van Doren et al., 1950), and they are surrounded by cutoff walls designed to collect the runoff and entrained sediment at a downstream outlet (Figure 3.20). Where the volume of runoff is too great to be collected in a tank (Figure 3.20A, B), a sampling device is usually employed to divert only a small proportion or aliquot sample of the outflow into the tank. The Geib multislot divisor (Geib, 1933) is an example of a fraction divider consisting of a metal box in which the water passes through a series of vertical slots of equal size. The outflow from one of these slots is collected, and a five-slot divisor will, therefore, collect 20 per cent of the total outflow (Figure 3.20C). Further divisor boxes may be incorporated to provide a smaller fraction. Flumes can be used to measure the volume of runoff from large plots, and these can be combined with Coshocton Wheel sampling devices (p. 166) which collect a small fraction of the total outflow from the flume (Figure 3.20D). Although the utilisation of artificial rainfall may extend the scope of an erosion plot study, careful thought must be given to the degree of simulation of water quality, drop size, terminal velocity and other rainfall parameters. Furthermore, a plot may not be large enough to reproduce the true erosive effects of overland flow and the concentrations of flow occurring in nature.

Radioactive tracers can also be applied to studies of sheet erosion, and De Ploey (1967) describes the use of ground glass sand labelled with Scandium 46 (^{46}Sc). In his study the labelled sand was placed on the ground surface, and the extent of sheet erosion was determined by measuring the residual concentration and the pattern of dispersion after a storm. Samples placed beneath a shield were also used in order to isolate the effects of sheetwash by eliminating rainsplash.

In some circumstances, subsurface movement of sediment, particularly colloids, through the soil and in piping systems may be an important process of sediment removal associated with sheet erosion. Throughflow troughs and trenches can be

Figure 3.20 Runoff and erosion plots
Small erosion plots of the type shown in A were used by Soons (1967) and Soons and Rainer (1968) to study runoff and sediment production in New Zealand. More permanent plots (B) are used by various research organisations and where the volume of water and sediment is too great to be collected in the tanks a series of divisor boxes is often used (C). H-flumes and Coshocton Wheel samplers can be used on large plots (D) to continuously record flow and to collect a small aliquot sample of the runoff.

modified to collect this sediment, and Schick (1967) has described a special Gerlach trough which may be used for this purpose.

An alternative approach to the assessment of sheet erosion involves measurement of the rate of lowering of the ground surface and calculation of the volumes removed. Values obtained in this way provide a useful check against direct measurements of removal rates. Erosion pins can consist of a 25 cm nail with a washer below the head, driven into the ground so that the washer is flush with the soil surface (Emmett, 1965). Subsequent measurements of the gap between the nail head and the washer, or accumulation above the nail head, provide an indication of surface lowering or accretion respectively. A large number of pins will be required to provide a representative result for an area. Where surface configuration changes markedly as a result of sheet erosion, periodic surveying using permanent benchmarks may also provide useful data.

3.6c   Rill and gully erosion

The assessment of rill and gully erosion can be approached in two ways. Firstly, measurements can be made of the material discharged by the stream at the outlet of a gully, and an attempt made to relate this transported material to the overall volume of erosion. Methods for studying the transport of material by a stream are considered in detail in section 3.7, and where check dams have been constructed, observations on rates of sedimentation can provide useful data. This technique is only really applicable to sizeable gullies with well defined stream courses. Alternatively, attempts may be made to document the development of the rill or gully system in both section and plan by periodic surveying and to calculate the amount of material removed (Tuckfield, 1964). Repeated aerial photographs may also prove valuable for demonstrating the development of these erosion phenomena (Stehlik, 1967). General studies of rill and gully growth can be supplemented by more intensive studies at individual sites using erosion pins.

3.6d   Mass movement

Mass movement must be included within a study of the slope phases of fluvial processes both for the purposes of comparison of its efficacy with sheet and gully erosion, and in its capacity as a supplier of material to the stream at the base of a slope. In many areas this source of supply will be unimportant, and Emmett (1965) has shown that in the south-western United States it constitutes less than 1 per cent of the amount contributed by sheet erosion. However, in areas with steep and unstable slopes such as occur in the Gisborne and East Coast area of New Zealand and in the San Gabriel Mountains of southern California, mass movement can provide a major sediment source (Campbell, 1955; Krammes, 1965). Furthermore, it is possible to view mass movement as a variant of water-borne movement where the proportion of water is very low. Many measuring techniques have been developed, the precise details varying with the type of movement and the site, and in Table 3.9 brief description is given of some of these techniques.

The application of these methods involves many problems including the disturbance of natural soil or regolith conditions caused by the apparatus and the relocating of

Table 3.9 Techniques for measuring mass movement

Type	Worker	Method
Rockfall	Rapp (1960)	Measurement of the accumulation on sack carpets and wire netting
	Prior et al. (1971)	Box traps (0·2 m²)
Snow avalanche	Luckman (1971)	Measurement of the accumulation on polythene sheets (2·33 m²) and on marked boulders
Debris movement	Krammes (1965)	Measurement of accumulation in troughs and behind barriers
Earthflow	Campbell (1966)	Continuous recording using a modified water stage recorder
Mudflow	Prior et al. (1971)	Movement of surface peglines
	Prior and Stevens (1971)	Continuous recording using a modified Munro water-level recorder
	Hutchinson (1970)	Movement of surface pegs, and deformation of vertical flexible tubes measured with an inclinometer
Solifluction	Rapp (1967)	Annual checking of downslope and transverse test lines marked by stakes and painted boulders
	Benedict (1970)	Movement of surface stakes and assessment of vertical velocity profile using cement rods
	Williams (1957)	Measurement of deformation of vertical plastic tubes by means of strain gauges
Soil creep	Miller and Leopold (1963)	Movement of metal stakes set along contours
	Hadley (1967)	Movement of glass beads embedded in an auger hole
	Rudberg (1967)	Movement of lines of painted rocks and deformation of plastic or wooden 'test pillars' placed vertically in the ground
	Kirkby (1967)	Movement of tilt bars or T-bars
	Young (1960)	Movement of pins set into the side of a re-excavated pit
	Emmett and Leopold (1967)	Movement of small aluminium plates set vertically in the side of a pit
	Selby (1968)	Movement of a wire attached to an aluminium 'creep cone' buried in the soil
	Everett (1962)	Observation of the movement of small aluminium pressure plates connected to potentiometers

markers or pits, and they are primarily suited to studies of long term movement. The problems of obtaining short term measurements have to some extent been overcome, in the case of soil creep, with the use of T-bars, and attempts at continuous recording of earth flow and mud flow movements have been described by Campbell (1966) and Prior and Stephens (1971). Detailed measurements of movement at depth can be made by observing the deformation of flexible tubes (e.g. Hutchinson, 1970).

3.6e   Deposition

In many cases, material eroded from one point on a slope may be deposited lower down that slope before reaching the vicinity of the stream channel. Values of net erosion must take into account measurements of both gross erosion and deposition on a slope. Many of the techniques described for documentation of sheet, rill and gully erosion can be extended to measurement of deposition. For example, accretion above the head of an erosion pin or above a survey profile can readily be measured.

3.7   Sediment in channels

Documentation of channel processes is in some respects easier than that of slope processes, because the channel provides a well defined site and the spatial sampling problems are, therefore, not as pronounced. A description of the techniques available is best approached using the traditional division into erosion, transportation and deposition.

3.7a   Channel erosion

Techniques for documentation of channel erosion are primarily based upon assessment of changes in channel form and calculation of the volume of material removed. The methods used are similar to those employed for monitoring changes in surface levels on slopes, and include erosion pins inserted into the banks, measurements to reference pegs and periodic surveys of both channel cross section and long profiles (Leopold, Emmett and Myrick, 1966). Long-term studies could be tentatively based upon evidence from maps and aerial photographs (for example, Daniel, 1970). Detailed studies can also be made of the development of individual features such as small waterfalls and meander scars. Local bed scour can be documented using erosion chains (Emmett, 1965) which consist of short lengths of chain buried vertically in the stream bed and indicating the depth of subsequent scour by the number of links exposed or laid over in a horizontal position. Laboratory studies of the relative erodibility of different bank and bed forming materials could be achieved using an artificial flume apparatus (Selby, 1970).

The effects of the forces of corrosion, corrasion and attrition on the bed material and the transported load are more difficult to assess. Newson (1971) has attempted to document these processes within karst subterranean drainage systems by studying the changes in size of transported sediment between swallet and resurgence and by using tablet-weighing experiments. Limestone tablets suspended in the channel and surrounded by nylon netting in order to exclude sand and gravel provided an indication of solution effects; unprotected limestone tablets were subject to both solution and abrasion, whilst sandstone tablets primarily demonstrated the effects of abrasion. These techniques could be extended to natural channels. Because abrasion is itself likely to remove any painted mark or index, it is difficult to monitor the effects of abrasion on labelled bed material, although if a naturally distinguishable source of bed material, such as quartz vein, can be located, downstream changes in roundness and size may be assessed. Less quantitative evidence can be provided by the general downstream increase in bed material roundness (Figure 2.11) exhibited by most stream channels (Gregory and Walling, 1971).

3.7b   Sediment transport

Assessment of the transport of sediment through a stream system requires quantitative data on the velocities (m/sec or cm/sec) and rates (kg/sec) of sediment movement, and the associated variations both within the channel cross section and through time. Attempts can be made to estimate this data, from theoretical considerations of the properties of the sediment load and the transport potential of the channel flow, but in the absence of a generally applicable and accurate theoretical approach, direct measurement must provide the major source of information. Furthermore, any theoretical procedure will require to be evaluated against field data. Although it is possible to measure the total sediment load of a stream as one component using a turbulence flume or turbulent river section where all transported load is in suspension (Benedict et al., 1955), this approach is limited in application, and it is more usual to measure the components of suspended load and bedload separately.

3.7c   Suspended sediment transport

It is difficult to calculate suspended sediment transport on a theoretical basis, because although the suspended bed-material fraction can possibly be viewed as a capacity load that is related to flow parameters (Einstein, 1950; Lane and Kalinske, 1941), the wash load fraction is essentially a non-capacity load. The magnitude of the wash load component will depend upon the supply of suitable sediment, and in most circumstances the transporting capacity will exceed this supply. A theoretical estimation procedure would need to consider conditions within the entire watershed. Catchment simulation models would seem to offer most promise, and the work of Negev (1967) and Fleming (1968) on the Stanford Sediment Model is noteworthy in this field.

Measurements of suspended sediment concentration are of primary importance because most assessments of suspended sediment transport are based upon values of sediment concentration (mg/l). Concentrations vary in a stream cross section, and therefore a sampling device is required to provide a representative sample of the concentration at a particular point or zone within the channel. Sampling instruments range in sophistication from rope-suspended buckets, through simple dip-bottles, to the sophisticated apparatus developed in the United States by the Federal Interagency Sedimentation Project (Witzigman, 1965). However, not all samplers will provide a representative sample, because three major requirements must be fulfilled. Firstly, flow velocity at the sampler intake should be the same as stream velocity; secondly, the presence of the sampler should cause a minimum disturbance effect on the flow

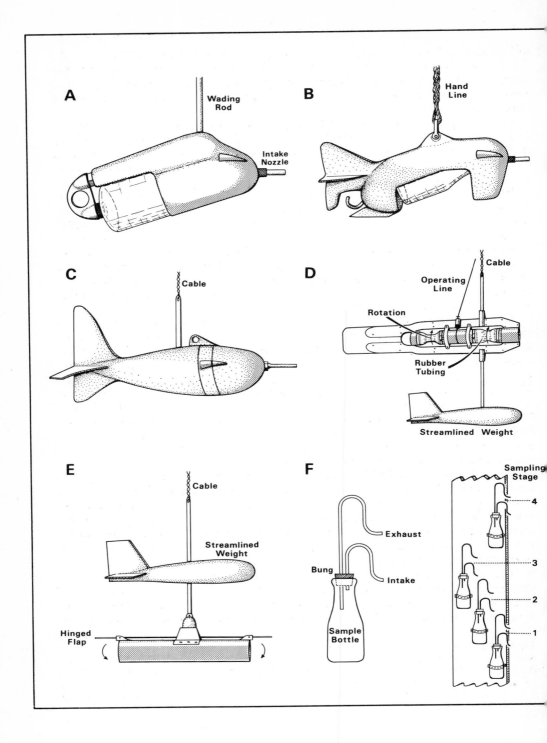

A

Wading
Rod

Intake
Nozzle

B

Hand
Line

C

Cable

D

Cable

Operating
Line

Rotation

Rubber
Tubing

Streamlined Weight

E

Cable

Streamlined
Weight

Hinged
Flap

F

Exhaust

Bung

Intake

Sample
Bottle

Sampling
Stage

4

3

2

1

at the sampling point; and thirdly, the sampler intake should be orientated into the flow in both the vertical and horizontal planes. Most purpose-designed sampler bodies are therefore streamlined or fish-shaped, and equipped with orientating fins. A major drawback with nearly all samplers is that they do not permit sampling close to the stream bed.

Many sampling instruments have been specially developed for river measurements and a distinction is generally made between instantaneous, point-integrating, and depth-integrating devices. Recent developments have also introduced pumping samplers for automatic operation and apparatus for continuous monitoring of suspended sediment concentrations.

Instantaneous samplers incorporate a horizontal tube which is lowered into the stream and aligned parallel to the flow by a fin or by the sinker weight (Figure 3.21E). The ends of the tube are then closed with spring flaps operated by an electrical solenoid catch, a trigger and string, a weighted messenger, or a dissolving capsule (Fleming, 1969). The Tait-Binckley sampler (Stichling, 1969) is a variant upon this type in which the tube is sealed by rotating it against rubber gaiters connected to either end (Figure 3.21D). The prime disadvantage of instantaneous samplers is that the sample obtained makes no allowance for the fluctuations in concentration associated with flow turbulence.

Point-integrating samplers overcome the major disadvantages of the instantaneous type in that the sampler intake nozzle can be opened at the required point in the channel and the sample container allowed to fill during a short time interval. The sample is therefore time-integrated. A solenoid valve is generally used to control the nozzle and provision must be made for a separate exhaust for air displaced from the sample bottle. The U.S. Federal Inter-Agency Sedimentation Project has developed a streamlined sampler design (Figure 3.21C) which includes the USP 61 sampler, a 48 kg bronze-bodied instrument 71 cm long (Plate 13), and the larger USP 63 model (91 kg) (Guy and Norman, 1970). The 90 kg Neyrpic sampler developed in France (Roche, 1963), is similar in design to the American instruments except that a continuous flow of compressed air is used to seal the intake nozzle.

The Delft sampler (N.E.D.E.C.O., 1959) used in some European countries operates on a different principle in that it collects a sample of the transported sediment rather than a representative water sample. The stream filament enters the nozzle and flows into a wide chamber where, as a result of the sharp decrease in velocity, the particles settle out. The water leaves the sampler at the tail, and it is claimed that the majority of the suspended load will have been trapped, although particles less than 0·05 mm in diameter may not be deposited.

Depth-integrating samplers are generally similar to point-integrating devices, except that a simple open nozzle is used, and the sampler is lowered to the stream bed and raised to the surface at a constant rate. The depth-integrated sample is therefore collected at a rate proportional to the flow velocity at a particular level in the vertical,

Figure 3.21 Suspended sediment samplers
Many types of suspended sediment samplers exist and this figure shows the USDH 48 depth integrating wading rod sampler (A), the USDH 59 depth integrating handline sampler (B), the USP 61 point integrating cableway sampler (C), the Tait-Binckley instantaneous horizontal sampler (D), a simple flap-type instantaneous horizontal sampler (E) and a series of rising stage sampling bottles (F).

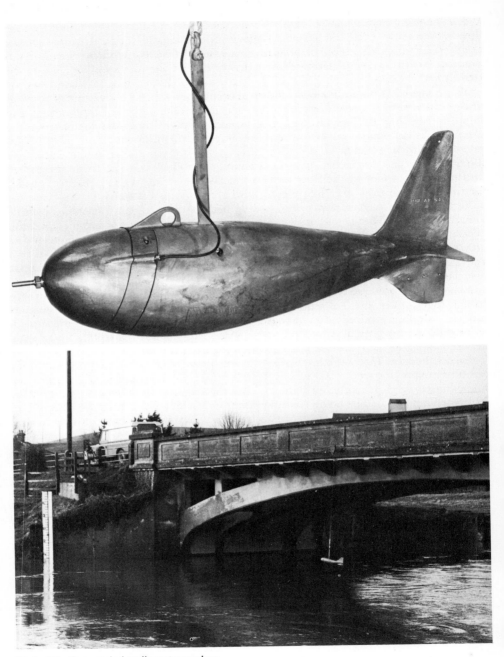

Plate 13 Suspended sediment samplers
Above, a USP-61 point-integrating suspended sediment sampler (weight 28 kg).
Below, a USD-49 depth-integrating sampler being used from a cableway over the River Exe at Thorverton.

and its concentration will represent the discharge-weighted mean concentration for that vertical. The U.S. Federal Inter-Agency Sedimentation Project has designed a range of 'fish-shaped' depth-integrating samplers including the USDH 48 wading rod sampler (Figure 3.21A), the USDH 59 handline sampler (Figure 3.21B) and the USD 49 cableway sampler (Guy and Norman, 1970) (Plate 13). All these samplers collect a volume of approximately 400 cc, although the USP 63 can collect 800 cc. A simple wading rod depth-integrating sampler can be made from a milk bottle and small bore copper tubing (Gregory and Walling, 1971). Particular care is required when using this type of sampler in determining the vertical transit rate. The transit rate is governed by the flow velocity and by the capacity of the sample bottle which should not become full before the end of its travel. In general it should not exceed 40 per cent of the maximum velocity in the vertical, and in deep streams the nozzle diameter can be reduced to slow the filling of the bottle.

Single stage samplers are essentially a type of point-integrating device which has been developed to meet the need for automatic sampling. They usually consist of a series of bottles mounted on a rack and which fill when the stream reaches a series of predetermined levels. The U.S. Federal Inter-Agency Sedimentation Project (F.I.A.S.P., 1961) has designed a simple arrangement of siphon-shaped air exhaust and intake tubes to fill bottles at successive levels of a rising stage (Figure 3.21F). A similar apparatus has been described by Schick (1967), and devices which sample on the falling stage could also be developed (for example, Knedlhans, 1971). These samplers provide valuable flood event data that might have been missed by a programme of manual sampling, but it is important to define the relationship between samples taken in this manner at the side of a stream and natural mid-stream concentrations.

Pumping samplers provide a further refinement in automatic sampling because they can be designed to collect samples at preset time intervals or when switched on by the rising water levels at the beginning of a flood event. The USXPS 62 sampler (F.I.A.S.P., 1962), a prototype pumping sampler, includes provision for three types of handling system for the pumped samples, firstly, individual sample bottling (145 bottles), secondly, an accumulative weight recording system, and thirdly, a volume recording apparatus utilising sedimentation tubes. Simple portable samplers have also been described by Doty (1970) and Walling and Teed (1971). In all cases the sampling inlet must be carefully designed and the concentrations of the pumped samples calibrated against the results of hand sampling.

Continuous monitoring apparatus can provide very valuable data on variations in sediment concentration through time at a particular point within the stream channel. Two major approaches have been followed; firstly, photo-electric turbidity measurements and, secondly, radio isotope measurements. The major problem involved is calibrating the monitor output in terms of suspended sediment concentration. The photo-electric devices are based upon the principle of measuring the reduction in light reaching a photo-cell due to the presence of solid particles suspended between the light source and the photo-cell. In detail, designs vary and Fleming (1969) has compared three instruments, the Southern Analytical Suspended Solids Monitor (Thorpe, 1964), the Davall (Instanter) Siltmeter (Jackson, 1964) and one designed by himself based on a design from the British Water Pollution Research Laboratory. The first instrument requires a pump to draw water from the stream and through the measuring cell, whereas the probes of the other two instruments can be placed directly

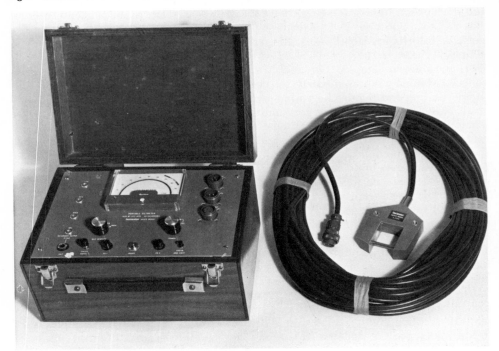

Plate 14   A turbidity meter
This Instanter photo-electric turbidity meter was developed by the British Transport Docks Board. The probe (*right*) which contains a light source and a photo-electric cell, can be mounted in a stream channel and readings of relative turbidity can be read from the panel meter or recorded on a separate recorder.

in the stream (for example, Plate 14). Algae growth on cell windows and ambient light can present problems with these instruments, and the calibration will be sensitive to changes in sediment particle size characteristics.

Radioisotope gauges operate on the principle of comparing the attenuation of X-rays passing through the ambient turbid water and a clear water standard. Cadmium (^{109}Cd) has been used as the radioactive source, and instruments have been described by Ziegler *et al.* (1967) and Florkowski and Cameron (1966). Field tests in New Zealand (Morrissey, 1970) and in the United States (Murphree *et al.*, 1968) have showed promising results and an instrument is now commercially available (Parametrics Inc., U.S.A.). However, cost, the need for replacement of the radioactive source at twelve-monthly intervals, and a general insensitivity to concentrations below 1000 mg/l limit the application of these devices.

Measurements of suspended sediment discharge for the overall channel may be required in addition to assessments of sediment concentration at a point or within a vertical. Because neither velocity nor sediment concentration is constant over the cross section a special technique is required for measurement and calculation of these values, and two widely used procedures have been developed for use with depth-integrating samplers (Guy and Norman, 1970). In the E.D.I. (Equal discharge increment) method, the channel cross section is divided into a series of segments of equal discharge, and depth-integrated samples are collected at their centroids. The sediment discharge is calculated for each segment as the product of discharge and concentration:

$$Qs = \frac{QCs}{1000}$$

where   Qs = sediment discharge in kg/sec
        Q = stream discharge in m³/s
        Cs = suspended sediment concentration in mg/l

and the sum of these values provides the total sediment discharge. The E.T.R. (Equal transit rate) technique does not require a previous knowledge of the velocity pattern within the channel and provides a discharge-weighted mean concentration for the overall channel. This is obtained by moving the sampler at a constant transit rate through a series of equally spaced verticals and determining the concentration of the compounded sample. In a small stream a single bottle may suffice for the entire sample. Both techniques are time-consuming and in many cases it may be sufficient to define the relationship between a single sample concentration and the mean concentration and to use a single sample and a correction factor for most observations. In addition, it should be recognised that most samplers will only operate to within 9–12 cm of the bed and will therefore leave an unsampled zone of high concentrations. This problem may be overcome by computing a correction (Colby and Hubbell, 1961), or by taking measurements in a natural or artificial turbulence flume where the concentration distribution in the vertical will be approximately uniform.

Laboratory analysis of suspended sediment samples is integrally associated with field sampling activities. The suspended sediment concentration is usually determined by filtration and weighing, using either Gooch crucibles with asbestos filter mats or, more simply, glass fibre, membrane or millipore filter circles (for example, Douglas, 1971). With higher concentrations where filtration is tedious, evaporation may be employed, although a correction is necessary for dissolved matter in the residue. Results are expressed in milligrammes per litre (mg/l) defined as:

One million times the ratio of the dry weight of sediment in grammes to the volume of the water/sediment mixture in cubic centimetres.

In some cases weight per weight units have been used (parts per million or ppm), although these are no longer recommended. The two units are not completely identical but values of mg/l can be derived from ppm values by multiplying the latter by a conversion constant which ranges from 1·0 (i.e., values are identical) for concentrations below about 15000 ppm, and increases to 1·5 for concentrations in excess of 529000 ppm (A.S.C.E., 1969). Because filtration is a time-consuming operation, there would seem to be considerable scope for application of photo-electric turbidity measurements and radio-active isotope techniques to the laboratory determination of sediment concentrations.

In addition to the total sediment concentration, data may be required on the particle size of the sediment. Standard methods for particle size analysis include sieving, the Visual Accumulation Tube, the Bottom Withdrawal Tube, the pipette method, the hydrometer technique and the sedimentation balance. Detailed descriptions of these procedures have been published by the British Standards Institution (1967), the U.S. Geological Survey (1969), and the American Society of Civil Engineers (1969) and in Table 3.10 the particle size ranges and the quantities of sediment

Table 3.10 Commonly used methods of particle-size analysis and their range of application

Method of analysis	Size range (mm)	Quantity of sediment required (gm)
Sieves	0·062–32·0	> 0·05
Visual accumulation tube	0·062– 2·0	0·05– 15·0
Pipette	0·002– 0·062	1·0 – 5·0
Bottom withdrawal tube	0·002– 0·062	0·5 – 1·8
Hydrometer	0·002– 0·062	20 –200
Sedimentation balance	0·002– 0·062	0·1 – 0·5

required are listed. More sophisticated instruments have also been developed recently, and the Coulter Counter (Fleming, 1967) and the Hydrophotometer (Jordan *et al.*, 1971) would seem to offer considerable promise.

3.7d   *Bedload transport*

The assessment of bedload transport is an extremely difficult task, and Hubbell (1964) aptly concluded a survey of the methods and techniques available thus:

> No single apparatus or procedure whether theoretical or empirical has been universally accepted as completely adequate for the determination of bedload discharge.

The problems involved in direct measurement include the alteration of the pattern of flow and transport by the presence of a sampling device, the difficulties of positioning the measuring apparatus on an uneven and variable channel floor, sampling efficiency, and the irregular movement of bed material. In view of these practical difficulties, much attention has been devoted to the derivation of theoretical formulae for the estimation of bedload transport. Because bedload can essentially be thought of as a capacity load, it is theoretically possible to predict the rate of transport from a knowledge of the character of the bed material and of the flow within the stream channel. However, as Hubbell implies, no theoretical formula would seem to offer a completely satisfactory estimating procedure, although in the absence of field measurement theoretical data may be the only data available. Tracers have also been employed in studies of bedload transport, but they primarily provide data on velocities of movement and travel paths rather than total load. In some circumstances where the bedload is dominantly sand sized a turbulence flume can be utilised for measurements on the assumption that the entire load will be in suspension and may be sampled using suspended sediment measuring apparatus. A brief consideration will be made of instruments for direct measurement, tracer studies, and theoretical formulae.

Bedload measuring apparatus can be grouped into several types according to their underlying principle, and these comprise slot traps and collecting basins, basket samplers, tray samplers, pressure difference samplers, and acoustic and pressure sensitive devices. The principle underlying a slot trap (Figure 3.22A) is largely self-explanatory. The slot may extend the full width or only part width of the stream, and this type of installation is particularly suited to small streams where the pit can contain a removable collecting box (Gregory and Walling, 1971). On larger streams a system of sluicing and pumping may be required for emptying the pit. Although simple in concept, the practical application of a slot trap involves considerable problems in terms of construction, emptying, and overfilling. A collecting basin is similar in concept, except that a structure is built across the stream to form a pool in which the bedload will be deposited. Calculation of the amount of bedload moved can be based on periodic surveying or emptying of the basin, making due allowance for its trap efficiency.

Where a permanent installation on the stream bed is impractical, a bedload sampling device can be lowered to the bed to collect a sample of the transported load. Basket samplers are the simplest form of sampling apparatus, and consist of a mesh box or basket lowered to the bed and orientated with the entrance upstream by means of a tail fin (Figure 3.22D). Because of the mesh construction, they are best suited to coarse bedload, and careful consideration must be given to the sampling efficiency. The Swiss Federal Authority Sampler (1939) and the Mühlhofer Sampler (Mühlhofer, 1933) are particular examples of this type. Pan or tray samplers consist of a flat pan- or tray-shaped apparatus which retains the bedload moving over it in collecting baffles. The basic design of the Polyakoff Sampler (Orlova, 1966) is shown in Figure 3.22B, and this type of sampler is best suited to low stream velocities and rates of bed load transport. A major problem associated with basket samplers is the decrease in flow velocity and associated reduction in bedload movement at the entrance, caused by the obstruction of the natural flow pattern. Pressure Difference devices have been specially designed to overcome this disadvantage. A pressure drop is induced at the inlet by a diverging downstream section, so that inlet velocities closely approximate ambient velocities. The load is collected in a mesh container or baffle system. The B.T.M.A. or Dutch Arnhem Sampler is illustrated in Figure 3.22C, and the Vuv Sampler (Novak, 1959) and Don Sampler (Orlova, 1966) can also be included in this category.

Attempts have also been made to continuously monitor the transport of bedload, and acoustic and pressure sensitive apparatus have been developed for this purpose, although they are more suited to laboratory than field measurement. Acoustic devices record the audible sound waves caused by bedload moving along the stream-bed as indicative of the magnitude of bedload movement (Bedeus and Ivicsics, 1964; Johnson and Muir, 1969). Because the microphone sensor can be housed in a stream-lined body suspended above the bed, this type of apparatus does not interfere with the natural pattern of movement. Prototype impact sensing devices in which the collision of particles with a receiving plate can be monitored have been described by Solovyev (1967).

Experimental studies using tracers and tagged bed material can provide useful and detailed information on the bedload transport processes within a particular stream reach and the techniques could also be applied to the suspended load. Fluorescent and radio-

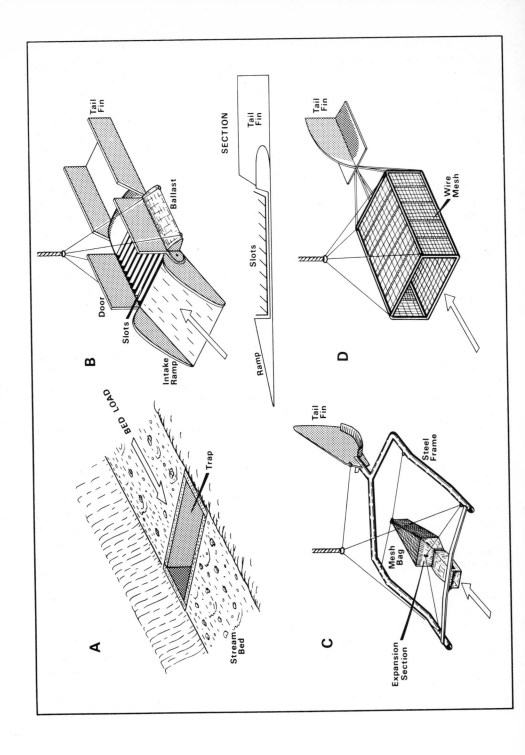

A

Stream Bed

BED LOAD

Trap

B

Tail Fin

Ballast

Door

Slots

Intake Ramp

SECTION

Tail Fin

Slots

Ramp

C

Tail Fin

Steel Frame

Mesh Bag

Expansion Section

D

Tail Fin

Wire Mesh

active tracers are the most commonly used, and radioactive techniques seem particularly promising due to the ease of detection using a scintillator counter which can, if
necessary, be towed over the streambed on a sledge. Radioactive labelling can be
achieved by irradiating natural or artificial particles, by coating particles with a
radioactive substance such as Iridium 192 (^{192}Ir) solution (Hubbell and Sayre, 1965)
or, in the case of larger particles, by sealing radioactive material into a hole (Kidson
and Carr, 1962). In all cases the half life of the radioactive substance should be related
to the duration of the transport phenomena under study. The use of simple painted or
numbered stones has also been described in Vigil Network experiments (Leopold,
Emmett and Myrick, 1966). Most of the information provided by tracer studies is
primarily semi-quantitative in nature and concerns direction of transport, areas
affected by transport and the extent of longitudinal and transverse scattering. However, attempts have been made by Crickmore and Lean (1962) and Rathbun and
Nordin (1971) to obtain more quantitative data by adapting dilution stream gauging
methods to sediment transport.

Bedload Formulae of many types exist for the computation of bedload transport, and
these vary in complexity from pure empirical approaches to attempts at a more theoretical modelling of the physical processes involved (for example, Einstein, 1950). The
general approach is that the magnitude of bedload transport per unit width is related
to bed shear stress. If the character of the bed material, the bed shear stress and the
critical shear stress required to initiate movement are known then the volume rate of
bedload transport can be calculated. In detail, however, theoretical evaluation of bedload transport involves complex considerations of open channel and loose boundary
hydraulics and fluid mechanics. The reader is referred to the treatment of this topic
by the American Society of Civil Engineers (1971) and Graf (1971) for a comprehensive coverage, and by Herbertson (1969) for a critical review. Three simple formulae
are listed in Table 3.11 and although these have to a large extent been superseded by
more detailed computation procedures they provide a useful indication of the theoretical background. Colby and Hubbell (1961) have described a very useful simplified method for calculating bedload transport using the Einstein procedure which
computes a transport function for individual size fractions.

An immediate difficulty encountered in the application of bedload formulae is that
different procedures can provide completely different results for a single channel
section (Figure 3.23A). Muir (1970) has evaluated several formulae for a section of
the River Tyne at Bywell and the calculated annual bedload transport rates varied
between 1290 and 15138 tonnes. Comparison with field measurements can indicate
further discrepancies (Figure 3.23B). Much depends on the similarity between conditions in the stream being studied and those of the channel for which a formula was
originally developed. In the case of the River Clyde (Figure 3.23B), the work of
Fleming (1969) has demonstrated that measured rates of bedload transport were
only about 0·1 per cent or less of those calculated by theoretical formulae, and would
seem to indicate that bedload cannot in every case be viewed as a capacity
load. Many bedload formulae have been derived from laboratory flume transport
experiments, and although such experiments are not directly concerned with field

Figure 3.22 Bedload samplers
The design and principle of operation of a pit trap (A), a Polyakoff-type tray sampler (B), an
Arnhem-type sampler (C) and a basket sampler (D).

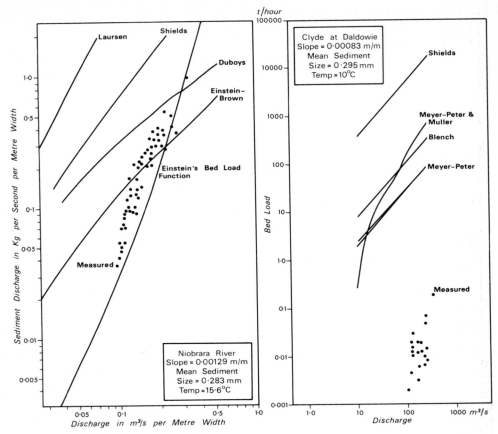

Figure 3.23   Bedload formulae

A comparison is made between computed and measured values of bedload transport for the Niobrara River, U.S.A. (based on Vanoni *et al.*, 1961, after Henderson, 1966) and the River Clyde, Scotland (based on Fleming, 1969).

measurement, they can provide much information of interest to the geomorphologist.

Detailed information on the character of the bed material is needed for many bed-load formulae, in order to calculate the critical shear stress or tractive force required to move that material. Consequently, several types of apparatus have been developed for sampling bed material. These include drag buckets and scoop samplers (Stichling, 1969), clamshell samplers (Figure 3.24) and two devices specially developed by the U.S. Federal Inter-Agency Sedimentation Project (F.I.A.S.P., 1963). The first is the

Table 3.11 Simple formulae for the calculation of bedload transport

1 DuBoys Formula

$$qs = Csr_o(r_o - r_c)$$

where   qs = bedload discharge
Cs = coefficient
r_o = bed shear stress
r_c = critical shear stress

2 Meyer Peter Formula

$$qs^{2/3} = 250 \, q^{2/3}S - 42.5 \, d_{50}$$

where   qs = bedload discharge
q = water discharge
S = slope of stream
d_{50} = median size of bed material

3 Shields Formula

$$qs = 10qS\frac{(r_o - r_c)}{\left(\dfrac{y_s}{y} - 1\right)^2} d_{50}$$

where (as in 1 and 2 above) and
ys = specific weight of the sediment particles
y = specific weight of the water

U.S. BMH 53 or hand-operated piston sampler which can be used to obtain a core from a sand bed stream (Figure 3.24), and the second is the U.S. BM 54 or 60 bucket sampler which incorporates a spring-loaded scoop operated by a slackening of the tension on the suspension cable when the sampler rests on the streambed (Figure 3.24).

3.7e   Channel deposition

The processes of erosion and transportation within a stream channel are complemented by processes of deposition. The geomorphologist is often criticised for paying too little attention to these latter processes (Allen, 1970) and any study of channel dynamics should attempt to document the occurrence of deposition. Detailed investigations are the realm of the sedimentologist, but there is a need for general study of channel and floodplain cross sections and long profiles, and for consideration of individual bedforms. Aerial photographs and periodic accurate surveying can provide useful data on changes in channel form and the associated deposition, through time (O'Loughlin, 1969), whilst stakes, and scour and fill chains can be used to study the complex interactions of erosion and deposition or scour and fill at individual points within the channel (Emmett and Leopold, 1965). Studies of the evolution and modification of such bedforms as sand ribbons, ripples, antidunes and linguoid bars can be based on field observation, but turbid water may make subsurface inspection during flood events very difficult so that detailed surveying is restricted to low water conditions. Laboratory flume studies can be used to study bedforms in greater detail (Guy, Simons, and Richardson, 1966).

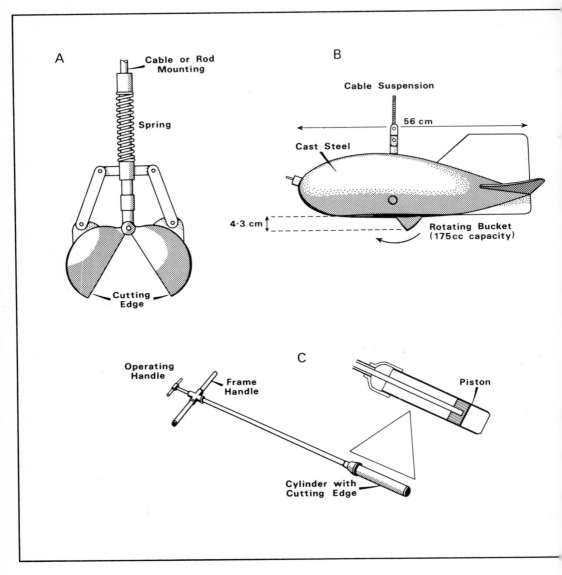

Figure 3.24 Bed material samplers
A clamshell-type sampler (A), the USBM 54 bucket sampler (B) and the USBMH 53 piston sampler (C).

3.8   Sediment yields

The sediment yield of a drainage basin (Tonnes/km^2 p.a.) is the resultant of the processes of erosion, transportation and deposition operating in that basin, and its magnitude reflects the sediment delivery ratio. It is possible to estimate the sediment yield by evaluating all the above processes, but in general direct measurements are made of the total sediment load transported by the stream at the outlet of the watershed, over a long period of time. Several techniques are available for assessing long-term loads and these include reservoir sedimentation studies, fraction collectors, and sediment sampling.

3.8a   Reservoir sedimentation studies

Reservoirs and artificial stock ponds can provide a very valuable source of data on sediment yields, because measurements of the rate of accretion of deposited sediment, suitably corrected for the trap efficiency of the water body, directly reflect the sediment yield of the tributary watershed (for example, Glymph, 1954; Roehl, 1962). Where possible a series of monumented sections should be established during or immediately after the construction of the dam, and these can be surveyed periodically in order to provide detailed data on rates of accretion. Sophisticated sounding techniques have been developed, utilising boats equipped with supersonic depth recorders to record depth configuration, and calibrated wheels moving along a taut wire to measure distance along a traverse (Shamblin, 1965). On larger water bodies the exact position of the boat on the traverse can be determined by the Raydist method involving phase comparison of radio waves from two transmitters, one on the shore and one in the boat (Shepherdson, 1965). Where the original bed level is unknown, the depth of accumulation can be assessed by using probes or a spud bar consisting of a weighted probe with a series of cups to indicate the depth to which it sinks into the sediment (Castle, 1965). In order to calculate values of sediment yield from volumes of accumulated sediment, data on the density of the deposits is required. Samples for laboratory analysis can be taken with core or pipe samplers comprising a sampling tube, a driving weight and a piston within the tube to retain the sample by creating a vacuum. Gamma probes have been used for *in situ* determination of density (Heinemann, 1962) and these are based upon measurement of the absorption by the deposit of the gamma rays emitted from a radium source in the probe. The count of photons reflected back to a detector in the probe is inversely proportional to the density of the surrounding sediment. A variant upon this principle is the dual probe or transmission apparatus (McHenry, 1964) which utilises two probes, one containing a radioactive source and the other a detector, connected 30 cm apart. This latter apparatus is particularly valuable for determining the density of thin layers. Reservoir surveying can be much simplified if the water body is drained or if the water levels are very low and normal ground-surveying techniques can be employed (Hall, 1967; Young, 1969). All results when converted to mass using density data must be corrected for the reservoir trap efficiency. This will depend on sediment characteristics, the shape and volume of the reservoir, the detention-storage time, the pattern of water and sediment discharge, and several other factors (Brune, 1953). Reservoir data on sediment yield does not distinguish between the suspended load and bedload components of the total yield and the results are essentially long term.

3.8b   Fraction collectors

Where a reservoir is not available for a sediment accretion survey or when more detailed and short term data is required, alternative techniques of assessing sediment yield must be employed. The most direct is an approximation of the reservoir principle and involves collecting the total sediment load transported by a stream. Bedload can be trapped in a debris trap or box, but with suspended load, the runoff will require collection for subsequent settling, unless a filter apparatus can be devised (for example, Blocker and Bower, 1963). Fraction collectors have been devised to overcome the practical problems of collecting the total discharge of water and suspended sediment by enabling a small but representative sample to be collected; nevertheless, the technique is only applicable to small catchments and erosion plots. If the sediment load of the fraction or aliquot is determined, the total yield can be computed from the ratio of the fraction collected to the total runoff volume. The many complex installations used for this purpose include multislot divisors (Geib, 1933), sharp-edged slot samplers (Barnes and Frevert, 1954) and the Coshocton Wheel Sampler (Parsons, 1955). The multislot divisor involves a settling box for coarse material and one or more divisor boxes consisting of several slots, the flow through only one of which is collected (Figure 3.20). A five-slot divisor will collect 20 per cent of the flow. Slot samplers can be installed in the nappe below a weir crest and will collect a very small proportion of the flow. Brown et al. (1970) describe installations using this type of device on watersheds in Beaver Creek, Arizona ranging from 120 to 520 hectares in size. The slot collects 1/600 of the flow and this fraction is further reduced by two splitters to collect a final aliquot of 1/60000 in a storage tank. This type of structure does, however, require careful calibration. The Coshocton Wheel sampler utilises a different principle and incorporates a horizontal water wheel which revolves when placed beneath the outflow of an H-Flume (Figure 3.20). As the wheel rotates, a slot positioned along the radius is carried intermittently into the flow and collects an aliquot of 1/100th of the total runoff of water and sediment.

3.8c   Sediment sampling procedures

Fraction collectors are specialised installations suitable only for small catchments. An alternative method of assessing sediment yield, and one suited also to large drainage basins, involves sediment sampling techniques. If frequent measurements of suspended load and bedload discharge are made at a streamflow gauging station in order to define the variations through time, estimates of sediment yield can be computed on a daily or longer basis. Apparatus for continuous monitoring of suspended load (p. 155) could be used to overcome the necessity for very frequent spot measurements, especially during periods of rapidly changing stream stage, and automatic pumping devices could also prove extremely valuable. Indeed, in a small catchment a pumping sampler or a continuous monitoring apparatus may be essential if the variations in sediment transport through time are to be recorded accurately.

In the absence of frequent samples, resort is usually made to the sediment rating curve technique of calculating sediment yields (Campbell and Bauder, 1940). This technique is primarily applicable to the suspended load and involves deriving a generalised relationship between sediment discharge or concentration and stream-flow, which can be used to predict the variations in sediment transport from a continuous record of stream discharge. The relationship or rating curve is usually presented as a straight line graph plot on logarithmic coordinates (see Figure 4.9). Several types of rating curve can be derived and these include instantaneous rating curves which relate sampled suspended sediment concentrations to the discharge at the time of sampling, daily rating curves which relate daily mean discharge to daily mean concentration, and rating curves for particular sediment sizes (Colby, 1956). Sediment sampling is therefore required only to derive and check the rating curve and it is possible to tentatively extend the relationship to periods when no sampling was carried out, but for which discharge records are available. The exact method of computing the estimate of sediment yield may involve streamflow duration curves and manual data processing (Miller, 1951) or a detailed digitized streamflow record and computer processing (Walling, 1971). Care is required if errors in the estimated yields are to be avoided, and Walling (1971) has shown that the use of daily mean discharge values instead of hourly point values, with an instantaneous rating curve, can lead to underestimation by as much as 50 per cent. Because suspended load is not a capacity load, the graphical relationship between sediment transport and discharge (Figure 4.9) often exhibits considerable scatter (Colby, 1956), and in some cases this scatter may be so marked as to render the technique of little value. Seasonal effects on the relationship can be overcome to some extent by constructing individual rating curves for different seasons (for example, Miller, 1951; Walling and Gregory, 1970). The Rating Curve technique was originally devised for suspended load, but it is equally applicable to bedload, and the theoretical and measured relationships between bedload discharge and streamflow shown in Figure 3.23 could be used as rating curves.

3.9   Measurement of solute dynamics

The assessment of the production and transport of solutes or dissolved load within a drainage basin is a field of measurement in which it may be extremely difficult to isolate the natural processes from such man-induced influences as municipal water pollution and the agricultural application of fertilisers. Furthermore, this topic of study overlaps with investigations by ecologists interested in nutrient cycling (Likens et al., 1967), and by public health authorities monitoring pollution and the quality of water supplies (Owens and Edwards, 1964) and with geochemical studies (MacKenzie and Garrels, 1966). Nevertheless, it is important that the geomorphologist should attempt to study the solute dynamics of a catchment as indicative of chemical erosion and as a complement to studies of suspended sediment and bedload production and transport. In some areas the dissolved load of a stream can be many times greater than the solid load (Jaworska, 1968), and form the major component of the denudation system. Laboratory analysis is necessarily of considerable importance, and the collection of field samples may only be the beginning of a lengthy and detailed chemical analysis of the water samples (for example, Rainwater and Thatcher, 1960; Hem, 1970; U.S. Geological Survey, 1970).

3.9a   Solute production

The dissolved load of a river can originate from several sources, notably precipitation, chemical weathering and erosion, atmospheric fallout, mineral springs and man-made pollution, but the geomorphologist is primarily concerned with documentation of the contribution of precipitation and of chemical weathering and erosion processes. The solute content of incoming precipitation can be assessed by collecting samples. Techniques of collection may involve raingauges and funnels or polythene sheets (for example, Egner and Eriksson, 1955), but in all cases the apparatus must be chemically clean and provision is often made for exclusion of leaves and bird droppings from the collecting vessel. The apparatus should be emptied regularly in order to avoid excessive contamination and to enable the chemical quality to be related to particular meteorological conditions. In addition, a distinction could be made between samples where the collecting vessel is only opened during storm periods and those where it remains open continuously and therefore also collects particles of atmospheric fallout. The combination of solutes and fallout is often termed bulk precipitation. Laboratory analysis can provide data on many parameters of chemical quality, but the concentration of total dissolved solids (T.D.S.) in milligrammes per litre (mg/l), specific conductance or conductivity (rmhos) and the concentrations of individual ions (mg/l) are of most interest.

On arrival at the ground surface, the precipitation may increase in solute content as a result of contact with litter, the soil surface, and the soil and rock; and the extent of any increase will primarily reflect the intensity of chemical weathering. Samples of soil water and groundwater can provide a useful indication of the associated phases of chemical weathering and leaching (Arrhenius, 1954), although the data must be evaluated in conjunction with analyses of the quality of incident precipitation. Groundwater samples can be obtained from wells or boreholes. If samples from different depths within the aquifer are required, specially designed sampling apparatus should be employed. An instantaneous vertical sampler could consist of a vertical tube, with provision for closing the ends when it has been lowered to the required depth. It is more difficult to collect a sample from the aeration zone or unsaturated soil, although a vacuum or suction device could be used to withdraw the sample (Parizek and Lane, 1970). Hendrickson and Krieger (1964) have analysed soil water samples obtained by leaching soil columns in the laboratory, but they noted considerable differences between the leachates derived from distilled water and those from natural rainwater. Movement of solutes into a stream can be documented by analysing the chemical quality of water collected in apparatus designed for assessing runoff production (p. 116), and in some cases the apparatus might be installed primarily for this purpose. Water samples from overland flow troughs, throughflow trenches, and from springs and seepages would indicate the variations in chemical content of runoff from different flow paths. In areas of limestone with subterranean drainage, useful comparisons can often be made between the chemical quality of swallet water, percolation water and rising water in order to determine the major sources of solutes (Newson, 1971).

3.9b   Transport of solutes in channels

Most studies of solute dynamics focus upon the transport of dissolved load by the stream; because the channel provides a convenient sampling site, and because the chemical quality of streamflow reflects the resultant of the many processes operating within the basin. It is usually assumed that solute concentrations are uniformly distributed throughout the channel cross section, that a simple dip sample using a chemically-clean bottle will provide a representative sample of dissolved load concentrations (mg/l), and that the simple product of concentration and discharge provides the value of dissolved load discharge (kg/sec). In theory, therefore, the assessment of dissolved load discharge is considerably simpler than that of suspended load where width- and depth-integration of samples is required to provide a discharge-weighted mean concentration (Section 3.7c). In practice, however, the streamflow may not always be homogeneous, and Anderson (1963) provides the example of the Susquehanna River at Harrisburg, U.S.A. where solute concentrations vary markedly across the bifurcated channel, probably as a result of the incomplete mixing of tributary inflows (Figure 3.25). Furthermore, Johnson (1971) suggests that whereas homogeneous chemical contents will exist in fast, turbulent channel sections and that variations in the cross section will generally be less than 3 per cent; in wide, deep sections with still pools, variations of up to 20 per cent may occur. The sampling site should be carefully selected to overcome these problems, but if incomplete mixing remains a problem, composite samples should be collected from several depth-integrated verticals. A programme of sample collection could be directed towards study of variations in solute transport in both time (for example, Durum, 1953) and space (Miller, 1961), and subsequent laboratory analysis might involve consideration of the concentrations of specific ions and other water quality parameters, in addition to total solute content.

When a study is being made of changes in solute concentrations through time, in response to such factors as rainfall and streamflow, very frequent sampling may be required to fully document the variations. The problem of frequent sampling has to some extent been overcome with recent developments of automatic pumping samplers and continuous monitoring instruments. Pumping samplers can be used to collect intermittent gulp samples (for example, Walling and Teed, 1971), samples withdrawn at equal flow volume increments (Claridge, 1970), individual period-integrated samples, or composite samples withdrawn either at a constant rate or at a rate proportional to the streamflow (for example, Fredriksen, 1969). Composite samples are useful for computation of solute discharge on a daily or similar time basis. Continuous monitoring has been developed largely in response to the control of effluent discharge and pollution, but a water quality station, as it is often termed, could be used to provide useful information for specialised geomorphological studies. In most installations, a pump is used to circulate water from the stream and through a system of pipes or chambers containing water quality probes and housed in a shelter on the bank; although some cells have been developed for direct immersion in the stream. Output from the probes or cells can be recorded on strip chart mechanisms or on magnetic tape data loggers (Hall, Prain and Hoer, 1968). The range of water quality parameters that can be monitored is limited at present and is primarily related to pollution and public health, but temperature, turbidity, conductivity and pH are of particular

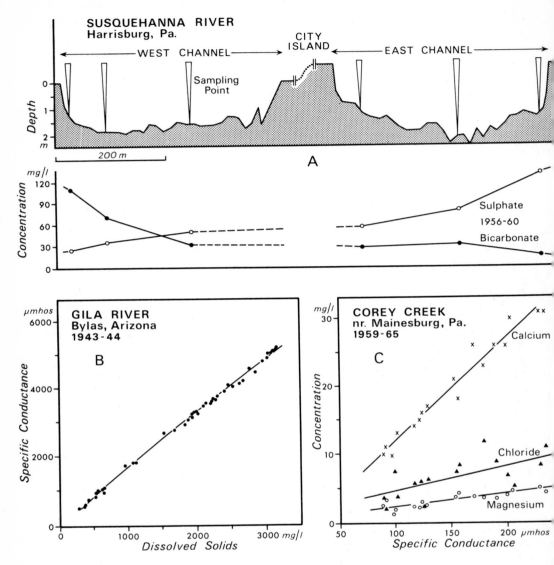

Figure 3.25 Measurement of solute transport
Solute concentrations are generally uniformly distributed throughout a stream cross-section but the Susquehanna River provides a clear example of incomplete mixing (A). In many streams, values of specific conductance can be used to estimate concentrations of dissolved solids (B) and individual ions (C) (based, A on Anderson (1963), B on Hem (1970), and C on data from U.S. Geological Survey Water Supply Papers).

relevance to the geomorphologist, and specific ion electrodes which are now being developed would seem to offer considerable scope for the future (Reynolds, 1971). Conductivity data is particularly valuable in studies of solute transport, because values are directly related to dissolved solids concentrations (Figure 3.25), and in many cases direct relationships can be found between conductivity and concentrations of individual ions (Figure 3.25), such that variations in the latter can be predicted from a continuous record of the former. Several direct reading conductivity meters incorporate facilities for continuous recording and these instruments could form the basis of a simple solute monitoring installation.

3.9c   Laboratory analysis

Although portable instruments can be used for the field determination of certain water quality parameters including conductivity, pH and, with the recent introduction of specific ion electrodes and meters, concentrations of individual ions, analysis of water samples to determine the character and magnitude of the solute content is primarily laboratory orientated, and the geomorphologist must lean heavily upon the techniques of the specialist water analyst (for example, American Water Works Association, 1966; Rainwater and Thatcher, 1960; The Institution of Water Engineers, 1960; Golterman and Clymo, 1969). Great care is required in the type of storage container used and the time elapsing before analysis, because changes can result from reactions both with the material of the container and within the sample itself, particularly if organic material is present. For these reasons, pH and conductivity are probably best determined in the field. The American Waterworks Association (1966) suggests that a limit of seventy-two hours should be placed on storage of samples of unpolluted water and that this period should be reduced with polluted water; and the results of experiments carried out by Johnson (1971) indicated that pH and conductivity should be determined immediately upon return to the laboratory and that the remaining analyses should be completed within six days and preferably within two or three.

The detailed choice of chemical parameters for determination by analysis, and the exact methods used will depend upon the purpose and scope of the study. In some cases the value of total solute concentration may suffice, while in others detail on individual constituents may be required. A pilot analysis is often worthwhile in order to determine the best course for subsequent work. The determination of the total dissolved solids content of a sample (T.D.S. in mg/l) is basic to most studies of solute dynamics, and this is achieved relatively easily by evaporation of a known volume of filtered sample over a steam bath or at 105 °C, and weighing the residue. Associated with values of T.D.S. are measurements of specific conductance or conductivity, which is the ability of the water to conduct an electrical current. The basic unit is the mho but the results are more commonly expressed as micromhos per centimetre (μmhos) and because the values increase with temperature, a reference temperature of 25 °C is generally used. Values may be corrected to 25 °C by using a correction of 2 per cent per degree centigrade. Values of conductivity reflect the T.D.S. content and they are often used as a rapid means of estimating the latter. In most cases the relationship is of the form

$$TDS = AK$$
where   TDS = total dissolved solids concentration (mg/l)
         K = conductivity (μmhos)
         A = conversion factor

(Figure 3.25) and the value of the conversion factor normally falls between 0·55 and 0·75 according to solute types. Methods for the determination of concentrations of individual anions and cations include titration, colorimetry and spectrophotometry (for example, Rainwater and Thatcher, 1960), flame emission and atomic absorption-spectrophotometry (see, Fishman and Downs, 1966) and the recently introduced specific ion electrodes which give a direct meter reading. Results are usually expressed in weight per volume units (mg/l) but weight per weight units (ppm) are sometimes used, and where Equivalent Weights are incorporated into the data to provide milli-gramme equivalents, the complementary units are milliequivalents (meq) and equiva-lents per million (epm) respectively.

Students of limestone solution have developed their own techniques of analysis and interpretation (for example, Douglas, 1963 and 1968) and most results are expressed in terms of water hardness, or the content of alkaline earth metals. Calcium and magnesium are the principal alkaline earths in natural waters, and the hardness due to their soluble salts is termed calcium hardness and magnesium hardness. Hardness is usually expressed in terms of an equivalent concentration of calcium carbonate (mg/l $CaCO_3$), but a distinction should be made between temporary or carbonate hardness which results from calcium and magnesium carbonates and bicarbonates in solution, and permanent or non-carbonate hardness which results from the other salts of the two metals (for example, $CaSO_4$, $CaCl_2$, $Ca(NO_3)_2$).

3.9d   Determination of dissolved load yields

Techniques for the assessment of the dissolved load yield of a drainage basin are similar in many respects to those employed for suspended sediment (Section 3.8). If a fraction collector device has been installed for measuring the suspended load, then the T.D.S. content of the aliquot can also be determined and used to calculate the total mass of dissolved material transported. A pumping device, which continuously with-draws a small fraction of the streamflow at a rate proportional to the discharge and which collects a composite sample, could be used to provide similar results. Frequent sampling of solute loads at a recording stream gauging station is probably the most generally accepted means of estimating short term dissolved load discharges and, therefore, long term yields. Because solute concentrations do not vary quite so markedly with discharge as suspended sediment concentrations, the problem of obtaining sufficient samples to adequately indicate the trend of the concentrations during storm runoff events is not so severe, but it may still be important in small catchments with flashy hydrographs. Pumping samplers can, therefore, often provide valuable assistance in a sampling programme, and a continuous record of specific conductance could be used either to estimate the continuous record of T.D.S. (Figure 3.25) or to interpolate the trend of the variations between individual samples. Where frequent sampling is impossible, dissolved load rating curves, relating solute concentration or discharge to streamflow, can be constructed (Figure 4.10) and used in conjunction with the appropriate discharge records to provide estimates of annual or longer term yields.

3.10 Rates of erosion or denudation

Values of sediment and solute yields from drainage basins are often used to calculate the rate of erosion or denudation within that catchment, expressed as the volume of material removed per unit area within a given time (m^3/km^2 yr). The basic calculation involves the conversion of the sediment and solute yields into volumes or depths of erosion per unit area, and therefore involves values of specific gravity or density, i.e.

$$\text{Denudation in } m^3/km^2 \cdot yr \quad \text{or} \quad mm/1000 \text{ yrs} = \frac{\text{total load (tonnes)}}{\text{area } (km^2) \times \text{specific gravity}}$$

Care must be taken to correct the dissolved load and to some extent the sediment yield for that portion not reflecting natural denudation, but originating as cyclic salts in the incident precipitation or as fertiliser or other man-induced pollution (Winkler, 1970). It is usual to convert the stream load to an equivalent volume of solid rock, and although Corbel (1964) has suggested that a general figure of 2·5 should be employed for the specific gravity, this value could be varied to suit different rock types. Furthermore, this approach may be misleading when dealing with soil erosion and the suspended load of a stream because the sediment would originate from the soil horizons which could have a specific gravity of less than 1·0 (Walling, 1971), and rates of erosion are possibly more logically expressed as depths or volumes of actual soil.

However derived, the values of denudation rates obtained in this manner are very generalised and care is required when attempting to extend them in space and time. The rates are expressed as an average for the entire watershed, and in reality the majority of the load may be derived from a small proportion of the catchment. In many catchments, most of the solid load is derived from the channel and the immediately surrounding areas and not from the watershed slopes. If denudation rates are extrapolated backwards in time to obtain estimates of rates of landscape development (for example, Young, 1969) allowances should be made for changes in climate and vegetation which could alter the loads and for the fact that present day yields are very much influenced by the effects of man on fluvial processes (Meade, 1969; Douglas, 1967).

3.11 Some logistics of drainage basin instrumentation

In the preceding sections, some of the many facets of drainage basin dynamics which may be documented, and the methods and instruments employed, have been described. The extent to which these various techniques are incorporated into a particular scheme of drainage basin instrumentation will depend upon the scope and objectives of the associated project. In some cases it may be intended to build up a detailed and comprehensive picture of the physical hydrology and the operation of fluvial processes, whilst in others the study may be directed towards a more specialised aspect of the dynamics. In neither case is it possible to rigidly define the required equipment because this must depend on the character of the watershed and the approach and means of the researchers involved, although the World Meteorological

Organization (1965) have specified minimum levels of instrumentation for general hydrometeorological measurements within an area, and Toebes and Ouryvaev (1970) have listed the minimum equipment necessary for basic hydrological observations within an instrumented catchment. The number and siting of specific instruments should be determined with reference to a network design which embodies considerations of statistical sampling techniques and which will produce high quality data for subsequent analysis. Much of the current philosophy of network design was presented at the joint International Association of Scientific Hydrology/World Meteorological Organization symposium on the design of hydrological networks held at Quebec, Canada in 1965 (International Association of Scientific Hydrology, 1965).

Several analytical frameworks for, or approaches to, the organisation of drainage basin studies have been suggested in recent years (for example, Ward, 1971), and the terms representative and experimental watershed and other associated nomenclature have figured large in hydrological literature. Although primarily hydrological in context, these considerations are closely connected with, and very relevant to, geomorphological studies within the drainage basin. The main differences between representative and experimental watersheds, as originally conceived within the I.H.D. (for example, Toebes and Ouryvaev, 1970), is that the former should have the minimum natural or artificial change during the study period and should be selected as representative of a hydrological region, whereas, in the latter, one or more of the catchment characteristics is deliberately modified. In general, representative basins range from 1–250 km² and experimental basins are less than 4 km² in area. In this context, representative basins are therefore used for detailed studies of the hydrological cycle and to provide an insight into the characteristics of the area which they represent, while experimental watersheds are used principally to study the effects of cultural changes on drainage basin dynamics. However, the term experimental has also been used to denote any small catchment which has been instrumented for a study of hydrological phenomena and investigation of principles, relationships and prediction methods (for example, American Geophysical Union, 1965; Ward, 1971). Under this latter interpretation, a representative basin takes the more specific role of representing a broad area to which the data can be transferred, and the definition given by the Australian Water Resources Council (1969) is in this vein, for example

> a catchment which contains within its boundaries a complex of land forms, geology, land use and vegetation which can be recognised in many other catchments of a similar size throughout a particular region.

The locations of the ninety-three representative basins originally selected for a coverage of Australia are shown in Figure 3.26.

Within the wider definition of a representative watershed, several specific variants have been distinguished. Benchmark catchments (Toebes and Ouryvaev, 1970; World Meteorological Organization, 1965) were conceived within the I.H.D. as

Figure 3.26 Representative and experimental catchments
This figure illustrates the series of representative basins selected for a coverage of Australia (based on Australian Water Resources Council, 1969) and the multiple watershed experiment initiated at the Moutere Soil Conservation Station in New Zealand (based on New Zealand Ministry of Works, 1968).

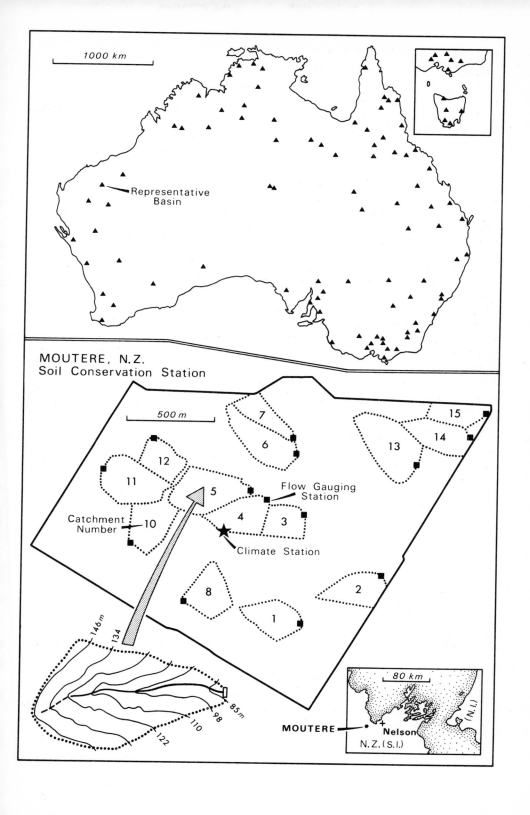

1000 km

Representative
Basin

MOUTERE, N.Z.
Soil Conservation Station

500 m

7

6

15

14

13

12

11

5

Flow Gauging
Station

4

3

Catchment
Number

10

Climate Station

8

2

1

146 m

134

85 m

98

110

122

80 km

MOUTERE

Nelson

N.Z. (S.I.)

(N.I.)

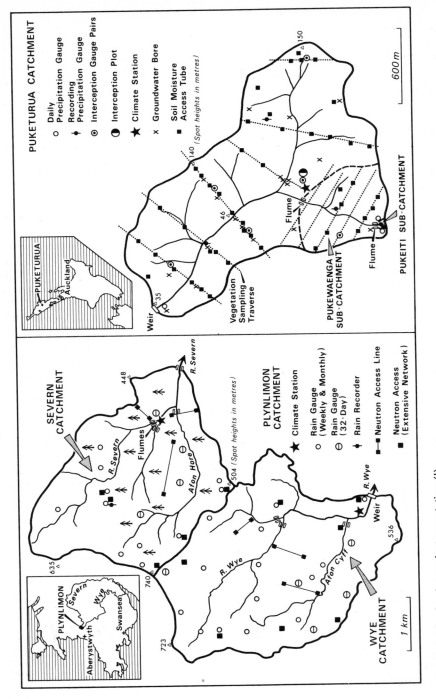

Figure 3.27 Catchment instrumentation (I)

Details of the instrumentation in the Plynlimon catchments established by the Institute of Hydrology and in the Puketurua experimental catchment set up by the New Zealand Ministry of Works.

representative basins which are still in their natural state and which have soil and vegetation conditions that are not expected to change for a long time, therefore allowing study of the interrelationships of climatic and hydrologic variables uninfluenced by the effects of human activity. The concept of vigil basins originally developed by the United States Geological Survey (Leopold, 1962) has also been incorporated within the I.H.D. to denote a type of representative basin similar to the benchmark, but not protected from artificial changes. The prime objective in this case is to carry out simple measurements of hydrological and geomorphological variables in order to watch (a vigil) changes with time in the landscape and to clarify the relationships between man, the land, and the hydrological cycle. The United States Forest Service has established a series of barometer catchments (Dortignac and Beattie, 1965) ranging from 200–600 km² in area in order to represent the broad climatic-physiographic regions of the country and to provide an inventory of hydrological data for use in assessing the effects of watershed management techniques in the National Forests.

Studies in representative basins are primarily observational, whereas those in experimental catchments, established for assessing cultural changes, involve both observation and a deliberate experimental procedure to introduce changes and to measure their effects (Boughton, 1968). The exact techniques used for evaluating the impact of the changes vary, but some form of calibration is inevitably required in order to determine the normal response of a catchment before the changes are introduced and to predict the pre-treatment response. In the single basin or 'before and after' technique, the catchment is calibrated for a number of years against climatic variables, the treatment is carried out, and the deviations from the predicted response are assessed. The paired catchment, comparative basin or control watershed procedure involves use of an untreated control catchment with similar characteristics to the experimental catchment which is to be treated. The two basins are initially calibrated for a period of years so that the behaviour of one can be predicted from the response of the other, and subsequent to calibration one is treated and the other left as a control. The effects of treatment are measured as departures from the predicted behaviour of the treated basin. Multiple catchment or multi-watershed experiments involve the use of a group of similar catchments which, being within a small area, are subject to similar climatic conditions. The catchments are subjected to several types of treatment, usually involving replicates, and by comparison with control basins and by statistical analysis the influence of individual treatments can be assessed. The study initiated at the Moutere Soil Conservation Station in New Zealand by the Water and Soil Division of the Ministry of Works (New Zealand Ministry of Works, 1968) provides an excellent example of a multiple catchment experiment designed to study the effects of different land management practices on the response of small watersheds (Figure 3.26). Twelve catchments are involved, and the planned treatments are shown in Table 3.12. Treatments 1–4 form a carefully designed 2 × 2 factorial experiment, each combination possessing two replicates.

In recent years, several workers have questioned the value of experimental watersheds in hydrological research (for example, Slivitzky and Hendler, 1965; Ackermann, 1966; Reynolds and Leyton, 1967). Factors of cost, leakage, unrepresentativeness, the difficulty of transferring results, and the difficulties of detecting changes have been raised as criticisms, and plot studies and models have been suggested as alternatives.

Table 3.12   Programme of catchment treatment, Moutere Soil Conservation Station, 1965

Treatment	Catchment numbers
1 Mob stocking plus contouring	2 and 6
2 Set stocking plus contouring	4 and 15
3 Mob stocking no contouring	5 and 14
4 Set stocking no contouring	1 and 3
5 Gorse	8 and 13
6 Cultivated and sown into crops	10 and 12

Source: New Zealand Ministry of Works (1968)

Although model studies, which include laboratory models (for example, Chow, 1967; Chery, 1966; Black, 1970; Roberts and Klingeman, 1970), analogue models (Riley *et al.*, 1967; Tinlin and Thames, 1969; Skibitzke, 1963) and digital models such as the Stanford Watershed Model IV (Crawford and Linsley, 1966), are undoubtedly of value in drainage basin studies, catchment instrumentation must always play a substantial part in studies of physical hydrology and fluvial geomorphology. Many of the criticisms listed above have been answered specifically by Hewlett, Lull and Reinhart (1969) and more generally by Ward (1971).

3.12   Examples

Figures 3.27 and 3.28 provide examples of schemes of instrumentation used in several different research projects. Research objectives in the Plynlimon catchments of the Institute of Hydrology (Institute of Hydrology, 1968) and in the Puketurua experimental catchment of the New Zealand Ministry of Works (New Zealand Ministry of Works, 1970) are primarily concerned with detailed assessment of the underlying physical hydrology (Figure 3.27). In the Plynlimon project a comparison is being made between the behaviour of the Wye catchment, which is primarily upland sheep pasture, and the Severn catchment, which is forested, in order to investigate the influence of forest cover on the runoff dynamics (Tinker, 1971). Instrumentation is therefore primarily concerned with the water balance, and in particular with accurate measurements of rainfall, streamflow, and soil moisture. In the Puketurua watershed a detailed assessment is being made of the hydrological characteristics of the soil/vegetation types contained within the catchment and of the effects of changes in land use and land management on the hydrological regimen. Particular emphasis is being placed on the development of mathematical models of catchment behaviour. The scheme of instrumentation of the three nested basins is more comprehensive than in the Plynlimon watershed and also includes interception gauges, an interception plot, suspended sediment measurements at the three gauging structures, and vegetation survey transects.

   The three examples of catchment instrumentation presented in Figure 3.28 are more concerned with detailed documentation of particular aspects of drainage basin dynamics. The illustrated portion of the Otutira experimental catchment has been

Figure 3.28   Catchment instrumentation (II)
Details of the instrumentation established for specialised studies within the Otutira Catchment, Slopewash Tributary and the Happy Valley Catchment.

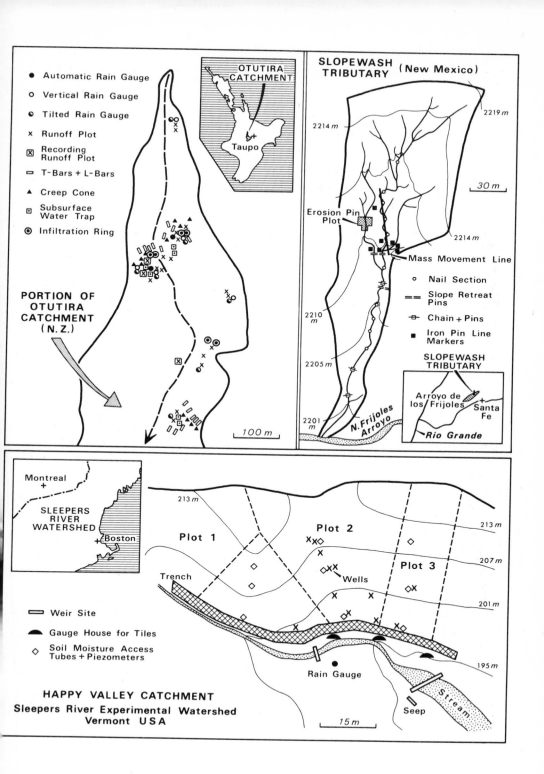

PORTION OF OTUTIRA CATCHMENT (N.Z.)

- ● Automatic Rain Gauge
- ○ Vertical Rain Gauge
- ◑ Tilted Rain Gauge
- × Runoff Plot
- ⊠ Recording Runoff Plot
- ▭ T-Bars + L-Bars
- ▲ Creep Cone
- ▫ Subsurface Water Trap
- ◎ Infiltration Ring

OTUTIRA CATCHMENT

Taupo

100 m

SLOPEWASH TRIBUTARY (New Mexico)

2219 m
2214 m
2214 m
2210 m
2205 m
2201 m

Erosion Pin Plot

Mass Movement Line

30 m

- ○ Nail Section
- ═ Slope Retreat Pins
- ⊟ Chain + Pins
- ■ Iron Pin Line Markers

SLOPEWASH TRIBUTARY

N. Frijoles Arroyo

Arroyo de los Frijoles

Santa Fe

Rio Grande

HAPPY VALLEY CATCHMENT
Sleepers River Experimental Watershed
Vermont USA

Montreal

SLEEPERS RIVER WATERSHED

Boston

213 m
213 m
207 m
201 m
195 m

Plot 1

Plot 2

Plot 3

Trench

Wells

Rain Gauge

Seep

Stream

- ▭ Weir Site
- ◣ Gauge House for Tiles
- ◇ Soil Moisture Access Tubes + Piezometers

15 m

selected by the University of Waikato, New Zealand for a detailed study of slope processes, and in particular runoff production and soil creep (New Zealand Ministry of Works, 1970). The instrumentation includes twenty small runoff plots ($4m^2$), three of which have provision for automatic recording, six subsurface water traps, six infiltration rings and a series of 64 creep cones, 113 L-bars and 46 T-bars for the measurement of creep. Slopewash Tributary was used by Leopold, Emmett, and Myrick (1966) for an intensive documentation of slope and channel processes within the general study of the Arroyo de Los Frijoles in New Mexico. Erosion pins, iron pins, mass movement stakes and scour chains formed the basis of this study, and provided valuable data on the rates of operation of the processes of slope and channel erosion. Certain drainage basin studies may involve intensive instrumentation of only a small portion of the watershed and the part of the Happy Valley Catchment illustrated in Figure 3.28 is a good example of such a study. This slope was instrumented by Dunne and Black (1970) for a detailed documentation of runoff production. The interceptor trench provided data on surface and subsurface runoff from three individual slope segments, one convex, one concave and one straight, and background information on moisture conditions within the slope and streamflow were provided by soil moisture neutron access tubes, piezometers and groundwater observation wells and stream gauging weirs.

Selected reading

A useful collection of short articles concerning techniques of measurement for geomorphological studies is contained in:

INTERNATIONAL GEOGRAPHICAL UNION, (Comp.) 1967: Field methods for the study of slope and fluvial processes. A contribution to the International Hydrological Decade: *Revue de Geomorphologie Dynamique*, **17**, 152–88.

Detailed descriptions of instrumentation techniques can be found in specialist manuals and guidebooks produced by several organisations, for example:

TOEBES, C., OURYVAEV, V. (eds.) 1970: Representative and experimental basins. An international guide for research and practice. *Unesco studies and reports in hydrology*, **4**.

WORLD METEOROLOGICAL ORGANISATION. 1964: *Guide to hydrometeorological practices*: W.M.O. **168**, TP 82.

U.S. GEOLOGICAL SURVEY. 1967 onwards: *Techniques of water-resources investigations of the United States Geological Survey*. (A series of manuals grouped under major subject headings called books and further subdivided into sections and chapters.)

The application of field measurements to specific problems is demonstrated by:

LEOPOLD, L. B.; EMMETT, W. M.; MYRICK, R. W. 1966: Channel and hillslope processes in a semi-arid area, New Mexico: *U.S. Geological Survey Professional Paper*, **352**-G.

Part B: Drainage Basin Analysis

4 Quantitative Evaluation of Drainage Basin Processes

When you can measure what you are speaking about and express it in numbers, you know something about it, but when you cannot express it in numbers your knowledge is of a meagre and unsatisfactory kind. *Lord Kelvin*

The measurement of processes is essential to the understanding of drainage basin dynamics, but measurement itself is often only the prelude to detailed processing, analysis and evaluation of the data obtained. A streamflow record consisting of stage data contained in a series of charts or a spool of paper tape is of limited value until it is converted to a record of discharge by using a stage/discharge rating relationship. Furthermore, it is only when this processed data is analysed for the purpose of extracting and determining its major characteristics, that it becomes valuable to the geomorphologist who seeks to understand catchment dynamics, the relationships between form and process, and variations in catchment response in both space and time. In this context, a consideration must be made of the derivation of parameters indicative of, or describing, the operation of individual processes and overall catchment response; and this must necessarily draw upon the specialised and rapidly developing fields of data processing, computer analysis, and statistical evaluation. Process interrelationships also merit attention because they can assist our understanding of catchment response and because they can themselves be used as a measure of drainage basin dynamics. In addition, developments in the field of mathematical modelling must be reviewed as involving the simulation of catchment processes and interrelationships for the purposes of prediction and understanding of drainage basin dynamics.

4.1 Catchment inputs

Analysis and evaluation of precipitation or more particularly rainfall records involves three major elements, namely, parameters of depth, intensity and spatial variation. In addition, distinction should be made between values derived for an individual station and those associated with an area to which point data has been extrapolated.

4.1a Depth characteristics

The depth characteristics of a precipitation record are relatively easy to assess, and annual precipitation totals provide a useful measure of the setting within which a drainage basin functions. Values of *mean annual precipitation* are most often cited in order to overcome the problems of unrepresentativeness of individual annual totals, and Binnie (1892) has shown that thirty or more years of record can be required for derivation of the long-term mean to within an accuracy of ± 2 per cent (Figure 4.1A). Cyclical trends in precipitation character can provide problems in this context. The degree to which individual annual values vary from the mean is relevant and the standard deviation (σ) and the dimensionless *coefficient of variation* $\left(\text{CV} = \frac{\sigma}{\bar{x}} \right)$ can be used to quantify the variability of precipitation from year to year (Barry, 1969). Variation of precipitation within the year according to season can be described by the use of *regime diagrams* essentially similar to those used for runoff (Figure 6.12). The precipitation ordinates are expressed either as absolute values or, when comparison between stations is required, in dimensionless units as a proportion of the annual total. Numerical indices of seasonal variability have also been devised and the $\frac{p^2}{P}$ index (where P = the mean annual precipitation and p = the precipitation in the wettest month) was used by Fournier (1960) as indicative of both the magnitude of annual rainfall and the degree of seasonal imbalance.

The incidence of short term periods of heavy rainfall or of drought can be described by use of *frequency or probability analysis*. Although this type of analysis is perhaps primarily concerned with prediction so that, for example, the magnitude of the daily precipitation total which is likely to be exceeded on average once during fifty years can be estimated for design purposes, these specific values and the associated data plots (Figure 4.1C) are equally valuable as indices reflecting the long-term character of the precipitation regime. The detailed background to frequency analysis is best consulted in a specialised text (e.g. Chow, 1964; World Meteorological Organisation, 1965), but a brief description can be given of the procedure that would be used to

Figure 4.1 Analysis and characterisation of precipitation data
This figure illustrates, (A) the effect of period of record upon the reliability of the associated values of mean annual rainfall (after Binnie, 1892), (B) the method of plotting rainfall probability or return period data using Gumbel paper, (C) rainfall probability plots for three contrasting stations (after Rodda, 1970), (D) a drought frequency plot (after Hall, 1967), (E and F) rainfall intensity–duration relationships for the River Ray Catchment (after Institute of Hydrology, 1967 and 1971), and (G) hypothetical depth-area curves (see text for explanation).

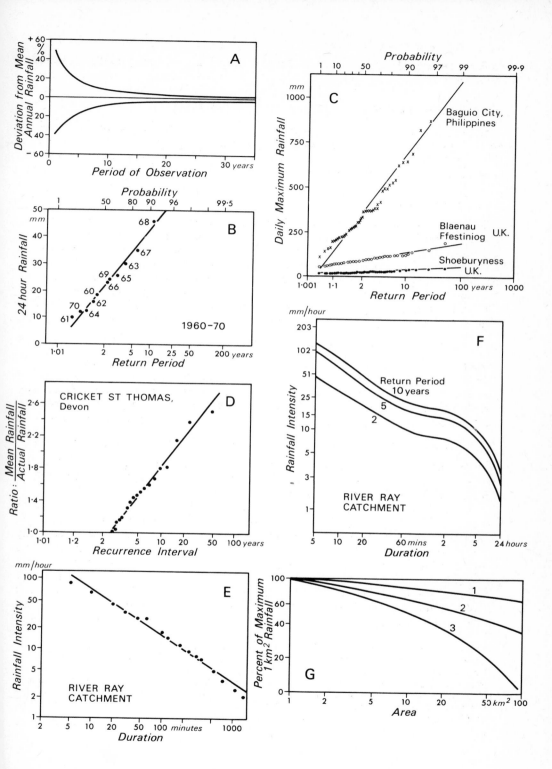

establish the frequency of occurrence of high magnitude 24 hour rainfall totals for a station with eleven years of record (1960–1970, Figure 4.1B). A distinction is usually made between analysis of a partial series and an annual series of data and the technique used for deriving the basic data comprises the major difference. In the former instance all values above a particular limit are extracted from the record, while in the latter case the maximum value occurring in each year is selected. The annual series is usually employed in this particular context and the simplest method of analysing an annual series is to employ a special probability plotting paper. The Gumbel Type I probability distribution is often used for this purpose (Rodda, 1970) in order to obtain a straight line relationship (Figure 4.1B). In this example the eleven values of rainfall depth selected from the record are plotted on the paper at positions determined by their respective return periods or probabilities (Table 4.1). The frequency plots for

Table 4.1   Computation of return periods for an annual rainfall series (*cf. Figure 4.1B*)

Year	Maximum 24 hr rainfall	Rank (m)	Return period $\dfrac{(n+1)}{(m)}$
1960	27 mm	7	1·72
1961	10	11	1·09
1962	22	8	1·5
1963	50	3	4·0
1964	15	9	1·33
1965	42	4	3·0
1966	36	6	2·0
1967	60	2	6·0
1968	82	1	12·0
1969	38	5	2·4
1970	14	10	1·2

Where: n = number of years of record

three stations are shown in Figure 4.1C and contrasts can be distinguished both in terms of the absolute magnitude of the rainfall depths for different return periods and in the slope of the individual relationships. This type of analysis could be extended to shorter period rainfall totals, for example 30 minute, so that a station could be characterised by values of its mean annual precipitation and the rainfall totals associated with different durations and different return periods (e.g. 10 year, 30 minute rainfall and 20 year, 24 hour rainfall). The maximum possible or probable maximum precipitation (P.M.P.) is the least upper bound of estimates representing the realistic upper limit of precipitation that can occur within a specified duration. Frequency analysis can similarly be applied to the drought extremes of a rainfall record and Figure 4.1D portrays a frequency plot of low annual rainfall totals expressed as a ratio to the mean annual rainfall total. Where detailed records are lacking, frequency data can often be extrapolated by establishing relationships between falls of a certain frequency and such general parameters as mean annual precipitation and the mean number of days per year with precipitation (Herschfield and Wilson, 1958; Rodda, 1967). An extremely useful rainfall atlas has been produced for the United States (Herschfield, 1961) which enables rainfall totals for several different durations and recurrence intervals to be estimated for any location.

Antecedent precipitation indices can also be included in an evaluation of the depth characteristics of a precipitation record. They provide an indication, for a specific moment in time, of how much precipitation has fallen in a preceding period, and are primarily concerned with an estimate of the associated soil moisture conditions. A decay function is normally employed in the calculation so that past rainfall exerts progressively less influence on the value of the index as the time span increased. Butler (1957) proposed an annual index, Pa, of the form

$$Pa = aP_0 + bP_1 + cP_2$$

where P_0, P_1, and P_2 = the annual rainfall totals of the current and preceding
two years
a, b, and c = weighting coefficients (a > b > c > o and
a+b+c = 1)

A similar system of weighting coefficients could be employed on a day to day or hourly basis; although a reciprocal or exponential decay function is often used to evaluate Pa for a rainfall total Pt occurring t days previously, for example

$$Pa = Pt \cdot 1/t$$

or

$$Pa = Pt \cdot K^t \quad (K = <1 \cdot o \text{ and usually } o \cdot 85 - o \cdot 98)$$

The exponential function is easily incorporated into a day to day accounting procedure since

$$Pa = (Pa_{d-1} \cdot K) + P$$

where Pa_{d-1} = index for preceding day
P = rainfall in preceding 24 hours

The precise value of K used in the exponential function may have to be derived by trial and error and although a constant value is usually employed it can be argued that it should vary from month to month in accordance with the magnitude of evapotranspiration losses.

4.1b   Intensity characteristics

It is difficult to separate consideration of the intensity characteristics of a precipitation record from those of depth, because values of depth are themselves inevitably qualified by a period of time and therefore indirectly reflect intensity. However, *depth-duration and intensity-duration analysis* is primarily concerned with these characteristics. An intensity-duration curve for an individual station displays the maximum recorded intensities for various durations (Figure 4.1E) while the depth-duration plot portrays the maximum depths recorded within various durations (Figure 6.13). Both provide a useful summary of the character of the record at that station and they will clearly indicate the contrasts between areas experiencing frontal rain and those subject to convectional rainfall. Intensity-duration and depth-duration analysis can also be combined with the frequency aspects of a rainfall record to produce plots for various frequencies or return periods (Figure 4.1F). Values of *average intensity, maximum intensity, maximum 30 minute intensity* and other similar indices are often used to characterise individual storms when attempting to relate the form of runoff hydrographs, or the magnitude of storm period sediment concentrations and yields, to the causative storm rainfall. Several specialised indices reflecting rainfall intensity have been developed for use in empirical equations for predicting soil erosion. At the level of the individual storm, Ellison (1945) expressed the energy characteristics of the rainfall in terms of three variables viz: (c.f. pp. 211)

$$V^{4.33}.d^{1.07}.I^{0.65}$$

where   V = velocity of the raindrops (m/s)
        d = diameter of the drops (mm)
        I = intensity of the rainfall (mm/hr)

On an annual basis, the rainfall erosion factor (R) developed by Wischmeier and Smith (1958) and used in the Universal Soil Loss Equation (p. 318) is derived as follows:

$$R = \frac{\Sigma EI}{100}$$

where   E = kinetic energy of the storm (Joules/cm.m²)
        I = maximum 30 minute intensity (cm/hr)

(The annual index is proved by the summation of the EI values for individual storms occurring within that year.) In the Musgrave Equation for predicting sheet erosion (Musgrave, 1947) the rainfall factor (P_{30}) is defined as the maximum 30 minute rainfall depth for a return period of 2 years, and Mircea (1970) suggested a similar index calculated as the product of the precipitation amount and the maximum intensity of a 15 minute storm.

4.1c   Spatial characteristics

Measures of spatial characteristics are particularly concerned with storm events and attempts to describe the areal variation of storm rainfall in terms of the degree of localisation or the rate of decrease in recorded fall depths away from the centre of the storm. Frontal rainfall will exhibit a much more gradual change with distance than convectional rain. A simple *distribution coefficient* applicable to the storm rainfall over a drainage basin could be calculated as

$$\frac{\text{Maximum point rainfall}}{\text{Average basin rainfall}}$$

More detailed information is provided by a *depth-area curve* which plots the decrease in average rainfall depth or maximum point rainfall, as a function of increasing area within a storm or distance from the storm centre. The precise shape of the curve will vary from storm to storm, particularly in response to storm duration, but most areas will demonstrate a characteristic general trend and the hypothetical curves shown in Figure 4.1G could refer to an area of dominantly frontal rain (1), an area of localised convectional rainfall (3) and an intermediate area (2). Empirical equations have also been developed to quantify the depth-area relation for a particular area or drainage basin (Table 4.2).

Table 4.2 Depth—area rainfall formulae (*after Court, 1961*)

Worker	Region	General formula
Horton (1924)	North-east U.S.A.	$p = m(1 - kA^{\frac{1}{2}})$
Light (1947)	Ohio, U.S.A.	$p = a + b \log A$
Huff and Stout (1952)	Illinois, U.S.A.	$p = a - bA^{\frac{1}{2}}$
Boyer (1957)	Ohio, U.S.A.	$p = 2mb^{-2} x^{-2}[1 - (1 + bx) \exp(-bx)]$

where p = average precipitation
A = area
x = distance from centre of the storm
m = precipitation at the centre of the storm
a, b, k = coefficients

4.2 The water balance

Although detailed study of the water balance is primarily the realm of the hydrologist (for example Edwards and Rodda, 1970), it can nevertheless provide useful background knowledge for the geomorphologist evaluating catchment response, by indicating the apportionment of precipitation input between runoff, evapotranspiration losses and storage, i.e. for a given drainage basin

$$P = Q + E \pm \Delta SMS \pm \Delta GWS \pm \Delta DS + GWO$$

or more simply

$$P = Q + E \pm \Delta S$$

where P = total precipitation input
Q = total streamflow
E = total evapotranspiration loss
ΔSMS = change in soil moisture storage
ΔGWS = change in groundwater storage
ΔDS = change in depression storage and snowpack
GWO = groundwater outflow at depth
S = storage

Furthermore, the water balance equation can provide a valuable check on the accuracy of measurement of the various components. It is usually evaluated on an annual basis, for a period beginning and ending with similar conditions of minimum natural storage, so that the storage terms can be largely ignored. The water year which runs from 1 October to 30 September has been designated to fulfil these requirements in most instances. Evaluation on a shorter time basis is more difficult, because changes in storage will have to be monitored or carefully estimated, but this

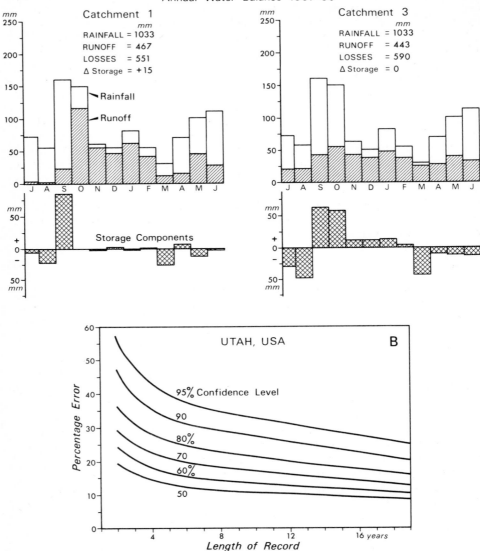

Figure 4.2 Water balance diagrams and the reliability of annual mean runoff values
Water balance diagrams for two east Devon catchments are shown in (A), and (B) demonstrates the influence of length of record upon the reliability of annual mean runoff values for catchments in Utah, U.S.A. (after Jeppson *et al.*, 1968).

would provide a clearer insight into the relative importance of the various storage volumes (see Figure 4.2A). Digital runoff models such as the Stanford IV Watershed Model (p. 229), which employs a continuous moisture accounting system and routes the precipitation input through several stores, could also be employed to demonstrate the magnitude and functioning of catchment storage.

A useful index of soil moisture conditions within a catchment at a particular instant in time, based upon the water balance concept, is that of the *soil moisture deficit* (for example, Grindley and Singleton, 1967). In the calculation procedure, a zero S.M.D. is assumed during late winter when the soil moisture storage is at field capacity, and subsequently the cumulative effect of rainfall accretion minus evapotranspiration loss is evaluated, preferably on a daily basis. Because the values of evapotranspiration will commonly be derived by the Penman Formula as potential values, careful considera-tion must be given to the need to allow for the reduction in the evapotranspiration rate below potential with increasing soil moisture deficits. The simplest procedure to allow for this occurrence is that suggested by Penman (1949) and which assigns a 'root constant' to different types of vegetation. When the soil moisture deficit increases beyond the root constant threshold, evapotranspiration rates fall below potential. Where different types of vegetation occur in a catchment, a multi-capacity basin accounting system, employing different root constants for the various land use pro-portions, must be employed.

4.3 Runoff response

Detailed evaluation of streamflow records necessitates a substantial amount of pro-cessing in order to convert the stage data provided by a chart or punched tape into values of instantaneous discharge, mean discharge and runoff volumes, by using a stage/discharge relationship. These values can in turn be further analysed to extract the major characteristics of the discharge record. The development of analogue-to-digital punched tape recorders has considerably assisted the task of processing because the digitized data can be quickly prepared for direct input into a computer. However quality control checking and correction procedures are of utmost importance with paper tape processing because there is no direct visual trace to indicate recorder failure or malfunctioning. The processing of chart records can be facilitated by the use of a chart reader or pencil follower which converts the trace into a digital record by accurately defining the hydrograph with a series of time and stage coordinates (Plate 15). The take-off interval can be either constant or varied according to the hydrograph shape, but should, nevertheless, be carefully selected with regard to the precise character of the record. The major characteristics of the runoff response of a catchment can conveniently be grouped under three headings, namely, general characteristics, flow components, and detailed hydrograph characteristics.

4.3a *General characteristics*

A graphical plot of a discharge record can be very revealing in terms of the pattern of response, particularly to the trained eye, but quantitative indices and simple graphical representation of its character are required if two or more drainage basins are to be accurately compared (for example, Walling, 1971) or if meaningful relation-

Plate 15 Automated data processing
Streamflow charts, in this case stored on microfilm for ease of storage and retrieval, can be digitised
using a d-mac pencil follower or chart reader. Output from the pencil follower can be in the form
of magnetic tape (*left*), paper tape and teleprinter (*right centre*) or punched cards (*right*), for
direct input into a computer (by kind permission of d-mac Ltd., Glasgow).

ships between form and process are to be established. Basic parameters include *daily,
monthly and annual mean flows*, the *total runoff volume* for a specified period, and the *runoff
percentage* which expresses the runoff as a proportion of the precipitation input for a
given period. Careful thought should be given to the reliability of estimates of long
term mean runoff derived from records of short duration. For example, in Utah,
U.S.A. it has been demonstrated that there is a 5 per cent chance that the mean of a
10 year record could deviate by as much as 33 per cent from the long term (50 year)
mean (Jeppson, *et al.*, 1968) (Figure 4.2B). The above parameters give little impression
of the variability within the record, and whereas values of *maximum and minimum
recorded flow* can be cited, more detailed indications are often required. Variability can
be usefully demonstrated by the *flow duration curve* (for example Searcy, 1959) which is
essentially a cumulative frequency diagram. It displays the duration or frequency
with which various magnitudes of flow are equalled or exceeded, and it may be thought
of as the hydrograph of the associate period with its flows arranged in order of
magnitude (Figure 4.3A).

The simplest technique for plotting a duration curve is illustrated in Figure 4.3A$_1$;
because both x and y coordinates are linear, the area beneath the curve is directly

proportional to the total discharge. However, detail concerning the low flows is almost obscured, and to overcome this particular disadvantage the discharge values are often plotted on a logarithmic scale (Figure 4.3A$_2$). Furthermore, normal probability scales are sometimes employed for the duration ordinate in order to provide detail on the extreme ends of the discharge range and because the duration curve will often approximate to a straight line if a logarithmic discharge scale is used (Figure 4.3A$_3$). In addition, the shape of a curve, and particularly its slope, will vary according to the time base of the discharge data used, for example daily, monthly or annual mean flows (Figure 4.3B). Indices have also been developed to quantify the shape or slope of a flow duration curve, and Hall (1967) used the simple *30/70 ratio* (the ratio of the flow level exceeded 30 per cent of the time to that exceeded 70 per cent of the duration) for this purpose, while Lane and Lei (1950) have proposed the *variability index*. This index is essentially the standard deviation of the flow data and is derived by reading off the values of discharge at 10 per cent duration intervals between 5 and 95 per cent and calculating the standard deviation of the logarithms of these values. The variability index for an average river is approximately 0·6, but Lane and Lei found a range between 0·14 and 1·17 in their study of rivers in the eastern United States. When a direct visual comparison of duration curve form is required, dimensionless curves can be constructed where the discharge values are expressed as a ratio to the mean (Figure 4.3D).

The chronology of the flows is completely masked in a duration curve because there is no indication of the season in which the maximum or minimum discharges occurred or whether they occurred as one continuous event or several separated minor events. The duration curve can, therefore, be usefully qualified by a simple *regime diagram* (for example Figure 6.12), which portrays the magnitude of the individual mean monthly flows either as absolute values, or in dimensionless units as a ratio to the annual mean monthly flow or as a proportion of the mean annual flow. Other measures of timing which are complimentary to the flow duration curve, are the *flow interval indices* used by several workers. Court (1962) defined the *half-flow interval* as the length of the period between the dates when one quarter and three quarters of the annual flow was attained, and this was subsequently modified by Satterlund and Eschner (1965) to refer to the shortest period during which half of the annual runoff occurred. Sopper and Lull (1967) developed the approach further by calculating *quarter- and half-flow intervals* defined as the shortest interval during which one half and one quarter of the runoff occurred and two *low-flow intervals* derived as the largest number of days required to discharge 1 per cent and 5 per cent of the annual runoff.

The extremes of high and low flow exhibited by a runoff record can be usefully qualified by frequency or probability analysis, in a similar fashion to rainfall data (p. 184). Again, the major application of flood and minimum flow frequency studies is probably for design and prediction purposes, but they nevertheless also provide a valuable means of summarising the record. (Figure 4.4A, B and C.) The return period or recurrence interval of a flood or drought is the average interval of time within which the given flow will be equalled or exceeded once, and the magnitude of the flood or low flow with a certain return period (for example 50 years) provides a useful index of the runoff dynamics of a catchment. Details of the computational procedures are best consulted in a specialist treatment of the topic (for example, Dalrymple, 1960; Nash, 1966), but the approach is basically similar to that used for rainfall data.

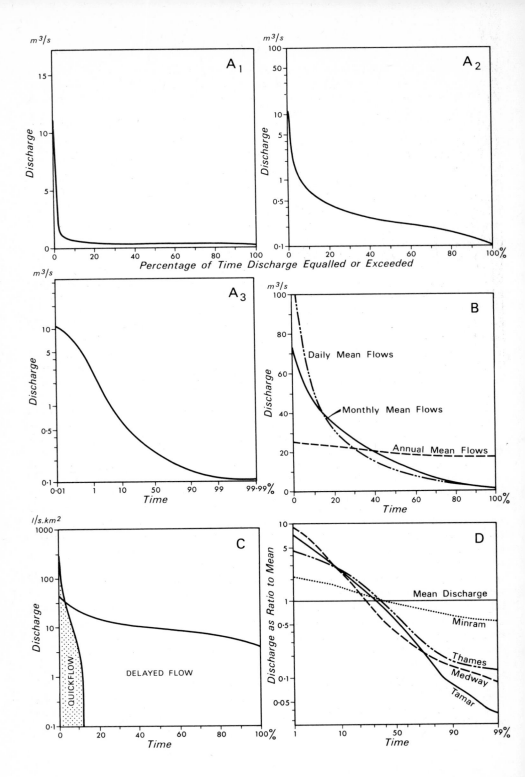

Data referring to the maximum floods or minimum drought flows are collected either as an annual series or a partial series, and the return period of each event is computed. There are several formulae available for calculating the return period Tr, but the simple evaluation of $Tr = \dfrac{n+1}{m}$ (n = number of years of record, and m = rank of a particular event) is generally applicable. Gumbel Type I plotting paper is often used for the graphical plot because the data usually conforms to a straight line (Figure 4.4A), although a normal probability scale and both logarithmic and linear discharge ordinates are also sometimes employed. A derivative from the Gumbel Type I extremal distribution is that the recurrence interval of the mean of the annual flood series will be 2·33 years and the discharge level with this return period is termed the *mean annual flood* ($Q_{2·33}$). This value is often abstracted from a frequency plot as a characteristic parameter.

In some instances the flood frequency plot does not conform to a straight line, but rather exhibits two straight line segments providing an upper and lower frequency curve (Figure 4.4B). The occurrence of a steeper upper frequency curve has been referred to by Potter *et al.* (1968) as the dog-leg phenomenon and from a study of runoff data for several hundred streams in the United States, they suggested that the boundary point between the two curves occurs at a return period of approximately 10 years. This phenomenon is probably the result of two different runoff populations, those forming the lower curve being governed by such variables as rainfall intensity, infiltration capacity and contributing area, while those forming the upper curve might represent runoff under saturated conditions from a maximum contributing area.

Frequency analysis can also be applied to the low flow extremes of a discharge record, and the data used could consist of instantaneous low flows or the mean flows for periods of different duration (see Figure 4.4C). Flood frequency analysis can also be extended to a regional level (Dalrymple, 1960; Cole, 1966) so that a generalised dimensionless flood frequency curve can be constructed for a hydrologically homogeneous region. The discharge ordinates on such a curve are usually expressed as a ratio to the mean annual flood (Figure 4.4E). The procedure for deriving a regional flood frequency curve and in particular for establishing the homogeneity of a group of flow stations has been described by Dalrymple (1960). The ratio of the 10 year to the 2·33 year flood is determined for each individual flow record and a mean ratio is computed. Estimates of the 10 year flood for each station are then derived by applying this mean ratio. The actual return periods of these estimates are computed and used in a homogeneity test which consists of confidence limits applied to a plot of computed return period against period of record (Figure 4.4D). If the computed return periods fall within these confidence limits which are based on two standard errors or 95 per cent probability (i.e. the differences between the stations could have arisen by chance), then the group of stations can be thought of as comprising a homogeneous region. The regional flood frequency curve is finally derived by calculating the mean ratio of

Figure 4.3 Streamflow duration curves
This figure illustrates several ways of plotting simple duration data (A_1, A_2 and A_3), the influence of the time base of the streamflow data on the precise form of the curve (B), duration curves constructed for individual streamflow components (C) and dimensionless curves (D, after Bannerman, 1966).

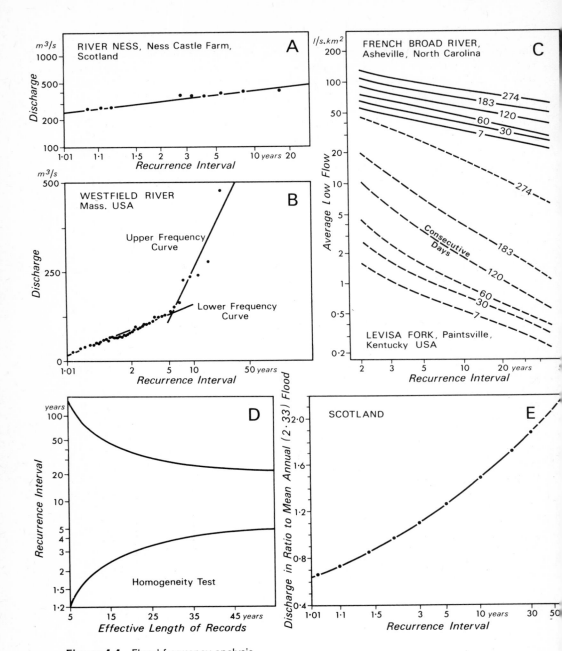

Figure 4.4 Flood frequency analysis
This figure illustrates, (A) a simple flood frequency plot (after Biswas and Fleming, 1966), (B) a frequency plot exhibiting the 'dog leg' effect (after Potter *et al.*, 1968), (C) low flow probability plots (after Schneider and Friel, 1965), (D) the homogeneity test used in regional flood frequency analysis, and (E) a regional flood frequency curve for Scotland (after Biswas and Fleming, 1966).

floods of several return periods to the mean annual flood. In some cases this type of regional analysis can be extended further by establishing a general relationship between mean annual flood and area for a particular region (Figure 6.4) so that a flood frequency curve could be synthesized for any catchment, providing its area was known.

In some instances it might be useful to cite *the maximum flood* experienced at a gauging station, but if comparisons between stations are to be made this value should be qualified by the period of record. Furthermore, it is difficult to compare catchments of different area because flood magnitude per unit area is inversely proportional to catchment area (Figure 5.9). For this reason, Pardé (1961) used an index of flood magnitude defined as

$$\frac{Q}{\sqrt{S}} \quad \text{(where } Q = \text{maximum recorded flood,} \\ S = \text{Catchment area)}$$

in his study of maximum floods throughout the world.

4.3b Runoff components

The techniques of hydrograph separation, described previously in the context of assessing the origins of runoff in channels (p. 119), can be extended to the annual hydrograph, in order to provide estimates of the relative proportions of total runoff originating from these sources. Arbitrary techniques of flow separation (Figure 3.9) are somewhat tedious to apply because a graphical plot of the entire record is required and the separation lines have to be constructed for each storm runoff event. It was in this way that Smith (1969) separated the hydrograph of the summer daily mean flows for the Bedburn and Greta catchments into three components, namely, storm runoff, basin storage discharge, and initial groundwater discharge. The time based separation advocated by Hibbert and Cunningham (1967) is more objective (Figure 3.9) and is suitable for direct computer processing of the digitized discharge values. Walling (1971) applied a computer routine of this nature to the hourly discharge values from several small catchments in east Devon, to evaluate the relative contribution of quickflow and delayed flow (Table 4.3). In this particular example the dif-

Table 4.3 Annual flow separation for five instrumented catchments in east Devon (1967–8)

Catchment no	Quickflow %	Delayed flow %	Response ratio %*
1	50·15	49·85	22·68
2	23·60	76·40	6·22
3	15·50	84·50	6·67
4	20·80	79·20	13·08
5	22·60	77·40	11·50

*The response ratio is defined as the ratio of the total quickflow volume to the total precipitation.

ference between the runoff response from catchment 1, which is relatively impermeable, and from catchment 3, which is underlain by more permeable rocks, is clearly shown. Chemical separation techniques are perhaps more meaningful in physical terms, but their application to long term hydrographs would be highly complex unless carried out on a fairly generalized basis (see Figure 3.10). Using such a generalised approach, Skakalskiy (1966) was able to evaluate the relative importance of different flow origins for the major natural regions of European Russia (Table 6.9). Values of the magnitude of the individual runoff components could be used to calculate *response factors*. Hewlett and Hibbert (1967) suggested two such factors, one defined as the ratio of the quickflow volume to the total precipitation input (Table 4.3) and the other as the ratio of the quickflow volume to the total runoff.

Hydrograph separation analysis could be extended further to incorporate study of the records of the individual components and these could be evaluated using many of the techniques described previously, for example duration curves (Figure 4.3C). This form of analysis is possibly most applicable to time-based separation because the associated computer routine can produce digitized records of the individual components for direct processing.

4.3c   Detailed hydrograph characteristics

None of the descriptive parameters and techniques discussed above provide detail on the precise shape of the streamflow hydrograph and methods of analysis of *storm hydrograph shapes* and *recession or depletion curve form* can be used for this purpose. The shape of the storm hydrograph produced by a catchment is influenced by the temporal and spatial distribution of the rainfall and by the catchment characteristics, and can itself be described by several indices (Figure 4.5A). More particularly, the *lag time* is the interval between the centre of gravity of the graph of effective rainfall and the hydrograph peak, the *time of rise* is the interval between the beginning of the rise and the associated peak, and the *base time* is the basal width of the hydrograph of storm runoff which itself depends on the technique of flow separation employed. A simple measure of the hydrograph peakedness can be derived as the *peak to mean ratio*, where the mean is the average level of storm runoff for the base time of the hydrograph. In some instances it is also of interest to establish the *runoff percentage* defined as the proportion of the storm rainfall occurring as storm runoff. The *effective rainfall* value is defined as the depth of rainfall equivalent to the volume of storm runoff.

Because hydrograph shape varies according to the magnitude and time distribution of the causative rainfall, an attempt must be made to standardise these controls if the characteristic hydrograph form of a catchment is to be isolated. This can be achieved by applying the *Unit Hydrograph* concept originally developed by Sherman (1932). A unit hydrograph of duration T, as originally defined, is the hydrograph of direct runoff resulting from 1 centimetre (or 1 inch) of effective rainfall generated uniformly in space and time over the catchment in unit-time, T. More recently, however, the definition has been altered slightly in that the term 'unit' is usually applied to the rainfall (i.e. unit depth—1 centimetre) rather than to the rainfall duration, although the basic concept remains identical. The duration T can be chosen arbitrarily and for example, 1 hour, 6 hour and 12 hour unit hydrographs could be constructed (Figure 4.5C).

A unit hydrograph can be derived from observed data either graphically or numerically (see Chow, 1964, Nash 1966). A simple graphical procedure (Figure 4.5B)

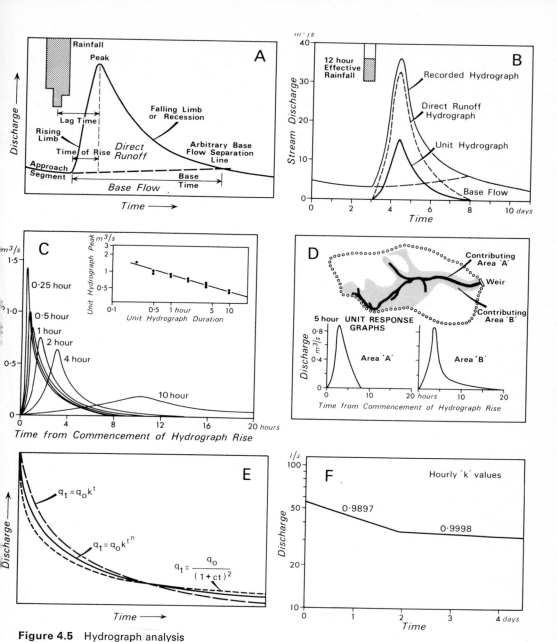

Figure 4.5 Hydrograph analysis

The various characteristics of a storm hydrograph are demonstrated in (A). The derivation of a 12-hour unit hydrograph is illustrated in (B) and examples of unit hydrographs of several durations constructed for the Rosebarn catchment are shown in (C). (D) demonstrates the influence of variable contributing area upon unit response graph form within a small east Devon watershed. Streamflow recession curves can be described by various mathematical functions (E) and individual segments conforming to the exponential function will plot as straight lines on semi-logarithmic coordinates and can be described by 'k' indices representing the slope (F).

would involve selecting an isolated short duration storm of uniform intensity and spatial distribution over the drainage basin, and its associated storm hydrograph. The hydrograph of direct runoff is separated from the observed stream flow hydrograph and the volume of direct runoff is measured and expressed as a depth of rainfall over the catchment. To construct the unit hydrograph it is necessary to adjust the ordinates of the direct runoff hydrograph by multiplying by $1/x$ (where x = the depth of rainfall equivalent to the volume of the direct runoff hydrograph). The resulting hydrograph is the unit hydrograph for the period of the duration of the effective rain and several examples should be constructed and superimposed to provide a generalised form.

A series of unit hydrographs of different durations can be derived (Figure 4.5C). For comparison of drainage basins of contrasting area, *dimensionless unit hydrographs* may be constructed in which the discharge ordinates are expressed as a ratio to the peak discharge, and the time ordinates as a ratio to the lag time. The *instantaneous unit hydrograph* is a mathematical abstraction produced when the duration of the effective precipitation becomes infinitesimally small. It is used primarily in the investigation of rainfall/runoff dynamics and its derivation is very complex, although computer techniques can be employed (for example Diskin, 1967; Johnson, 1970).

Although there is a certain controversy over the linear or non-linear basis of the runoff process and its relation to the unit hydrograph, and many of the assumptions involved are based essentially on the Horton infiltration approach to run off (p. 28) rather than a partial area or variable source area model, the unit hydrograph is nevertheless valuable as a means of studying hydrograph form. The technique of standardising hydrograph form by adjusting the direct runoff hydrograph to represent 1 cm (or 1 inch) of effective rainfall should, however, perhaps be referred to as the *unit response graph* approach (Walling, 1971) to avoid the wider implication of traditional unit hydrographs. One should not expect completely similar response graphs to be produced by storms of a certain duration, because variations in the extent of the contributing area will also influence the precise form of the hydrograph. In Figure 4.5D the unit response graphs derived from two storms of 5 hours duration over a small catchment in east Devon have been reproduced. The estimated contributing areas were completely different for these two events and this factor is reflected in the response graph form. The unit response graph associated with the small contributing area exhibits a rapid rise and recession while that associated with the larger area rises more slowly and possesses a more protracted recession.

The other major element of the runoff record for which shape analysis can be undertaken is the recession limb or depletion curve which comprises the gradually falling trace of streamflow following a storm event when unaffected by rainfall. Several investigators have suggested that either the entire recession curve, or segments of it, can be expressed as a mathematical function (Figure 4.5E). The simplest function is the basic exponential equation of the form

$$q_t = q_0 e^{-at}$$
$$\text{or} \quad q_t = q_0 k^t$$

where q_t = discharge at time t
q_0 = initial discharge

e = base of the natural logarithms
a = constant
t = time interval
k = constant representing (e^{-a})

which can also be written as

$$\log q_t = t \log k + \log q_o$$

and will plot as a straight line on semi-logarithmic graph paper (Figure 4.5F). Other equations reviewed by Toebes and Strang (1964) include the double exponential—

$$q_t = q_0 e^{-btn}$$

or

$$q_t = q_0 k^{tn}$$

where   n = constant
k = (e^{-b}) = constant

and the hyperbola—

$$q_t = \frac{q_0}{(1+ct)^2}$$

where   c = constant.

If the streamflow recession curve can be fitted by one of these equations then its form can be simply described by the value of the constant or constants. The exponential function is often used for this purpose because it can be described by the k value which in turn represents the slope of a semi-logarithmic plot (Figure 4.5E). In many cases it is also possible to derive a master recession curve from the streamflow record for a catchment by combining the characteristics of the various recession records. Methods available for this purpose include the correlation method (Langbein, 1940), the strip method (Wisler and Brater, 1959) and the tabulating technique (Johnson and Dils, 1956). Different master curves may be required for the summer and winter seasons because of the greater importance of evapotranspiration loss from storage during the summer months.

4.4   Sediment and solute production

The major characteristics of sediment and solute production and transport within a drainage basin can be summarised in terms of magnitude, timing, and duration and frequency aspects. The detail with which these facets can be studied necessarily depends on the nature of the data available; whether it is derived from continuous monitoring, frequent sampling, fraction collectors, estimation using a rating curve procedure, or from reservoir surveys. Furthermore, data of similar detail may not be available on all components of the load. Data processing will in most cases involve use of discharge records in order to convert concentration data into values of sediment and solute discharge (kg/sec, kg/day etc.) and computer routines can often be profitably employed.

Table 4.4 Sediment production statistics

A Variation in annual suspended sediment load

River	Year	Load (tonnes)
Kura River, Tiflis, Poland[2]	1928	16,253
	1929	9,467
	1930	8,295
	1931	28,939
Paria River, Lees Ferry,[3] Arizona, U.S.A.	1949–50[1]	1,460,000
	1950–51	1,546,000
	1951–52	2,007,000
	1952–53	4,626,000
	1953–54	2,337,000
	1954–55	4,384,000
	1955–56	1,058,000
	1956–57	3,249,000
	1957–58	11,412,000
	1958–59	2,823,000

Source: [1]Water Year
[2]Jarocki (1957)
[3]Mundorff (1968)

B Relative importance of sediment and solute loads

River	% of Total load	
	Sediment	Solutes
Colorado River, Grand Canyon, Arizona[1]	94	6
Green River, Green River, Utah[1]	90	10
San Juan River, Bluff, Utah[1]	97	3
Wind River, Riverton, Wyoming[1]	73	27
Saline River, Russell, Kansas[1]	87	13
Iowa River, Iowa City, Iowa[1]	83	17
Corey Creek, Pennsylvania[1]	56	44
Pond Branch, Maryland[2]	16	84
Volga, U.S.S.R.[3]	36	64
Don, U.S.S.R.[3]	45	55
Amur, U.S.S.R.[3]	73	27
Ob, U.S.S.R.[3]	33	67
Wieprz, Poland[4]	5	95
Pilica, Poland[4]	7	93
Tyne, Bywell, U.K.[5]	65	35
Derwent, Eddysbridge, U.K.[5]	70	30
East Devon Catchments[6]		
Catchment 1	23	77
Catchment 2	40	60
Catchment 3	55	45
Catchment 4	36	64
Catchment 5	45	55

Source: [1]Wolman and Miller (1960)
[2]Cleaves *et al.* (1970)
[3]Strakhov (1967)
[4]Jaworska (1968)
[5]Hall (1967)
[6]Walling (1971)

C   Relative contribution of bedload and suspended load to total sediment load

River	Proportion of total sediment load (%)	
	Bedload	Suspended load
Upper Niger, Baro[1]	6·5	93·5
Lower Niger, Shintaku[1]	5	95
Benue, Yola[1]	6	94
Alpine Mountain rivers[2]	70	30
Central Asian rivers[2]		
a Mountainous	15–23	77–85
b Hilly	5–15	85–95
c Lowland	1–3	97–99
Volga, U.S.S.R.[2]	0·3–2·0	98–99·7
Mississippi, U.S.A.[2]	0·3–10·0	90–99·7
Tyne, Bywell, U.K.[3]	13	87
East Devon Catchments[4]		
Catchment 1	11	89
Catchment 2	1·3	98·7
Catchment 3	1·8	98·2
Catchment 4	2·8	97·2
Catchment 5	2·2	97·8

Source: [1]N.E.D.E.C.O. (1959) (values of suspended load include both wash load and suspended load totals)
[2]Jarocki (1957)
[3]Hall (1967)
[4]Walling (1971)

4.4a   Magnitude and general characteristics

Because of inherent inaccuracies in the various methods that could be used to calculate values of *annual sediment and solute yield* and associated data, these figures should be qualified by an indication of the technique used (cf. pp. 165–7). The problem of assessing long-term mean annual loads, particularly sediment loads, is probably greater than in the case of rainfall and runoff (Figures 4.1A and 4.2B) because the occurrence of a flood of high return period could move more sediment in a few days than in the previous several years, and furthermore there are few accurate long period records against which to evaluate short-term estimates. An indication of the greater range to be found in annual sediment loads than in annual runoff is provided by four years of record from the Kura River in Poland and ten years of record from the Paria River, Utah, U.S.A. (Table 4.4A). This year to year variation could be qualified by use of the coefficient of variation of the annual totals. The values of mean annual yield (tonnes/year or m³/year) can be qualified by data on mean concentrations to incorporate the runoff volume.

It is also worthwhile to subdivide the total load of a stream into estimates of the proportions transported as sediment load and solute load and to further subdivide the sediment load into the proportion carried as bedload and suspended load (Table 4.4B and C). In general, sediment load exceeds solute load and suspended sediment load exceeds bedload. In addition, total load can also be subdivided according to its sources. In the case of sediment load the relative importance of sheet erosion, gully erosion, mass movement and channel erosion could be evaluated and Table 4.5 lists

Table 4.5 Relative importance of erosion types in the southeastern United States

Major land resource area	Range in relative importance as % of total erosion	
	Sheet erosion	Channel erosion
Southern Coastal Plain	100–69	0–31
Highland Rim and Pennyroyal	100–99	0–1
Cumberland Mountains	100–100	0–0
Southern Piedmont	100–66	0–34
Southern Mississippi Valley Loess	84–77	16–23
Alabama–Mississippi Blackland Prairie	100–77	0–23
Sand Mountain Area	100–100	0–0
Southern Appalachian Ridges and Valleys	100–99	0–1
Blue Ridge	99–86	1–14

Source: Roehl (1962)

the range in relative importance of the two major sediment sources in the south-eastern United States. In this particular region, sheet erosion is in all cases consider-ably more important than channel erosion, but information cited by Gottschalk (1962) suggests that channel erosion is dominant in the semi-arid and arid areas of the United States and in the mountainous areas of the central and southern Pacific coast regions. The total solute load of a stream can be subdivided according to the contri-bution made by cyclic salts or atmospheric sources and that made by the soil and rock of the drainage basin. The basin contribution could also be qualified as to the pro-portion attributable to the various flow origins. For example, Durum (1953) carried out an investigation on the Saline River in Kansas, U.S.A. and estimated that surface runoff comprised approximately 65 per cent of the total runoff for the years 1947 and 1948 and produced 24 per cent of the annual solute load, while groundwater runoff contributed 35 per cent of the total runoff and 76 per cent of the annual dissolved load. *The sediment delivery ratio* is a further index applicable to the total sediment yield of a catchment which expresses the ratio of the sediment yield to the gross erosion or sediment production within the drainage basin. Roehl (1962) cited values of the sedi-ment delivery ratio between in excess of 90 per cent and less than three per cent, for catchments in southern and south-eastern United States.

Information on the levels of instantaneous and daily mean concentrations at a measuring station can be provided by a listing of the maximum and minimum levels or more precisely by the associated values of the mean and standard deviation. In general the range of suspended sediment concentrations at a station considerably exceeds that of solute concentrations and whereas the range of the former could be as much as five orders of magnitude, that of the latter is usually less than one order of magnitude. *Rating curves* for suspended sediment and solute production which show the relationship between concentration or sediment and solute discharge and stream-flow (Figures 4.9 and 4.10) can also be used to provide an index of levels of concentra-tion, and in New Zealand suspended sediment rating curves are often used as a mea-sure of catchment condition (Campbell, 1962), in that the curves reflect the intensity of soil erosion. A more precise use of rating curve form is that proposed by Bauer and Tille (1967) who studied the variation in the parameters m and b in the suspended sediment rating curve equation i.e.

log Qs = log b+m log Q

where   Qs = suspended sediment discharge

Q = stream discharge

b+m = constants representing the intercept and slope of the rating plot respectively.

Within eleven catchments in Thuringia, German Democratic Republic, the value of parameter b was found to be related to mean annual discharge while the parameter m which ranged between 1·11 and 2·90 was found to be closely associated with geological contrasts between the catchments. The *erosion rate coefficient* proposed by Campbell and Caddie (1963) is also based on the suspended sediment rating curve and is derived for a drainage basin by calculating the discharge level with a return period of 2 years, $Q_{2\cdot0}$, and reading from the rating curve the sediment discharge corresponding to $0\cdot6Q_{2\cdot0}$. This value is divided by the catchment area (km²) to produce the erosion rate coefficient (tonnes/day.km²). Values of this coefficient cited for New Zealand ranged between 12 for the Wanganui River at Te Maire and 3320 for the Waipaoa River at Waipaoa Station.

4.4b   Time distribution characteristics

Variation of sediment and solute production within a year can be portrayed by extending the use of *regime diagrams* to sediment and solute data (for example, Figure 6.18). They provide a very graphic means of demonstrating the seasonal imbalance of sediment production in such areas as the Arctic. Although detailed studies have been undertaken on the relationship of sediment and solute concentration to discharge (Section 4.5b), very little work has been carried out on the detailed form of sediment or solute hydrographs. One notable exception is the concept of the *distribution graph* proposed by Johnson (1943), and which can be thought of as essentially analogous to the use of unit hydrographs or unit response graphs for analysing the shape characteristics of streamflow hydrographs. He studied the detailed shape of the sediment load graph for individual storm events (Figure 4.6A) and found that this appeared to be a function of the time of rise of the streamflow hydrograph, which in turn reflects the intensity and duration of the storm rainfall. By redrawing the sediment load graph with dimensionless load ordinates expressed as the percentage of the total load transported within a specified duration (for example, 1 hour), the graphs for a particular period of rise could be superimposed and a generalised master form produced (Figure 4.6B). Characteristic and contrasting shapes could subsequently be derived for different times of rise (Figure 4.6C). Therefore, although concentrations vary from storm to storm in response to many influencing factors (Section 4.5b) it would appear that the shape of the sediment load graph demonstrates a consistent pattern related to the time of rise of the discharge hydrograph. Within the East Fork of the Deep River, North Carolina, studied by Johnson, a relationship between the time of lead of the peak of the sediment concentration over the peak of the stream discharge, and maximum discharge was also established and he was able to use the distribution graph concept to generate estimated sediment load graphs for individual storms based upon one measurement of sediment concentration during the storm event. Little further work has been carried out on this particular theme, although the

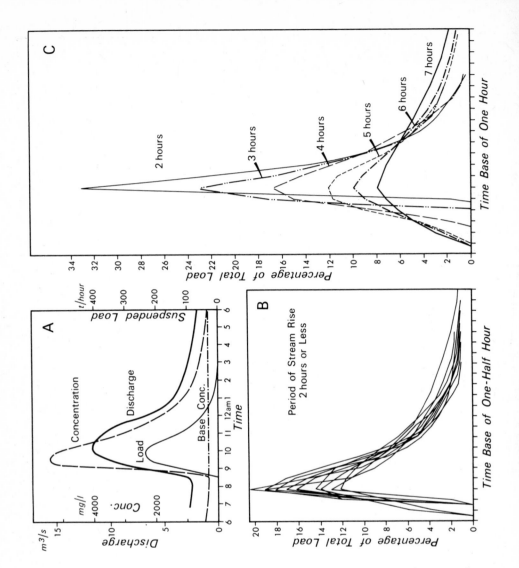

Figure 4.6 Suspended sediment distribution graphs

Graphs of streamflow, suspended sediment concentration, and suspended load for a typical storm hydrograph on East Fork of Deep River, North Carolina, (A). Superimposed distribution graphs for ten rises on the same river, in which the time of rise was two hours or less (B). Average distribution graphs for various times of rise on the same river (C). (after Johnson, 1943).

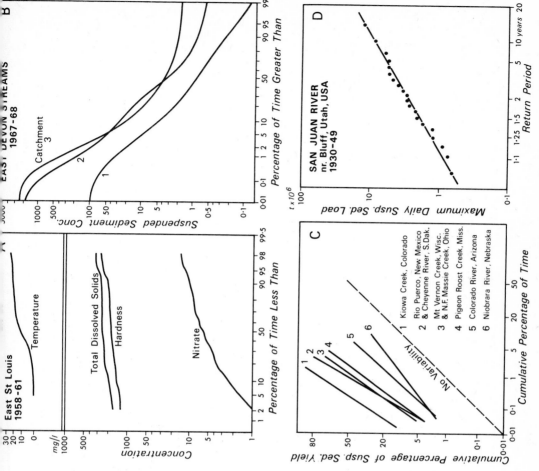

B

EAST DEVON STREAMS
1967-68

Catchment
3

2

1

Suspended Sediment Conc.

Percentage of Time Greater Than

1000
500
100
50
10
5
1
0.5
0.1

0.01 0.1 1 2 5 10 50 90 95 99

D

SAN JUAN RIVER
nr. Bluff, Utah, USA
1930-49

Maximum Daily Susp. Sed. Load

Return Period

$t \times 10^6$
100
10
1
0.1

1.1 1.25 1.5 2 5 10 years 20

East St Louis
1958-61

Temperature

Total Dissolved Solids

Hardness

Nitrate

Concentration

mg/l
1000
500
100
50
10
5
1

Percentage of Time Less Than

1 2 5 10 50 90 95 98 99.5

30
20
10
0

C

1 Kiowa Creek, Colorado
2 Rio Puerco, New Mexico
  & Cheyenne River, S.Dak.
3 Mt Vernon Creek, Wisc.
  & N.F. Massie Creek, Ohio
4 Pigeon Roost Creek, Miss.
5 Colorado River, Arizona
6 Niobrara River, Nebraska

1
2
3
4
5
6

No Variability

Cumulative Percentage of Susp. Sed. Yield

Cumulative Percentage of Time

80
50
20
5
1
0.1
0.01
0.001

0.01 0.1 1 5 20 50

Duration curves can be constructed
for various water quality parameters
(A, based on Harmeson and Larson,
1969, and B). (C) illustrates the
duration aspects of suspended sediment yield for several rivers in the
United States (after A.S.C.E., 1970),
and a sediment load probability plot
is shown in (D).

recent advent of continuous monitoring devices for suspended sediment concentration (p. 155) would seem to offer scope for future research.

4.4c Frequency and duration aspects

Many of the techniques of analysis applied to streamflow records in order to evaluate duration and frequency characteristics can also be applied to records of sediment and solute transport. Duration curves of solute concentration and load, the concentration of individual ions, and other water quality parameters can be produced (Figure 4.7A). Similarly suspended sediment data can be reproduced in duration curve form (Figure 4.7B) to demonstrate that high levels of concentration and sediment discharge only occur during a very small proportion of the period of record. This aspect of duration analysis can also be logically extended to evaluate the relative proportions of the total sediment and solute loads moved within specified time durations. For example, the results obtained by Walling (1971) for five small catchments in east Devon (Table 4.6) indicate that whereas approximately 78 per cent of the annual suspended sediment load was carried by flows equalled or exceeded 1 per cent of the time, or within an overall time period of about 3·5 days, the time taken to transport a similar proportion of the dissolved load was between 70 and 180 days. In these particular catchments suspended sediment production is dominated by the few major storm events which are associated with increased sediment concentration and

Table 4.6 Frequency characteristics of sediment and solute production from 5 instrumented catchments in east Devon

A Suspended load

| Flow frequency (%) | % total load moved by flows equalled or exceeded with the stated frequency | | | | |
	Catchment 1	Catchment 2	Catchment 3	Catchment 4	Catchment 5
0·01	2·50	8·00	3·20	4·50	4·90
0·10	20·50	46·00	28·00	30·30	31·90
1·00	73·70	83·00	81·90	77·30	77·60
5·00	93·20	92·80	94·90	88·40	91·20
10·00	96·70	95·60	96·80	91·80	94·50
20·00	98·80	98·10	98·00	95·30	97·00
50·00	99·80	99·70	99·30	98·90	99·10

B Dissolved load

| Flow frequency (%) | % total load moved by flows equalled or exceeded with the stated frequency | | | | |
	Catchment 1	Catchment 2	Catchment 3	Catchment 4	Catchment 5
0·01	0·65	0·37	0·26	0·43	0·45
0·10	5·78	3·00	2·50	3·60	3·60
1·00	30·50	12·70	13·70	16·40	16·00
5·00	56·80	28·50	29·30	29·00	30·80
10·00	69·70	40·10	38·50	38·20	39·70
20·00	79·90	57·40	52·90	51·30	52·50
50·00	93·30	83·40	75·10	77·20	76·00

Table 4.7 The importance of storm magnitude in suspended sediment production within the United States

Watershed	Catchment area (km²)	Proportion of average annual sediment yield contributed by large, moderate and small storms (%)		
		Large	Moderate	Small
Kiowa Creek, Colo.	287	36	16	48
Scantic River, Conn.	255·0	17	3	80
Tarkio River, Iowa	518	12	7	81
Elm Creek, Kans.	58·8	44	22	34
Pigeon Roost Creek, Miss.	303	10	7	83
West Tarkio Creek, Mo.	272	11	7	82
Fox Creek, Nebr.	199	42	16	42
Little Miami River, Ohio	334	14	8	78
Stillwater Creek, Okla.	427	20	12	68
Corey Creek, Pa.	31·6	24	14	62
South Tyger River, S.C.	451	6	4	90
Deer Creek, Tex.	211·9	29	13	58
Coon Creek, Wis.	199·9	25	13	62

Source: Piest (1965)

increased stream discharge, whereas solute production is a much more uniform process because solute concentrations tend to decrease or only increase marginally with increasing discharge and remain high during all levels of flow (Figure 4.10B). A similar tendency for the majority of the sediment load to be transported within a small proportion of the time is exhibited by most other rivers (for example, Figure 4.7C). If the data available on suspended sediment transport is sufficiently detailed, analysis can be extended to individual storms in order to evaluate the relative importance of storms of varying magnitude in sediment production (for example, Piest, 1965). Table 4.7 is based on the work of Piest and lists the proportion of the annual average sediment yield contributed by large, moderate and small storms within several drainage basins in the United States. This data shows that small storms which will occur on average several times a year are generally most significant, although the proportion contributed by large storms would seem to increase in the semi-arid regions of the country.

If suitable data is available, frequency analysis of a similar kind to that used for rainfall and streamflow data, can be applied to sediment and solute transport. The return periods of, for example, values of annual yield, maximum daily yield or maximum flood yields could be evaluated (for example, Figure 4.7D) and the resultant plots could be used to characterise the dynamics of a particular catchment.

4.5   Process interrelationships

Analyses of process interrelationships are valuable for two major reasons. Firstly, they can increase our knowledge and understanding of drainage basin dynamics and provide possibilities for predicting catchment response from information on a number of controlling variables. Secondly, they can in many cases be used to characterise the dynamics of a particular watershed to enable meaningful comparisons to be made between watersheds. For example, the suspended sediment concentration/stream discharge relationship or sediment rating curve for a catchment provides information on the sediment production dynamics of a catchment, enables estimates of concentration to be made for times when only discharge records are available, and the precise form of the relationship can be used as an index for comparing the response of several catchments (Figure 5.14B). Many different interrelationships could be considered, but an indication of their nature and importance is provided by selecting several examples central to this drainage basin theme and introducing input/output analyses and interrelationships between adjacent process subsystems.

4.5a   Input/output analyses

Simple input/output analysis can be applied to the process of suspended sediment production within a drainage basin, in that the output of sediment is primarily governed by the character of the precipitation input. A detailed study of sheet erosion carried out by Ellison (1945) attempted to relate the amount of soil splash to rainfall factors. Using small plots established on the Muskingum silt-loam soil at Coshocton, Ohio, and artificial rainfall simulators, he found that the weight of soil trapped in the splash samples was closely correlated with the drop size, velocity and intensity characteristics of the rainfall i.e.

$$E = K \cdot V^{4 \cdot 33} d^{1 \cdot 07} I^{0 \cdot 65}$$

where E = the weight of soil intercepted in the splash samplers during a
30-minute period (gm)

V = velocity of the raindrops (ft/sec)

d = diameter of the raindrops (mm)

I = rainfall intensity (in/hr)

K = constant, depending primarily upon soil type.

Surface runoff can loosen and remove soil particles from a slope, and should also be thought of as an input into the slope subsystem in the sheet erosion process. Ellison (1945) established that the soil loss caused by surface runoff or overland flow alone was related to the square of the flow velocities (Figure 4.8A). Furthermore, soil splash is in itself relatively ineffective in removing material from a slope, and surface runoff is required to transport the loosened particles. Because the incidence of both soil splash and surface runoff are associated with similar rainfall controls, relationships can be established between total soil loss output and rainfall input. A study by Wischmeier and Smith (1958) related total soil loss from plots on Shelby soil to a storm rainfall parameter derived as the product of the total storm energy E and the maximum 30-minute rainfall intensity (cf. pp. 318) (Figure 4.8B). A multiple regression equation was also developed to incorporate other variables, i.e.

$$Ls = 0 \cdot 0026E + 0 \cdot 0024E \cdot I_{30} + 0 \cdot 0517C + 0 \cdot 2330API - 2 \cdot 93$$

where Ls = total storm soil loss (ton/acre)

E = total storm energy (ft ton/acre)

I_{30} = maximum 30-minute storm intensity

C = accumulated rainfall energy since last tillage

API = antecedent precipitation index.

In this equation, the C factor provides a measure of the degree of compaction of the ground in terms of preceding heavy rainfall, and the API variable provides for the increased depth of surface runoff associated with wet conditions or high API values. Similar findings were obtained by Dragoun (1962) at the level of the small natural watershed. He studied two small catchment near Hastings, Nebraska and derived the relationship

$$Ls \propto E \cdot I_{30}(1 + Pa - Qa)$$

where Pa = an API index for the preceding 5 day period (in)

Qa = antecedent runoff for the preceding 5 day period (in)

in which the Pa and Qa variables incorporate an allowance for the soil moisture conditions within the catchment.

When studying the suspended sediment output from a larger basin, allowance must be made for sediment sources other than sheet erosion, namely, rill, gully and channel erosion. These are primarily controlled by the magnitude of the flow through the rill, gully or channel system which in turn closely depends on the rainfall input. Therefore, it should again be possible to establish relationships between storm period sediment yields and precipitation input variables and some of the results obtained in a classic study by Guy (1964) can be used as an example. In the Brandywine Creek catchment, Delaware, U.S.A., he related storm-period sediment yields, defined by the discharge-weighted mean concentration and the total sediment discharge, to variables representing rainfall input (Rq, Ri) and catchment condition and moisture status (Mt, Ta, Qb) (Table 4.8).

Table 4.8 Storm-period suspended sediment yields related to controlling variables; Brandywine Creek, Wilmington, Delaware, U.S.A.

Log C = 0·886 +0·644 log Qw −0·001 55 Mt +0·003 85 Ta −0·195 log Qb − 0·365 log 10 Rq +0·338 log 100 Ri

Log Qs = −1·634 +1·633 log Qw −0·001 42 Mt +0·003 65 Ta −0·199 log Qb − 0·354 log 10 Rq +0·328 log 100 Ri

(all regression coefficients significant at 95 per cent level or above)

where C = the water-discharge-weighted storm event mean sediment concentration (ppm)
   Qs = the storm event sediment discharge (ton)
   Qw = the net surface runoff (cfs day)
   Mt = the consecutive number of months during the period of record, beginning with May 1951
   Ta = the long-term mean air temperature for the given time of year (°F)
   Qb = the groundwater runoff at the beginning of the storm event (cfs)
   Rq = the mean precipitation for the basin (in)
   Ri = the mean precipitation intensity for the basin (in/hr).

Source: Guy (1964)

Storm rainfall/runoff analysis is in certain respects more complex than sediment production, because the runoff process involves both surface and subsurface components. However, the relationship is more direct as considerations of availability, dislodgement and transport of sediment, interposed between rainfall input and sediment output, are not involved. Furthermore runoff relationships have been studied extensively by hydrologists, particularly in the context of flood prediction. Most simple relationships established between storm rainfall and runoff involve measures of rainfall depth and intensity and a variable to account for moisture conditions within the basin at the time of the storm. Typical relationships are shown in Table 4.9. Those for the five small catchments in east Devon (cf. Figure 6.1C) were established by Walling (1971) and were able to adequately account for the two measures of storm runoff, namely, runoff volume and hydrograph rise, in terms of rainfall depth and an index of the soil moisture deficit. The equations were derived by stepwise multiple regression analysis and although independent variables representing rainfall amount, duration, maximum 15 minute intensity and soil moisture deficit were included in the original regression data, only rainfall amount and soil moisture deficit proved significant. The non-appearance of rainfall intensity and duration in the final equations is in some ways surprising because they are central to the traditional Horton concept of runoff production, in which overland flow occurs when rainfall intensity

Figure 4.8 Process interrelationships
Relationships established by Ellison (1945) between flow velocity and soil loss from plot experiments (A), and by Wischmeier and Smith (1958) between soil loss from plots and storm rainfall character (B). (C) illustrates the relationships between storm rainfall and storm runoff and hydrograph rise at two levels of soil moisture deficit for five small catchments in east Devon.

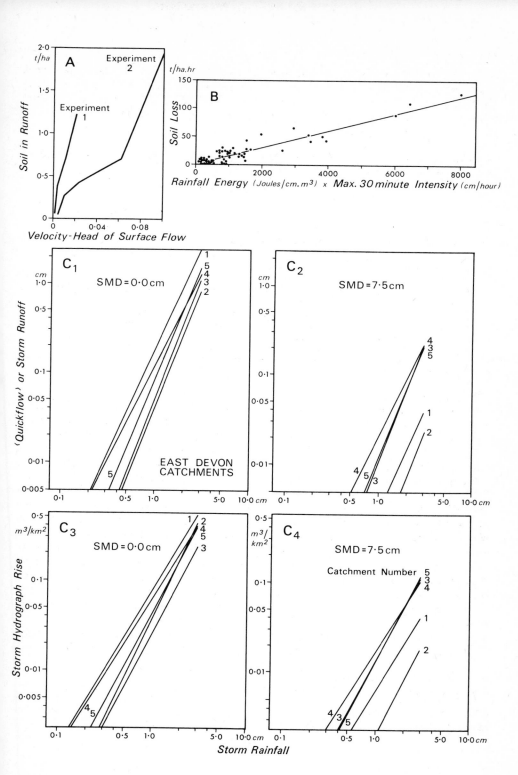

Table 4.9 Rainfall-runoff relationships

A East Devon Catchments

Catchment no	Relationship	Correlation coefficient
1	$\text{Log } Ra = -0.1751 + 2.3473 \log R - 0.6116 \text{ SMD}$	0.91
	$\text{Log } Hr = 1.5515 + 1.7580 \log R - 0.3739 \text{ SMD}$	0.92
2	$\text{Log } Ra = -0.7046 + 2.6668 \log R - 0.5217 \text{ SMD}$	0.90
	$\text{Log } Hr = 1.2185 + 1.9867 \log R - 0.3830 \text{ SMD}$	0.89
3	$\text{Log } Ra = -0.5638 + 2.7620 \log R - 0.2290 \text{ SMD}$	0.93
	$\text{Log } Hr = 1.1720 + 1.9930 \log R - 0.0990 \text{ SMD}$	0.90
4	$\text{Log } Ra = -0.4426 + 2.1141 \log R - 0.2593 \text{ SMD}$	0.90
	$\text{Log } Hr = 1.4388 + 1.6877 \log R - 0.1915 \text{ SMD}$	0.89
5	$\text{Log } Ra = -0.3937 + 2.6139 \log R - 0.2890 \text{ SMD}$	0.90
	$\text{Log } Hr = 1.4042 + 2.0076 \log R - 0.1745 \text{ SMD}$	0.91

where
Ra = storm runoff or quickflow depth (in)
Hr = hydrograph rise (cfs mi^2)
R = storm rainfall depth (in)
SMD = soil moisture deficit (in)

B Arizona and New Mexico Catchments (*Osborn and Lane, 1969*)
(*i*) *General relationships*

Watershed no	Relationship
LH-3	$Q = 0.48\, P_{tot} - 0.004\, D_{pd} - 0.076$
	$Qpr = 2.4\, P_{15} + 0.003\, D_{pd} - 0.65$
LH-4	$Q = 0.33\, P_{tot} - 0.24\, P_{ac} - 0.10$
	$Qpr = 1.7\, P_{tot} - 1.4\, P_{ac} - 0.50$
LH-5	$Q = 0.41\, P_{20} - 0.073$
	$Qpr = 3.7\, P_{10} - 0.50$
LH-6	$Q = 0.33\, P_{tot} - 0.003\, D_{pd} - 0.057$
	$Qpr = 2.8\, P_{10} - 0.45$

(All variables significant to the 1% level)

(*ii*) *Threshold prediction equations*

Watershed no	Equation	Threshold (in)	(mm)
LH-3	$Q = 0.40\, P_{tot} - 0.10$	0.26	6.6
LH-4	$Q = 0.32\, P_{tot} - 0.12$	0.38	9.7
LH-6	$Q = 0.25\, P_{tot} - 0.08$	0.32	8.1

where Q = total volume of runoff (in)
Qpr = peak rate of runoff (in/hr)
P_{tot} = total depth of precipitation (in)
P_{10} = maximum 10-minute depth of precipitation (in)
P_{15} = maximum 15-minute depth of precipitation (in)
P_{20} = maximum 20-minute depth of precipitation (in)
P_{ac} = antecedent precipitation index (exponential recession) (in)
D_{pd} = duration of runoff producing precipitation (intensity $\geqslant$ 0.4 in/hr) (min)

exceeds the infiltration capacity and where the infiltration capacity declines with increasing storm duration. However, it was suggested that in the alternative variable source area model, involving throughflow and saturated overland flow, the total rainfall amount is more important than intensity and duration. The form of the relationships for the individual catchments can be used as a basis for comparison (Figure 4.8C) and it can be seen that they appear to function very differently under wet and dry conditions. For example, catchment 1 produces the maximum values of runoff and hydrograph rise under wet conditions (SMD = 0) but amongst the lowest values under dry conditions (SMD = 7·5 cm). This could be explained in terms of the extent of the variable source areas at different moisture levels.

The relationships for the four small unit source watersheds (0·22–4·4 hectares) in the Walnut Gulch Experimental watershed in south-eastern Arizona (Table 4.9B) were established by Osborn and Lane (1969) and were similarly derived by stepwise multiple regression analysis. The resultant equations contrast with those for the Devon catchments in that rainfall intensity and duration variables are important for several of the catchments and in certain cases rainfall amount does not figure in the regression equations. This is perhaps to be expected in a semi-arid area where the infiltration approach to runoff production is essentially valid. Osborn and Lane also calculated the simple relationships between runoff volume and rainfall amount (Table 4.9B) and these were used to establish the threshold levels of rainfall required for runoff production (i.e. the minimum amount of rainfall required to initiate runoff) (Table 4.9B). This threshold level averaged 0·81 cm indicating that under semi-arid conditions only storms of medium magnitude and above produce a storm runoff response.

4.5b Interrelationships between subsystems

The relationship between suspended sediment transport and stream discharge provides an interesting example of the interaction of two adjacent process subsystems. Although bedload transport is essentially a capacity load controlled by stream discharge, the suspended sediment load, particularly the wash load component, is a non-capacity load and the rate of transport depends upon the supply conditions and involves the interaction of the sediment production subsystem and the streamflow subsystem. Both subsystems are nevertheless causally related to a common input, rainfall. The general relationship between suspended sediment concentration (or discharge) and stream discharge is usually presented as a rating plot (Figure 4.9A). A distinction could be made between instantaneous and daily, monthly, and annual mean curves depending on the nature of the basic data, although the instantaneous curve is most valuable in this context. Furthermore, for statistical reasons it is more meaningful to consider sediment concentration rather than sediment discharge. Because sediment discharge (Q_S) is by definition the product of stream discharge (Q) and sediment concentration (C), the relationship between it (Q_S) and stream discharge involves a common variable, Q, and this can give rise to correlations of a spurious nature.

The graph of concentration versus discharge usually plots as a straight line on logarithmic coordinates (Figure 4.9A) and therefore conforms to the equation $C = aQ^b$. In the experience of the authors working on streams in east Devon, the

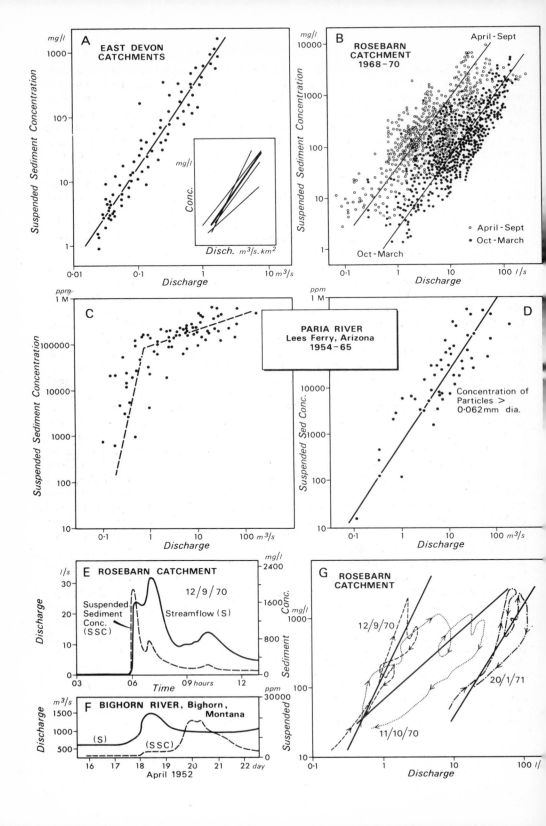

A **EAST DEVON CATCHMENTS**

Suspended Sediment Concentration (mg/l)
Discharge (m³/s)

Conc. (mg/l)
Disch. m³/s.km²

B **ROSEBARN CATCHMENT 1968-70**

April - Sept
Oct - March

Suspended Sediment Concentration (mg/l)
Discharge (l/s)

○ April - Sept
● Oct - March

C

Suspended Sediment Concentration (ppm)
Discharge (m³/s)

PARIA RIVER
Lees Ferry, Arizona
1954-65

D

Suspended Sed Conc. (ppm)
Discharge (m³/s)

Concentration of Particles > 0·062 mm dia.

E **ROSEBARN CATCHMENT**

12/9/70

Discharge (l/s)
Sediment Conc. (mg/l)

Suspended Sediment Conc. (SSC)
Streamflow (S)

Time (hours)

F **BIGHORN RIVER**, Bighorn, Montana

Discharge (m³/s)
Suspended Sediment (ppm)

(S)
(SSC)

April 1952 (day)

G **ROSEBARN CATCHMENT**

12/9/70
20/1/71
11/10/70

Suspended Sediment (mg/l)
Discharge (l/s)

value of the exponent b usually approximates to 2·0 (Figure 4.9A) although in other areas of Devon values as low as 0·80 have been found. However, studies by Campbell and Bauder (1940) and by Leopold and Maddock (1953) on rivers in the western United States found that the exponent in the sediment discharge/stream discharge relationship (i.e. $Qs = aQ^b$) usually assumed a value of 2·0 or between 2·0 and 3·0 respectively. The exponents in the associated concentration relationships would equal these values minus one (i.e. if $Qs = aQ^b$ then $C = aQ^{b-1}$ because $Qs = Q \times C$) and it may be that the concentration exponent generally lies between 1·0 and 2·0.

More detailed study of the rating plot often reveals considerable scatter about the straight line relationship (Figure 4.9B), indicating, as would be expected, that other variables besides stream discharge influence the level of concentration. Seasonal effects are often important, and work in Great Britain by Hall (1967) and Walling and Teed (1971) and in the eastern United States by Guy (1964) has demonstrated that for a given discharge, concentrations are generally greater in summer than in winter and separate rating curves are often constructed for these two seasons (Figure 4.9B). This occurrence can be partly accounted for in terms of the lower baseflows during the summer, meaning that a given discharge represents a greater proportion of storm runoff; the drier surface conditions during the summer; and the tendency, in Great Britain at least, for summer storms to be more intense and therefore more likely to erode material. In addition, many rating plots exhibit a dog leg effect with a tendency for the upper segment of the rating curve to flatten off or even approach horizontal (Figure 4.9C). This can be explained by the fact that above a certain limit sediment concentrations do not continue to increase with discharge at a uniform rate but may even remain constant when the supply potential of the catchment has been fully achieved. This tendency is clearly shown by the rating plot for the Paria River, Arizona (Figure 4.9C), a river well known for the record concentrations that have been reported (p. 329). Rating curves can also be constructed for individual size fractions of the suspended sediment load and this has been done for the Paria River by plotting the concentrations of particles >0·062 mm diameter for the same samples (Figure 4.9D). The transport of this coarser material is more closely controlled by the tractive force of the stream and there is therefore a closer single relationship with discharge than shown by the overall concentrations (4.9C). A further important cause of rating plot scatter is that the peak of the concentration graph rarely coincides with the peak of the streamflow hydrograph and maximum concentration will therefore be associated with levels of discharge less than the peak flow. On some rivers the peak of the sediment concentration can lag significantly behind the flood peak. Lewis (1921) cites the example of a flood on the Tigris River at Amara, Iraq, in which the maximum sediment concentration occurred 3 days after the maximum discharge, whilst at Baghdad, more than 560 km upstream, the two peaks coincided. A more

Figure 4.9 Suspended sediment concentration/discharge relationships
A clearly defined rating relationship is exhibited by a small catchment in east Devon and the rating curves from adjacent basins exhibit similar forms (A). A pronounced seasonal effect is shown by the rating plot for the Rosebarn catchment (B) and the rating plot for the Paria River, Arizona, demonstrates the tendency for near-constant concentrations to occur at high discharges (C). Rating curves can also be constructed for individual size fractions (D). (E) provides an example of where the peak suspended sediment concentration occurs before the streamflow peak and (F, after Heidel, 1956) exhibits a lag effect. The hysteresis effect in the rating relationship is demonstrated by rating plots for three individual storm hydrographs from the Rosebarn catchment (G).

detailed study of this phenomena was carried out by Heidel (1956) on the Bighorn River in Montana and Wyoming U.S.A., (for example, Figure 4.9F), and he established that the sediment concentration peak moved more slowly than the water discharge peak and that the sediment peak progressively lagged behind the flood peak in a downstream direction. Average velocities of the sediment and flood peaks were cited as 3·2 and 6·1 km/hr respectively and on this particular river the maximum sediment concentration of a flood passing Bighorn, Montana could occur 48 hours after the maximum flood discharge (Figure 4.9F). On other rivers, however, the peak sediment concentration occurs before the streamflow peak and this situation is related to conditions in which the maximum sediment production or availability occurs in the early stages of a storm event and then declines, even though storm discharges continue to increase (Figure 4.9E). In this particular example, for a small stream, the first period of intense rainfall removed most of the available sediment and the second period, although marked by a higher discharge peak, caused only a small increase in concentration. Similar patterns can occur on large rivers and Jarocki (1957) cites the case of the Vistula River at Torun, Poland where the rising limb of the concentration graph is displaced about 12 hours in advance of that of streamflow. The tendency for the supply of sediment to decline during a storm event can also often be distinguished on a storm to storm basis when a series of closely spaced and similar storm runoff events exhibit progressively lower sediment concentrations.

Even in cases where the peaks of water discharge and sediment concentration are coincident, the relationship between concentration and streamflow is further complicated by a hysteresis effect. More particularly, sediment concentrations on the rising stage of a streamflow hydrograph are usually greater for a given level of discharge than on the falling stage and the rating plot for an individual storm event will assume the form of a clockwise loop rather than a straight line (Figure 4.9G). This feature can be explained in terms of the cessation or reduction of the erosive effect of the rainfall on the falling limb and the increased volume of subsurface runoff contributing to the recession flow.

The interaction of the sediment production and streamflow subsystems necessarily means that the relationship between suspended sediment concentration and discharge is complex and multivariate in nature and for this reason the use of multivariate analysis should shed further light upon the relationship. Walling (1971) analysed a series of 1200 suspended sediment samples obtained during storm runoff events from a pumping sampler installed within the Rosebarn catchment (Figure 1.10) and attempted to explain the levels of concentration in terms of discharge and other independent variables. A stepwise multiple regression computer program was used to obtain the relationship listed in Table 4.10. The exact influence of each of these variables is better seen by considering the standardised regression coefficients or beta coefficients as suggested by Yevdjevich (1964) and the values obtained are listed in Table 4.10. These beta coefficients are dimensionless parameters and measure the effect of a particular independent variable on the variation of the dependent variable, and as they are dimensionless, they may be directly compared. From these values it can be seen that the timing of the sample during the storm hydrograph and the stormflow discharge are most important in accounting for the variation in levels of concentration. The signs of these coefficients exhibit physical significance because the maximum values of the X_1 variable occur on the falling stage of a streamflow

Table 4.10 The relationship between storm event suspended sediment concentrations and characteristics of the streamflow record for the Rosebarn catchment

(*i*) *Multiple regression equation*

$$\text{Log Conc.} = 2 \cdot 2648 - 0 \cdot 0020\ X_1 + 0 \cdot 4380 \log X_2 - 0 \cdot 2684 \log X_3 + 0 \cdot 0947 \log X_4$$

where Conc. = suspended sediment concentration (mg/l)
X_1 = time relation of sample to hydrograph peak: varying from negative on the rising stage to positive on the falling stage (min)
X_2 = stormflow discharge at time of sampling (cusec)
X_3 = flow level preceding storm hydrograph (cusec)
X_4 = index of flood intensity defined as $\dfrac{\text{Hydrograph peak} - X_3}{\text{Time of rise}}$

(*ii*) *Beta coefficients*

Variable	Beta coefficient	(B = b(Si/Sd))
X_1	−0·5307	
Log X_2	0·5275	
Log X_3	−0·4310	
Log X_4	0·1079	

where B = the beta coefficient
b = the regression coefficient
Si = the standard deviation of the independent variable
Sd = the standard deviation of the dependent variable

hydrograph where the values of concentration would be expected to be lowest, and an increase in concentration is associated with an increase in stormflow discharge (X_2). The index of previous flow level, variable X_3, is also important in explaining the variations in concentration, with reduced concentrations associated with the maximum value of this variable. However, it is more difficult to ascribe a precise physical meaning to this variable, since it will reflect moisture status within the catchment and the dryness of the soil surface as well as the season of the year. The flood intensity index, variable X_4, is less important than the first three, but indicates a tendency for high concentrations to be associated with sharp peaked storm hydrographs which themselves probably reflect intense rainfall.

The relationship between dissolved solids concentrations and discharge within a stream provides a further example of the interaction of two adjacent process subsystems, in this case runoff production and solute production, and reflects both the surface and subsurface dynamics of a catchment. The relationship between total dissolved solids concentration (C) and streamflow (Q) is usually inverse in form (for example, Figure 4.10 B_2 and C) and often exhibits a straight line plot on logarithmic coordinates conforming to the equation $C = aQ^{-b}$. The inverse character of the relationship is due primarily to a dilution effect, with increased flows marking an increased contribution from storm runoff which possesses a lower solute content than baseflow. In most cases the slope of the logarithmic plot (−b) is less than −1·0, indicating that whereas concentrations decrease with increasing discharge, total load increases. The rating relationships for some streams exhibit flattening at the upper and lower ends (for example, Figure 4.10A) and this occurrence can again be

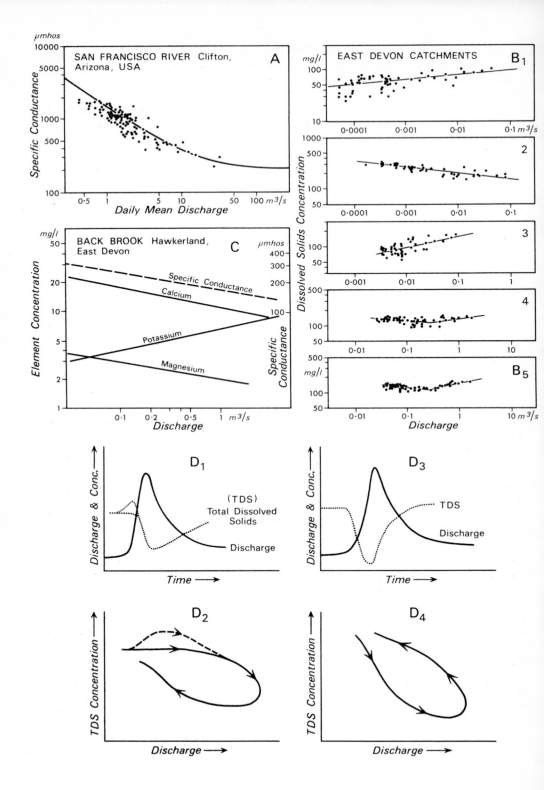

explained in terms of the dilution of a solute-rich baseflow component by storm runoff with a lower solute content. Under dry conditions, when baseflow is the only source of runoff and is therefore undiluted, there is a tendency for a constant maximum concentration value to be attained at low discharges. At very high discharges the dilution effect becomes progressively less significant and concentrations are dominated by the near-constant solute content of the storm runoff component.

In streams where solute sources and runoff dynamics are complex, different forms of the rating relationship may appear, and in certain cases the relationship may be very poorly defined if not non-existent. Figure 4.10B demonstrates the variety of relationships exhibited by five adjacent small catchments in east Devon (Walling 1971). In catchments 1 and 3 concentrations were found to increase with increasing discharge and this tendency can be tentatively explained in terms of the source areas of runoff and solutes. Low flows are sustained predominantly from one rock type, whereas at higher discharge levels the contributing area expands to include other rock types which supply runoff with a higher solute content. Catchment 2 is dominated by a single and different rock type which produces higher solute concentrations, and this watershed exhibits the classic inverse rating form. Catchments 4 and 5, which are larger in size, contain component watersheds exhibiting the varying characters of the three previously described basins and the rating relationships are expectedly complex. Elements of both inverse and direct relationships combine to provide 'U' shaped rating plots.

A study of dissolved load rating relationships for 99 streamflow stations throughout the United States made by Gunnerson (1967) found the inverse hyperbolic form for this relationship (i.e. $C = aQ^{-b}$) to be most generally applicable, although inverted 'U' shaped ratings were also found. The Salt River at Shepherdsville, Kentucky, was cited as unique in that the highest concentrations occurred at peak discharges and although no causal mechanism was suggested, the occurrence of cavernous limestone and connate brines was advanced as a likely explanation. Gunnerson further classified the inverse hyperbolic relationships into three categories; firstly those where the slope of the relationship was less than −0·1, secondly those where the slope was between −0·1 and −0·33 and thirdly those where the value was in excess of −0·33. The distribution of the three types within the United States showed no readily discernible pattern, although those ratings in the first group were found to be primarily associated with stations below reservoirs.

Solute rating relationships can also be established for the concentrations of individual ions (Figure 4.10C) and although in many cases they are similar in slope to those for specific conductance or T.D.S., some ions may exhibit very different or even reverse relationships. For example, in Figure 4.10C the ratings for calcium and magnesium are almost identical in slope to that for specific conductance, and concentrations of these two elements could probably be approximated fairly accurately from the conductivity record. Conversely, the rating for potassium shows a marked

Figure 4.10 Dissolved load concentration/discharge relationships
A typical dissolved load rating relationship is shown in (A, after Hem, 1970). However, many catchments exhibit other rating curve forms (B), and the rating plots for individual constituent elements can vary (C). In some catchments the relationship between solute concentrations and discharge during individual storm events exhibit a clockwise hysteretic effect while in others an anti-clockwise hysteretic effect is found (D).

positive trend. Further examples of the varying response of individual ion concentrations to fluctuations in streamflow are provided by Figures 3.9 and 3.10. In addition, certain ions maintain near constant levels of concentration over the range of flows and this occurrence may be the result of chemical buffering mechanisms within the system. Johnson et al. (1969) studied the variations in major ion concentrations in the stream water of the Hubbard Brook Experimental Watershed, and distinguished three major groups of ions; those that diluted with increasing stream discharge, those that concentrated, and those that exhibited limited extremes and were possibly subject to chemical buffering mechanisms (Table 4.11). In this particular drainage basin,

Table 4.11   General behaviour of major ions in stream water of the Hubbard Brook Experimental Watersheds 1963–7

Ion	Dilutes with stream discharge	Concentration with stream discharge	Limited extremes (chemically buffered)
Na	+ +	—	+
SiO_2	+ +	—	+
Mg	+	—	+ +
SO_4	+	—	+ +
Cl	—	—	+ +
Ca	+	+	+
Al	—	+ +	+
H	—	+	+
NO_3	—	+ +	+
K	—	+	+

+ irregular occurrence
+ + consistent occurrence
— not evident

Source: Johnson *et al.* (1969)

aluminium concentrations increased with increasing discharge, and this was explained by the fact that this element was wholly acquired from the soil and was dissolved by the initial reaction of rainwater with the upper soil (i.e. under storm runoff conditions). Chloride concentrations showed little variation, and because there was no dilution effect the influence of rainfall must have been compensated by a rapid reaction within the soil.

Understanding of solute concentration/discharge relationships can be furthered by the application of simple mixing models or mass balance equations in an attempt to approximate the physical processes involved (for example, Hall, 1970; Hem, 1970; Johnson et al., 1969). A very simple model could be built around the situation where a volume of water (Q_1) with a known solute concentration (C_1) enters the drainage network from an upstream spring and is diluted by the remaining runoff (Q_2) of a lower concentration (C_2) to produce the observed downstream concentration (C_3). The solute balance equation will take the form:

$$C_1 Q_1 + C_2 Q_2 = (Q_1 + Q_2) C_3$$

If the volume and concentration of the inflow (Q_1, C_1) are constant and the particular solute is absent in the diluting water (i.e. $C_2 = 0$) then:

$$C_3 = \frac{C_1 Q_1}{Q_1 + Q_2}$$

and because the $C_1 Q_1$ term is constant the solute rating plot will conform to a straight line with a slope of $-1\cdot0$ on logarithmic coordinates. In nature, it is more likely that the diluting water will also contain the solute concerned and in this case

$$C_3 = \frac{C_1 Q_1 + C_2 Q_2}{Q_1 + Q_2}$$

and because it is unlikely that the $C_2 Q_2$ term will be constant the rating will be curved when plotted on logarithmic coordinates. Furthermore, it is also unlikely that the springflow ($C_1 Q_1$) will remain constant, although the equation is still applicable. The mass balance equation could be extended to cover the case of inflow from several runoff components i.e.

$$C_n = \frac{C_1 Q_1 + C_2 Q_2 + \ldots C_n Q_n}{Q_1 + Q_2 + \ldots Q_n}$$

Hem (1970) describes the case of the San Francisco River at Clifton, Arizona where the rating plot (Figure 4.10A), which did not conform to a simple straight line, could be evaluated in terms of three major runoff components. These were, firstly, saline inflow from the Clifton Hot Spring which contributed about 60 per cent of the total dissolved load; secondly, storm runoff; and thirdly, the general baseflow of the river. The spring inflow exhibited a constant flow of 56 l/s and a conductivity of 16000 micromhos, the baseflow a conductivity of 500 micromhos and a maximum flow of $2\cdot8$ m³/s, and the storm runoff comprised all flows above $2\cdot86$ m³/s and was characterised by a specific conductance of 200 micromhos. Substitution of these values in an equation of the form

$$C_3 = \frac{C_1 Q_1 + C_2 Q_2 + C_3 Q_3}{Q_1 + Q_2 + Q_3}$$

yielded the line superimposed on the plot in Figure 4.10A and which adequately describes this particular dissolved load rating relationship. Hall (1970, 1971) further developed the use of mixing models in studies of solute concentration/discharge relationships. He established six major mixing equations and defined the form of the various associated rating plots. By using a linear transformation of the various model equations and linear regression techniques, or alternatively a non-linear optimisation procedure it is possible to determine the equation most applicable to a particular set of discharge and concentration data.

Whichever mathematical relationship between solute concentrations and discharge is adopted, the rating plot will in most cases exhibit scatter about the line describing the relationship (for example, Figure 4.10A or B). Several reasons for this scatter could be advanced. Sampling, instrumental and analytical errors should be carefully evaluated and plotting discrepancies should not be overlooked, because Hem (1970) has noted that instantaneous solute concentrations are often erroneously related to values of daily mean discharge. However, the detailed dynamics of solute and runoff production themselves also give rise to scatter about any idealised relationship. Hysteretic effects similar to those noted for suspended sediment rating plots (p. 218) also occur with solute rating relationships. For a given discharge, solute concentrations will often vary according to whether they are associated with the rising or falling stage of the streamflow hydrograph. This occurrence was described in detail for the Salt River at Shepherdsville, Kentucky by Hendrickson and Krieger (1960), who distinguished a cyclic relationship between streamflow and dissolved load concentrations for individual storm runoff events. Concentrations were greater on the rising than on the falling stage and the rating plot exhibited a clockwise hysteretic loop. The reverse occurrence, an anticlockwise hysteretic loop was described by Toler (1965) for Spring Creek, Georgia, where solute concentrations were greater on the falling stage.

Idealised examples of clockwise and anticlockwise hysteretic loops are presented in Figure 4.10D. In the first case (Figure 4.10D$_1$, D$_2$), the flood cycle can be divided into three phases. During the first stage, discharge increases rapidly but dissolved load concentration changes little and may even increase, as a result of the flushing of readily soluble material, especially evaporative deposits, from the soil and ground surface, which counters or even reverses the expected dilution effect of the storm runoff. In the second phase, discharge reaches a maximum and begins to recede and solute concentrations fall rapidly due to the 'fresh' character of the storm runoff and the high stream stages which prevent groundwater seepage from entering the stream. The third phase representing the recession limb is marked by gradually increasing concentrations, as a result of a reduction in the proportion of storm runoff and an increase in baseflow. The clockwise hysteretic form is therefore dominated by the effect of high concentrations of dissolved solids in the early runoff from the land surface. In the reverse case (Figure 4.10D$_3$, D$_4$) the major control is exerted by the dynamics of groundwater flow, and this occurrence is associated with areas where groundwater levels respond rapidly to infiltration and where good hydraulic connections exist between the groundwater body and the stream. Three phases can again be recognised. In the first phase of rapidly rising stage, inflow of groundwater into the stream is reduced, because river levels rise faster than groundwater levels and storm water may even move into bank storage. However, by the time the flood peak occurs, groundwater levels have responded to infiltration input and movement of groundwater into the stream occurs. In the third phase, after the peak, this movement of groundwater increases and for a given discharge level the contribution from groundwater and, therefore, the total dissolved load concentration is greater than on the rising stage. Modification and refinement of these two basic hysteretic effects is necessary to incorporate the precise details of runoff production associated with the variable source area concept in catchments of varying characteristics.

Seasonal effects may also be responsible for details of the rating plot. Hendrickson and Krieger (1960) noted that concentrations for a given discharge were lower during spring and early summer than in late summer and autumn, and these latter higher concentrations were explained by the accumulation of soluble salts near the surface of the soil during the summer and by the long storage duration since the winter recharge of the groundwater reservoir. A seasonal cyclic relation, analogous to that described for individual storms by Hendrickson and Krieger (1960), was demonstrated by

Gunnerson (1967) for several streams in the Columbia River basin. In these cases, the rating plots were described as 'elliptical doughnuts' and they exhibited an annual clockwise hysteretic effect. Maximum concentrations occurred during the low flow months of late summer, autumn and early winter. Then during late winter and early spring, increased runoff from the onset of the rainy season caused concentrations to decrease in response to the dilution effect, although this decrease was partly offset by the flushing out of solutes accumulated during the dry months. Minimum concentrations occurred at the time of maximum discharges in May and June and, as flows subsequently fell, concentrations began to rise, although for a given discharge, concentrations were less than during the period of increasing discharge in spring because of the absence of the flushing effect. Biological effects should also be acknowledged as capable of causing seasonal variations in dissolved loads, particularly in the concentrations of individual ions. The production and movement of solutes in a drainage basin is closely associated with the processes of nutrient cycling and the use of nutrients by the biota of the ecosystem. Johnson et al. (1969) demonstrated that within the Hubbard Brook Experimental Forest, New Hampshire, nitrate concentrations of the streamflow were lower for a given discharge in summer than in winter, because of the biologic uptake of the element during the summer. The annual rates of biological uptake of certain nutrient elements within this forest can be compared with the quantities transported by the stream (Table 4.12). Ions such as potassium and nitrogen

Table 4.12   Flux of nutrient elements through the Hubbard Brook watershed ecosystem

Element	A Mean annual transport by stream	B Annual rate of biologic uptake	A/BX100 (%)
Na	6·1 kg/ha. yr	1·6 kg/ha. yr	380
S	12·4	4·0	310
Cl	3·6	1·6	220
Al	1·9	1·6	120
Ca	10·6	32	33
Mg	2·5	16	16
K	1·5	32	5
N	1·9	44	4

Source: Johnson et al. (1969)

can be seen to be in critical supply and their concentrations would be expected to vary according to biological demand.

4.6   Modelling of catchment response

This consideration of the quantitative evaluation of drainage basin processes can be usefully concluded with a brief discussion of the application and use of models in geomorphological studies of the drainage basin. Models are viewed by some workers as alternatives to catchments for research purposes (cf. p. 176), but they should necessarily be thought of as complementary. Indeed, the Canadian National Committee for the I.H.D. (1966) stressed that 'The development of conceptual models, which explain the interrelationship of the various hydrological phenomena, is the end objective of all basin research studies.' Three major spheres of application of modelling techniques can be distinguished. Firstly, echoing the statement cited above, models can be used to increase our understanding of drainage basin response and the operation of the various processes, and the interrelationships of process and form (p. 264). Secondly, they can be used to extend the scope of watershed studies by providing the opportunity for reconstructing particular palaeohydrological situations and for prediction of catchment response. Thirdly, they may be employed to evaluate the effects of changes in catchment characteristics, particularly those occasioned by man's activity.

The literature abounds with references to many different types of model and several classifications have been proposed. However, following Ward (1971), three major types may be distinguished, namely, physical models, analogue models, and digital models. *Physical models* (for example, Eagleson, 1969) are essentially scaled-down models of watersheds, interpreted according to laws of dynamic similarity, and the laboratory catchment, where both input conditions and even basin form can be varied (pp. 264–5), can provide considerable scope for geomorphological research. A distinction is often drawn between studies involving physical models of actual catchments (for example, Chery, 1966) and those involving experimental prototype models (for example, Black, 1970). *Analogue* models constitute mechanical or electrical devices that possess functional characteristics equivalent to the characteristics of the system under investigation. The most widely applied analogy is that between the flow of electrical current and the flow of water. Tinlin (1969) described a passive direct electric analogue of a catchment which incorporated electronic circuits to represent the processes of interception, surface storage and runoff, and infiltration and subsurface storage. A constant current pulse was applied to the circuit (rainfall input), and this was progressively modified so that the final pulse resembled the streamflow hydrograph. The scope of this approach is considerably broadened by the possibilities presented in the use of analogue computers (for example, Riley and Narayana, 1969) and of electronic circuit network analysis programs for digital computers (Tinlin and Thames, 1971). *Digital models* are integrally associated with digital computers which can manipulate and process vast quantities of data in incredibly short periods of time. Any classification of digital models must be somewhat arbitrary, but a threefold division, on the basis of information scale, into Deterministic, Stochastic and Parametric models (for example, Snyder, 1971) can be usefully applied.

Deterministic models (for example, Woolhiser, 1971) are essentially physically-based and incorporate a theoretical structure based on the laws of continuity or conservation of mass, energy and momentum, and involving a series of ordinary or partial-differential equations. However, lack of knowledge of the precise nature and functioning of the various processes operating within a drainage basin and the complexity of the detail of natural systems means that, at this point in time, the scope for their application is limited, although, in certain instances, they may form part of a larger parametric model. The best example of their application is probably that of overland and open-channel flow, which can be modelled according to theories of unsteady free-surface flow using, for example, kinematic wave equations. A similar approach can be adopted in the context of infiltration processes, using the theory of single-phase flow in a porous media. Deterministic models can perhaps be thought of as the ultimate goal of watershed modelling, where the researcher possesses near perfect information on catchment dynamics. *Stochastic models* (for example, De Coursey, 1971) can conversely be applied when the researcher possesses little or no information on catchment dynamics and when individual processes must be viewed as stochastic occurrences. Stochastic analysis is used to generate synthetic sequences of hydrological or geomorphological data, by determining the statistical distribution characteristics of the sample historic data and by using a random number generator to extend the data sequence. For example, monthly runoff or sediment yield at a gauging station could be synthesised by a stochastic model based on the mean, standard deviation and serial correlation of the historic data. Stochastic processes used for this purpose commonly include Markov processes and Monte Carlo procedures. In the true sense, therefore, absolutely nothing is assumed as to the internal structure of the drainage basin, although stochastic techniques can also be used to generate input sequences for both deterministic and parametric models.

Parametric modelling techniques (for example, Snyder, 1971) involve the development and analysis of relationships between numerically defined characteristics of drainage basin form and process. Parameters are defined as the mathematical terms in the functional relationships between the variables. This group of models can be thought of as lying between the two former groups on the information scale. Sufficient information concerning the internal structure of the drainage basin and the associated processes is available, so that a stochastic approach is unnecessary, but the information is insufficient for the development of rigorous deterministic models. A consideration of parametric modelling techniques is particularly valid in the context of this treatment of drainage basin form and process, because these variables form the basis of the relationships involved in the models. Techniques used to establish these relationships and to model drainage basin processes vary in complexity from the black box approach, through grey box analysis, to the white box or systems synthesis approach.

The black box approach involves the fitting of input (for example, rainfall) to output (for example, runoff or sediment production) by some simple mathematical function (see pp. 210–15) without attempting to incorporate relationships or parameters representing the internal functioning of the drainage basin system. This method is particularly suited to computer optimisation procedures, because the functional parameters can be optimised until the predicted and observed output closely correspond, but the function obtained is applicable only within the limits of the data involved and should not be extended to new sets of data or to other drainage basins. Some of the objections levelled against the black box can be overcome by use of grey box techniques, especially correlation analysis. In these models, the internal functioning of the basin is represented by variables known to be important in the processes interposed between input and utput. These may represent either catchment characteristics (for example, slope, soil character, etc.) or measures of catchment condition at a particular point in time (for example, antecedent precipitation, temperature, etc.). The functional relationships are usually established using multiple regression analysis (see Tables 4.10, 5.11, 6.1, and 6.6) and problems involved include those of interdependence of the variables, and the physical reality of the statistical relationships. White box analysis attempts to model the actual processes operating in the catchment, within the limits posed by the analytical techniques and the precise knowledge of drainage basin dynamics. Two well known examples may be cited, firstly, a runoff model and secondly, a model of sediment production.

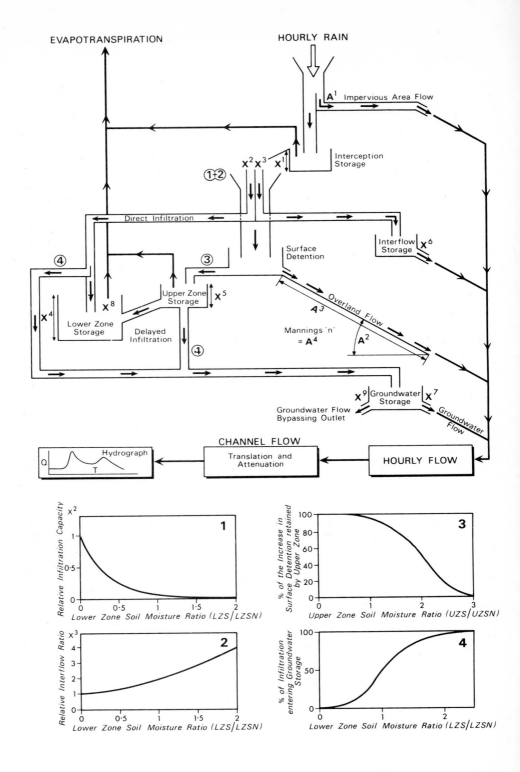

EVAPOTRANSPIRATION

HOURLY RAIN

A^1 Impervious Area Flow

Interception
Storage

X^2 X^3 X^1

①÷②

Direct Infiltration

Surface
Detention

Interflow
Storage X^6

④

③

Upper Zone
Storage X^5

X^8

A^3 Overland Flow

Mannings 'n'
= A^4

A^2

X^4

Lower Zone
Storage

Delayed
Infiltration

④

X^9 Groundwater X^7
Storage

Groundwater Flow
Bypassing Outlet

Groundwater
Flow

CHANNEL FLOW

Hydrograph

Q

T

Translation and
Attenuation

HOURLY FLOW

Graph 1 — X-axis: Lower Zone Soil Moisture Ratio (LZS/LZSN), Y-axis: Relative Infiltration Capacity X^2

Graph 3 — X-axis: Upper Zone Soil Moisture Ratio (UZS/UZSN), Y-axis: % of the Increase in Surface Detention retained by Upper Zone

Graph 2 — X-axis: Lower Zone Soil Moisture Ratio (LZS/LZSN), Y-axis: Relative Interflow Ratio X^3

Graph 4 — X-axis: Lower Zone Soil Moisture Ratio (LZS/LZSN), Y-axis: % of Infiltration entering Groundwater Storage

4.6a A digital runoff model

The Stanford IV Watershed Model (Figure 4.11) was proposed by Crawford and Linsley (1966) and has been successfully applied in several areas of the world (for example, Fleming, 1970 (Scotland); Wood and Sutherland, 1970 (New Zealand)). In this runoff model, the drainage basin is represented by five major storage containers (Figure 4.11), and data on rainfall (usually hourly) and evapotranspiration (usually daily) are used to calculate the changes in moisture content of each container and the resultant production of runoff. A continuous streamflow hydrograph can be simulated (Figure 4.12) and the basic conceptual framework is as follows. Rain falling on any impervious area does not enter the system but is diverted directly to the stream channel as impervious area flow, while the remaining precipitation is subject to interception and, if it fills the interception storage container, the overflow is available as input to the ground surface. The infiltration capacity determines the proportions of this rainfall that are diverted to surface detention and infiltration. The infiltrating water is subdivided between interflow storage, lower zone storage and groundwater storage, and the increment to surface detention either contributes to overland flow or enters the upper zone storage, which represents depression storage and storage in highly permeable surface soils. Moisture is lost from the upper zone storage by percolation to the lower zone and to groundwater storage. Interflow storage and groundwater storage are in turn depleted by outflow to the stream, and evapotranspiration acts upon interception storage, and upper and lower zone storage. The input of overland flow, interflow and groundwater flow into the stream channel is translated and routed to generate the output hydrograph of total runoff. Provision for variation in model response according to catchment characteristics is provided for by 13 parameters ($A^1 - A^4$ and $X^1 - X^9$, Figure 4.11) which represent the physical characteristics of the drainage basin under consideration (Table 4.13). Variations in response according to moisture conditions within a catchment is accommodated by four functions (Figure 4.11, 1, 2, 3 and 4), which vary the uptake and release of water according to the relative contents of the storage containers (the ratio of the actual content of the store to a nominal storage level within that store).

This particular runoff model could perhaps be faulted in that it does not make full allowance for the partial area and variable source area concepts of runoff production, but until these concepts are adequately translated into a generally applicable simulation model, it must remain a very useful tool for use by geomorphologists as well as hydrologists. Analysis of the size and operation of the various stores and the proportions of runoff originating from the different stores can provide insight into the functioning of the drainage basin system and study of the relationship of model parameters to measurable drainage basin characteristics can clarify interrelationships between form and process. Furthermore, the model could be used for palaeohydrological reconstruction and future prediction by modifying the input and the various parameters to represent historic or future conditions. James (1965) utilised the model to predict the effects of urbanisation on catchment response and Moore *et al.* (1969) used a

Figure 4.11 The Stanford IV watershed model
See text for explanation (based partly on Crawford and Linsley, 1966 and Wood and Sutherland, 1970).

Table 4.13 Parameters controlling the functioning of the Stanford IV Watershed Model

Parameter	Explanation
A^1	Percentage of impervious area
A^2	Overland flow slope
A^3	Overland flow length
A^4	Manning's n for overland flow
X^1	Interception storage capacity
X^2	Infiltration index
X^3	Interflow index
X^4	Nominal lower zone storage
X^5	Nominal upper zone storage
X^6	Interflow recession constant
X^7	Groundwater recession constant
X^8	Lower soil evaporation factor
X^9	Groundwater bypass fraction

'A' parameters can be estimated directly from the physical characteristics of the watershed. 'X' parameters require initial estimates which are subsequently optimised by comparison of actual and predicted streamflow data. *Based on:* Wood and Sutherland (1970)

modified version to estimate the natural pattern of runoff from a watershed which had been substantially modified by the construction of floodwater reservoirs, in order to evaluate the precise effects of the structures. Figure 4.12 demonstrates how a simplified form of the model has been used to simulate the natural pattern of runoff from the Rosebarn catchment. The model, calibrated against the pre-urbanisation period, will be used to estimate the response of the basin in the absence of the effects of urbanisation.

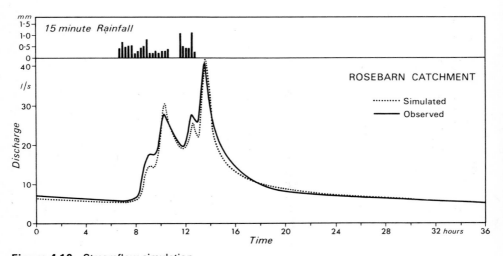

Figure 4.12 Streamflow simulation
The Stanford IV watershed model (15-minute time base) was used to obtain this close simulation of an individual storm hydrograph from the Rosebarn catchment.

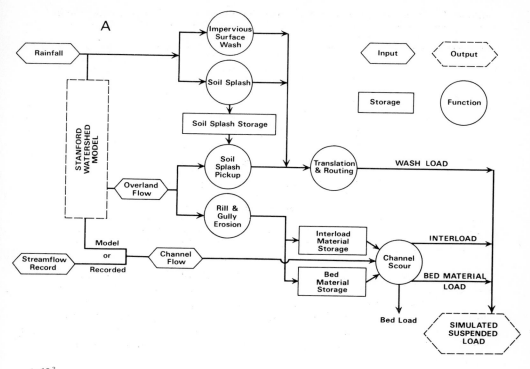

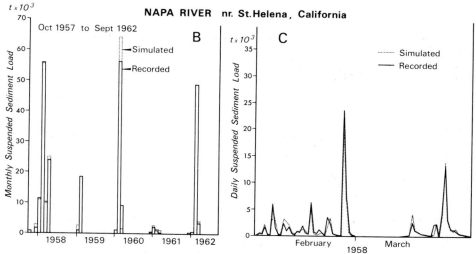

Figure 4.13 The Stanford sediment model
This digital sediment model (A) has been used successfully to simulate suspended sediment production and transport (B and C) (based on Negev, 1967).

4.6b   A digital sediment model

Little attention has as yet been given to the possibilities of white box modelling of sediment and solute production processes, although the Stanford Sediment Model developed by Negev (1967) stands as a notable exception. This model (Figure 4.13A) is essentially a development of the Stanford runoff model to include the processes of suspended sediment production and transport, and these processes are necessarily viewed in the context of the runoff dynamics underlying the parent model. Input to the model is provided by values of hourly rainfall, overland flow calculated by the runoff model, and total flow either as recorded or as calculated. A distinction is made between the land surface and stream channel in the internal functioning of the model and the output of suspended sediment is divided into wash load (very fine) produced by sheet erosion and interload (fine), and bed material load (coarse) produced by rill, gully and channel erosion. The precise operation of the model is controlled by several constants representing the characteristics of the drainage basin and which are established either by estimation or by optimisation procedures in a similar way to the runoff model. The results of the application of the model to simulation of suspended sediment transport in the Napa River, California is demonstrated in Figure 4.13B and C. In this example, the model successfully reproduced annual, monthly and daily sediment loads, and the theoretical division of the load into three size fractions showed a close correspondence to the actual grain size composition of the suspended sediment loads. The model has also been applied to the River Clyde in Scotland by Fleming (1968). Considerable scope for improvement remains, because such factors as flood plains deposition and seasonal effects in the availability of sediment should be included. Nevertheless the approach provides the geomorphologist with a useful technique for reconstructing and predicting the production of sediment and the progress of erosion within a catchment and for furthering study of the dynamics of sediment production. A similar approach could possibly be adopted to the simulation of dissolved load transport within a catchment, in that the various runoff components of the watershed model could be used as a basis for evaluating a mixing model such as those described on p. 223.

Selected reading

Techniques of data analysis are discussed by:

V. T. CHOW (1964); V. M. YEVDJEVICH (1964); D. R. DAWDY and N. C. MATALAS (1964), in: Statistical and Probability Analysis of Hydrologic Data, Section 8 in V. T. CHOW (ed.), *Handbook of Applied Hydrology*, 8-1—8-90,

UNITED STATES GEOLOGICAL SURVEY, *Techniques of Water Resources Investigations of the United States Geological Survey*, Book **4** Section A, Statistical Analysis.

A more detailed treatment of the frequency aspects of geomorphological processes is provided by:

M. G. WOLMAN and J. P. MILLER (1960), Magnitude and Frequency of Forces in Geomorphic Processes, *J. Geology*, **68**, 54–74.

Process interrelationships are considered by:

H. P. GUY 1964: An analysis of some storm-period variables affecting stream sediment transport, *United States Geological Survey Professional Paper*, **462**-E.

A mathematical approach to the study of processes and their interrelationships is presented in:

P. S. EAGLESON 1970: *Dynamic Hydrology*.

The use of models is discussed by:

R. J. MORE 1967: Hydrologic models and geography. In R. J. CHORLEY and P. HAGGETT (eds.), *Models in Geography*, 145–85.

5 Drainage basin form and process

> Though the main drive in American research in the two decades since Horton's classic work has been to quantify the descriptions of landforms and drainage systems, there are enough valuable instances of the correlation of the morphometric parameters with streamflow and land erosion to warrant the hopeful prediction that more and more will be done and learned. (*S. Shulits, 1967*)

5.1   **Water and sediment in river channels**
5.2   **Channel cross section: hydraulic geometry**
5.3   **The channel reach:** *5.3a Meandering channels; 5.3b Braided channels; 5.3c Alluvial fans and deltas; 5.3d Flood plains*
5.4   **The drainage basin:** *5.4a Topographic characteristics; 5.4b Rock type; 5.4c Soil; 5.4d Vegetation and land use*
5.5   **Drainage basin mechanics**

Relationships between drainage basin form and drainage basin process are of two kinds. Firstly, the character and magnitude of processes can be influenced by form or characteristics of the drainage basin and, secondly, the processes operating can be responsible for fashioning the form of the basin and therefore for determining the drainage basin characteristics. Which of these two relations applies depends upon the time scale being considered and also upon the scale of attention. The drainage network and pattern will influence the rate at which fluvial processes operate in the drainage basin but also on a longer time scale the processes may cause adjustment of the network and the channel pattern in quasi-equilibrium with the processes operating. Feedback mechanisms may, therefore, inevitably occur and, just as channel form can influence stream channel process, in turn the processes can bring about a modification of the channel form. Feedback mechanisms can be both positive and negative (Table 1.3) but the latter may predominate in drainage basin situations, whereas positive ones are more apparent in coastal systems (King, 1970). Many of the equilibrium notions developed in the context of the drainage basin (Table 1.3) have emerged from studies of the relationships between drainage basin characteristics and processes.

Studies of form—process relationships have been achieved in three main ways. Firstly, early work on the movement of water and sediment in stream channels in the nineteenth century led to the development of physical relationships; these were later complemented by empirical and laboratory studies of water and sediment movement; and they culminated in studies of the hydraulic geometry of stream channels in the mid-twentieth century. Secondly, work on river patterns has been based upon field, experimental and theoretical investigations; it has been prominent since the work of Inglis (1949) on meanders and the researches of the Vicksburg Waterways Experimental Station; and some geomorphologists have identified a branch of study denoted as alluvial morphology which is concerned with the depositional characteristics of flood plains. Thirdly, at the level of the drainage basin the work of Horton (1932, 1945) was foreshadowed by the work of engineers who devised empirical relations between drainage basin parameters and streamflow and sediment yield. Since Horton's work hydrologists, engineers, and geomorphologists have investigated drainage basin form and process and these studies have been catalysed by the needs to understand how runoff rates and yields of sediment vary, in association with soil erosion, changing vegetation cover, and eventual urbanisation. These three foci, hydraulic geometry, alluvial morphology and drainage basin process—response systems, provide a convenient sequence for attention but they must be preceded by some consideration of the controls upon water and sediment in channels.

5.1 Water and sediment in river channels

The necessity to understand the principles governing the movement of water and sediment in stream channels is emphasised specifically by Leopold, Wolman and Miller (1964) and more generally by Carson (1971) who contended that 'one suspects that geomorphology will emerge as a reputable discipline only when its students have become well-versed in the established principles of natural science'. Water flowing in channels is influenced by gravity but there are forces operating in a direction opposite to that of the flow, collectively termed shear resistance, which influence the flow through the frictional effect of the bed and banks and, to a lesser degree, at the air-water interface above. For this reason the velocity profile shows the highest velocities at the greatest distances from the bed and banks. When velocity is considered in successive cross sections (average velocities v_1, v_2, v_3 . . .) then, if density is constant, discharge in successive cross sections (area a_1, a_2, a_3 . . .) will be expressed as Discharge $= a_1v_1 = a_2v_2 = a_3v_3$ . . ., which is the continuity equation for steady flow. Steady flow exists if velocity at a point remains constant with respect to time. Two types of flow can be distinguished, namely, laminar and turbulent, although a combination of both may occur. Laminar flow exists if the flow of a fluid past a boundary, or another separate body of fluid which acts as a fixed or moving boundary, entails momentum transfer by molecular action only. Particles in the fluid thus tend to move in smooth, definite paths with uniform velocity and there is no significant transverse mixing during movement along the channel. Whereas viscosity forces dominate during laminar flow, as inertia forces increase instability develops, the flow becomes turbulent, and particles then move in irregular paths which are neither smooth nor fixed, and diffusion takes place by finite groups of molecules termed eddies and this is termed eddy viscosity. Laminar and turbulent flow may be distinguished by the Reynold's Number (R_N) which is given by:

$$R_N = \frac{\rho VL}{\mu} = \frac{VL}{v}$$

where V = velocity of flow
  L = characteristic length or measure of size of obstacle in terms of length
  ρ = density
  μ = dynamic viscosity
  v = kinematic viscosity (a measure of interference between adjacent layers of fluid during flow)

If the hydraulic radius (cross sectional area of flow/wetted perimeter) is used for L, laminar flow will occur when R is less than 500, turbulent flow when R is greater than 2500, and at intermediate values either type can occur.

Other types of flow may be distinguished and steady flow obtains when the velocity at a point does not change with time, whereas in unsteady flow the velocity may change in magnitude or direction, during the passage of a flood wave, for example. Flow may also be classified as tranquil (or streaming) and rapid (or shooting). The flow is tranquil when obstacles have a damming effect and disturbances are transmitted upstream, but rapid flow occurs when the motion upstream is not affected by the obstacle. Where flow changes from tranquil to rapid, the water level falls evenly but the transition from rapid to tranquil may often be abrupt and marked by a turbulent surface roller or hydraulic jump. The existence of tranquil or rapid flow depends upon the Froude number (F) where

$$F = \frac{V}{\sqrt{\dfrac{\Delta\gamma L}{\rho}}} = \frac{V}{\sqrt{gD}}$$

where V = average velocity of flow
  L = characteristic length dimension
  $\Delta\gamma$ = difference in specific weight across air-water interface
  ρ = density of fluid
  D = depth

The Froude number describes flow conditions by relating the inertia force to the gravity force and it distinguishes tranquil flow when $F < 1$, rapid flow when $F > 1$, and when $F = 1$ the flow is described as critical. These types of flow may be combined and Sundborg (1956) recognised four different states of flow (Figure 5.1a) indicating that laminar flow occurs only for slow currents or for shallow depths.

Velocity is thus a significant parameter and for more than two centuries flow equations have been available to relate flow velocity to character of the channel. The Chezy equation (1769) relates velocity (V) to hydraulic radius (R) and slope of the channel (S) in V = C/RS where C is the Chezy discharge coefficient. The Manning equation was developed for natural and artificial channel data and gives

$$V = K \frac{R^{2/3}S^{\frac{1}{2}}}{n}$$

where n is the Manning roughness coefficient (pp. 127–9). Illustrations of the Manning n have been published (Barnes, 1967) and some selected values are included in

Table 3.4. In wide shallow channels the hydraulic radius is approximately equal to mean depth and so, from the Chezy equation, velocity in such channels is proportional to the square root of the depth multiplied by the slope. The roughness of channels is controlled by a number of factors including depth, slope and physical size of bed material. Whereas in channels which are rigid the roughness is independent of flow, in natural river channels the roughness of the channel will vary with the flow and this is expressed in alluvial channels by a definite sequence of bed forms (Figure 5.1(c)).

The amount of material transported as suspended sediment and as bed load (including contact and saltation load) will depend upon the relationship of particle size and flow velocity. A critical erosion velocity or component velocity has been employed by several workers (for example, Sundborg, 1956) as the minimum current capable of initiating movement of grains of a given diameter. Although sediment will vary in distribution and concentration throughout the cross profile of a river channel in sympathy with the velocity distribution, Hjulstrom (1935) produced a series of curves in a single diagram which identified fields where erosion, deposition and transport occurred. These curves were modified by Sundborg (1956) to include suspended sediment as shown in Figure 5.1(b). Such curves will indicate for a particular cross section whether the channel is experiencing erosion or deposition. Channels were distinguished as stable or unstable by Leliavsky (1955). In stable channels the particles are in constant movement but the general form of the bed is unchanged because when a particle is moved forward it is immediately replaced by another from upstream. Although a bar may be swept away during high discharge it will be replaced during the subsequent low flow. In unstable channels this quasi-equilibrium is not maintained and bars and banks which do appear are soon removed.

Forms in alluvial channels include sand waves which are ridges on the stream bed, formed by the movement of bed material usually approximately normal to the flow direction, and shaped like a water wave; dunes which move downstream and are sand waves of approximately triangular cross section with a gentle upstream slope and a steep downstream slope; and antidunes which usually move upstream and are indicated on the water surface by a regular undulating wave. Simons and Richardson (1961) have demonstrated a sequence of bed forms which are associated with increasing intensity of flow (Figure 5.1(c)). Stages A to C (Figure 5.1(c)) are associated with the lower regime whereas stages E to H are typical of the upper flow regime, and D is transitional. When bed forms change, the roughness of the channel also changes. The effect of size of bed material and Froude number on form of bed roughness and on Manning n (Figure 5.1(d)) has been deduced for a range of flow conditions in a 2·44 m (8 ft) wide flume (Simons and Richardson, 1962). Langbein and Leopold (1968) have shown that the interaction among the individual units moving in a continuous flow pattern causes the velocity of these units to vary with their distance apart. Spacings of the units have properties which are described by the continuity equation and by a velocity relation and are termed kinematic. With these properties it is unlikely that particles can remain uniformly distributed along the flow path, for random influences lead to the formation of dunes in sand channels and to pools and riffles in gravel channels.

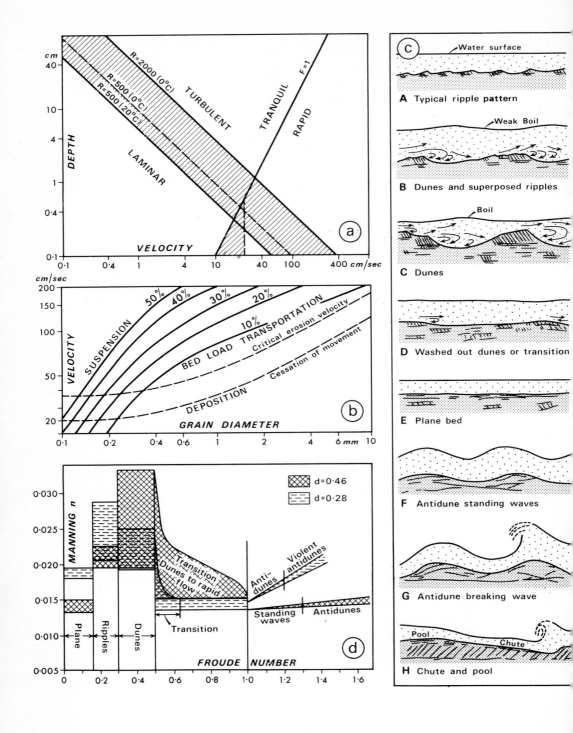

(a)

cm
40

10

4

1

0·4

0·1

DEPTH

VELOCITY

R = 2000 (0°C)   TURBULENT

R = 500 (0°C)

R = 500 (20°C)

LAMINAR

TRANQUIL   F = 1

RAPID

0·1   0·4   1   4   10   40   100   400 cm/sec

(b)

cm/sec
200
150

100

50

20

VELOCITY

50%   40%   30%   20%

10%

SUSPENSION

BED LOAD TRANSPORTATION
Critical erosion velocity

DEPOSITION   Cessation of movement

GRAIN DIAMETER

0·1   0·2   0·4 0·6   1   2   4   6 mm 10

(d)

0·030

0·025

0·020

0·015

0·010

0·005

MANNING n

d = 0·46
d = 0·28

Plane   Ripples   Dunes

Transition

Dunes to rapid
flow

Transition

Standing
waves

Anti-
dunes

Violent
antidunes

Antidunes

FROUDE NUMBER

0   0·2   0·4   0·6   0·8   1·0   1·2   1·4   1·6

(c)   Water surface

A Typical ripple pattern

Weak Boil

B Dunes and superposed ripples

Boil

C Dunes

D Washed out dunes or transition

E Plane bed

F Antidune standing waves

G Antidune breaking wave

Pool   Chute

H Chute and pool

5.2 Channel cross section: hydraulic geometry

From empirical, experimental and theoretical studies it is apparent that relationships should exist in natural channels between the measures of channel form and the processes operating in the channel. These relationships were studied, as hydraulic geometry of stream channels, by Leopold and Maddock (1953), by Wolman (1955) and by Leopold and Miller (1956). In these and subsequent studies mean velocity (v), mean depth (d) and width (w) of the flowing water were related to discharge (Q) in power functions of the form:

$$w = aQ^b$$
$$d = cQ^f$$
$$v = kQ^m$$

where a, c, k, b, f, m, are numerical coefficients
As cross sectional area $A = w \cdot d$ and $w \cdot d \cdot v = Q$
then $aQ^b \cdot cQ^f \cdot kQ^m = Q$ so that $a \cdot c \cdot k = 1 \cdot 0$ and $b+f+m = 1 \cdot 0$

Leopold and Maddock (1953) found that in the midwestern United States average relations gave values of b = 0·26, f = 0·40, m = 0·34, whereas Leopold, Wolman and Miller (1964) quote values of b = 0·29, f = 0·36, m = 0·34 for ephemeral streams in the semi-arid U.S.A. This indicates that in the case of the semi-arid ephemeral channels studied, width increases more rapidly with discharge than it does in the humid areas. This example outlines one value of such studies in that they provide a method of comparing stream channel geometry from one area to another. The relations obtained were the basis for an average hydraulic geometry of stream channels (Figure 5.2). This shows that, in addition to the values quoted above, suspended sediment load, slope, channel roughness represented by a parameter of the Manning

Figure 5.1 Characteristics of water and sediment movement in channels
a Regimes of flow in a broad open channel based upon Sundborg (1956, Figure 1). Using hydraulic radius = $\left(\dfrac{\text{Cross-sectional area of channel}}{\text{Wetted perimeter}}\right)$ as L in the Reynolds Number flow is laminar when R < 500 and turbulent when R > 2000. The shaded area indicates the intermediate range where either type can occur.
The four types of flow defined are

Laminar		Turbulent	
R < 500–2500		R > 500–2500	
Tranquil	Rapid	Tranquil	Rapid
F < 1	F > 1	F < 1	F > 1

b Relation between flow velocity, grain size and state of sediment (after Sundborg, 1956). When flow velocities are less than the velocity for cessation of movement no sediment can be moved along the channel bed. In the band between cessation of movement and critical erosion velocity (dashed lines) there is transport of bedload but no erosion, whereas the zone between the two dotted bands indicates transport of bedload and possible erosion.
c Roughness of sand bed channels. Based upon Guy (1970) and Simons and Richardson (1966). A to C are representative of the lower flow regime when F < 0·4, and E to H of the upper flow regime when F > 0·7.
d Form of bed roughness and Manning n in relation to size of bed material and Froude number with sands of 0·28 and 0·46 mm median diameter (d) in an 8-ft wide flume. After Guy (1970) and Simons and Richardson (1962).

type, and bed material size can also be related to discharge for at-a-station and down-stream relations (Figure 5.2).

The size and dimensions of the river channel should be related to the nature of the sediment in the bed and banks. Studies of stream channels on the Great Plains by Schumm (1963) demonstrated a relation between channel shape, expressed as the

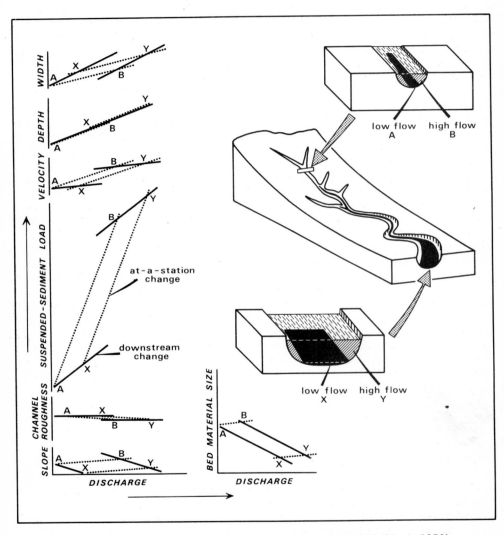

Figure 5.2 Average hydraulic geometry relations (after Leopold and Maddock, 1953)
The channel sections indicate high- and low-flow situations at two positions along the stream channel and the general graphical relationships are indicated for values at-a-station (dotted lines) and downstream values (solid lines).

width/depth ratio (F) and the percentage of sediment finer than 0·074 mm (per cent silt clay, M, see p. 71) in the form:

$$F = 255 \ M^{-1·08}$$

Although initially this relationship (Figure 2.10D) might be interpreted as indicating that more cohesive banks characterise the channels with higher silt-clay content, this interpretation is not completely satisfactory because the correlation is between channel shape and composition of sediment in the banks and also in the bed. Schumm (1963) therefore concluded that the type of load, and therefore the type of deposits, are associated with a particular channel shape. This is supported by the fact that wide channels have often been noted to have higher bed load transport than deep channels. The relation was the basis of a tentative classification of alluvial river channels in which the classification was based upon the type of load (bed load, mixed load, suspended load) and channel stability (eroding, depositing, stable) (Table 5.1). This classification acknowledges the tendencies of channels to be aggrading, eroding or stable and this is indicated by the deviation of specific instances from the general relation between channel shape and per cent silt clay (Figure 2.10E).

Relationships have, therefore, been demonstrated between the form of stream channels and the processes which operate within them as expressed by water discharge and sediment characteristics. A recurrent problem, however, is that some of these relations may obtain for a particular discharge frequency and this raises the questions of whether the form of the stream channel is adjusted to a stream discharge of a particular frequency. Although many studies have indicated that the bankfull discharge, with a recurrence interval often of 1·5 years, can be the significant process value which is related to channel width and to channel depth (Figure 5.3), some empirical studies have revealed a situation more complex than this. In a study of three streams in southern England, for example, Harvey (1969) demonstrated that on streams with an important peak-flow component in the hydrograph, as on the Ter draining the boulder clay of Essex and the upper Nar on the boulder clay of Norfolk, the balance between channel capacity and discharge is maintained approximately at the level of the mean annual flood. In contrast, on small streams with a dominant baseflow component, illustrated by the Wallop Brook draining the Chalk of Hampshire, the channel capacity was apparently related to less frequent events.

Variation in hydraulic geometry thus appears to be a complex response to the water and sediment moving through the channel and also to the character of the material in the bed and banks. It is necessary to consider the magnitude, frequency and duration of stream and sediment discharge as well as particular values of stream flow frequency. Hydraulic geometry studies are further complicated by the problems of channel definition, by the effects of local factors including vegetation, and by the basin factors which influence the supply of water and sediment to the channels. Definition of the stream channel, and particularly of the bankfull stage (pp. 57–8) is seldom easy and is complicated in some cases where the cross section is compound. In south-east Australia Woodyer (1968) identified three channel benches in addition to the flood plain and of these the lowest is visible only at low flow, the middle one appears to be associated with a bankfull frequency of 1·02 to 1·21 years, and the high bench, which is equivalent to the flood plain level, corresponds to a discharge with a

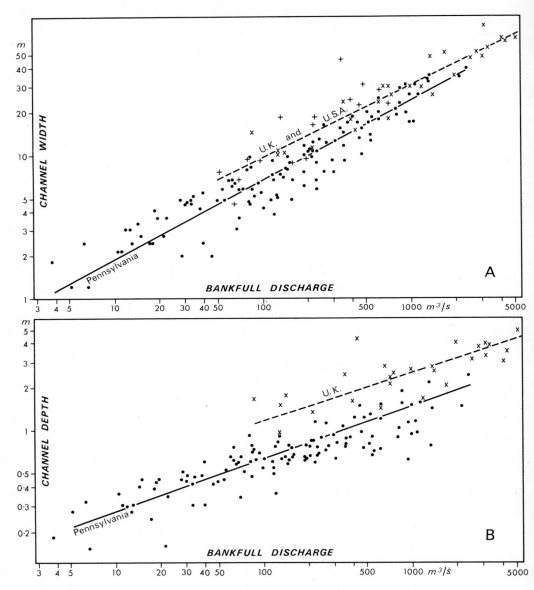

Figure 5.3 Channel width and depth related to bankfull discharge
Based upon data for Pennsylvania by Brush (1961) and for the U.K. and U.S.A. by Nixon (1955).

Table 5.1 Tentative classification of alluvial river channels (*after Schumm, 1963*)

	Stable	Depositing	Eroding	Suspended Load 100%	Bedload 0	M 100%
Suspended load channels	W/D ratio < 7 Sinuosity > 2·1 Gradient gentle	Major deposition on banks → narrowing	Dominant bed erosion. Widening minor	85	15	30
Mixed-load channels	W/D ratio 7–25 Sinuosity 2·1 → 1·5 Gradient moderate	Initial major deposition on banks followed by deposition on bed	Initial bed erosion followed by channel widening	65	35	8
Bedload channels	W/D ratio > 25 Sinuosity < 1·5 Gradient moderate	Bed deposition + island formation	Widening dominant. little bed erosion	30	70	0

frequency ranging from 1·24 to 2·69 years. At 34 sites in the Piedmont of the south-east U.S.A. Kilpatrick and Barnes (1964) described valley benches at different eleva-tions in relation to slope which appears to be a major factor determining the character of the benches. The low benches are all associated with floods of recurrence interval less than fifty years but, whereas the mean annual flood can be contained within a channel on a steep slope, along reaches which have a lower slope a comparable dis-charge may inundate the highest bench to a considerable depth.

Local factors, including rock type, superficial deposits and soils and vegetation cover will naturally influence channel form although the extent of their influence may depend upon the size of the channel. Thus where channels are narrower than 4 m, vegetation can exercise a binding effect on the banks, and vegetation may encroach on the channel. Contrasts will, therefore, occur between areas according to such local factors and width/depth ratios of channels on grassland are generally lower than the ratios under forest cover. This may reflect the fact that grass behaves more like consolidated sediment than bank material which is bound by tree roots. Sampled reaches of the Sleepers River basin in northern Vermont allowed Zimmerman, Goodlett and Comer (1967) to assess the influence of vegetation upon stream channel morphology. In this area they showed that channel width does not increase con-sistently downstream where drainage areas are less than 0·26–2·06 km^2 due to the encroaching vegetation which eliminates the effect of the downstream increase in discharge; that along a single stream variability in width reaches a maximum where the drainage area is approximately 5·2 km^2; and that on some streams least variability in channel width occurs where the drainage area is approximately 15·5 km^2. They therefore demonstrated the existence of two thresholds in that the smallest channels (catchment area less than 0·26–2·06 km^2) have their form and size primarily in-fluenced by vegetation factors, the intermediate channels (2·06–15·5 km^2) show a downstream increase in width but mean width and range of width are very much influenced by vegetation, and the largest channels (drainage areas greater than 10·3–15·5 km^2) are affected only to a marginal extent by vegetation. In a study of stream channels of the Mato Grosso, Brazil, Thornes (1970) demonstrated that although the main channels conform to the established relations between discharge and width, depth and velocity, the smaller head water channels show very different relationships.

The efficacy of studies of hydraulic geometry is established but further studies are required to demonstrate the ways in which channel characteristics vary with climate and with basin characteristics. Variations in the hydraulic geometry of stream channels have been demonstrated for twelve areas in the humid regions of the U.S.A. (Stall and Chih Ted Yang, 1970) (Table 5.2) but further studies are still needed. Such studies may be useful because it may be possible to utilise channel measure-ments as an indication of flow magnitude. Herdman (1970) showed that for streams in California mean annual runoff could be related to width (w) and depth (d) by rela-tions of the form $Q = 186w^{1\cdot54}d^{0\cdot88}$ (perennial channels) and $Q = 2\cdot58w^{0\cdot80}d^{0\cdot60}$ (ephemeral channels).

Understanding of the controls upon the hydraulic geometry of stream channels is important because future changes in basin processes, instigated by modifications of climate or of basin characteristics, could have implications for the size and shape of river channels. Thus Rango (1970) attempted to assess the effects that precipitation modification might have upon stream channel geometry. Small watersheds in

Table 5.2A Hydraulic geometry relations for three river basins in the humid areas of the United States (*after Stall and Yang, 1970*)

Roanoke Basin (335 mm annual runoff)	Susquehanna Basin (1052 mm)	Sangamon Basin (203 mm)
$\log_e Q = 0\cdot47 - 2\cdot35F + 1\cdot05 \log_e Ad$	$\log_e Q = 1\cdot48 - 3\cdot97F + 1\cdot05 \log_e Ad$	$\log_e Q = 0\cdot30 - 5\cdot39F + 1\cdot10 \log_e Ad$
$\log_e A = 0\cdot54 - 1\cdot46F + 0\cdot92 \log_e Ad$	$\log_e A = 1\cdot12 - 2\cdot39F + 0\cdot91 \log_e Ad$	$\log_e A = 1\cdot19 - 4\cdot20F + 0\cdot87 \log_e Ad$
$\log_e V = 0\cdot08 - 0\cdot88F + 0\cdot13 \log_e Ad$	$\log_e V = 0\cdot37 - 1\cdot57F + 0\cdot14 \log_e Ad$	$\log_e V = 0\cdot89 - 1\cdot18F + 0\cdot23 \log_e Ad$
$\log_e W = 1\cdot52 - 0\cdot34F + 0\cdot54 \log_e Ad$	$\log_e W = 1\cdot79 - 0\cdot92F + 0\cdot59 \log_e Ad$	$\log_e W = 1\cdot45 - 1\cdot51F + 0\cdot54 \log_e Ad$
$\log_e D = 0\cdot98 - 1\cdot13F + 0\cdot38 \log_e Ad$	$\log_e D = 0\cdot66 - 1\cdot47F + 0\cdot32 \log_e Ad$	$\log_e D = 0\cdot26 - 2\cdot69F + 0\cdot ee \log_e Ad$

where Q = discharge, cubic feet per second
A = cross-sectional area of stream channel in square feet
V = velocity in feet per second
W = channel width in feet
D = channel depth in feet

F = frequency of occurrence from 10 to 90% of days
Ad = drainage area in square miles

Table 5.2B Mean values of exponents for the same three basins (*after Stall and Yang, 1970*)

	Exponents		
	Width W (W = aQ^b)	Depth D (D = cQ^f)	Velocity V (V = KQ^m)
Roanoke	b = 0·12	f = 0·47	m = 0·41
Susquehanna	= 0·28	= 0·49	= 0·23
Sangamon	= 0·23	= 0·37	= 0·40
158 stations in U.S.A.	= 0·12	= 0·45	= 0·43
14 mountain stations	= 0·11	= 0·48	= 0·42

Colorado and South Dakota were classed into three groups according to area and soil type, and for each group regression equations were established to relate stream channel geometry parameters to per cent silt clay in bed and banks (M) and to mean annual discharge (Q). From these regression equations the average geometry was estimated for each of the three area-soil type groups (Table 5.3). Using discharge values (Q) estimated by assuming a 20 per cent precipitation increase in the same regression equations, the likely changes in stream channel geometry, assuming that M remained constant, could be estimated (Table 5.3). All three areas should see modifi-

Table 5.3 Stream channel geometry parameters predicted from regression equations and their likely change following a simulated 20 per cent precipitation increase (*after Rango, 1970*)

GEOMETRY	PREDICTED VALUES			ESTIMATED VALUES AFTER PRECIPITATION INCREASE (significant at more than 90 per cent shown as per cent in brackets)		
	Group A (Badger wash)	Group B (Newell clay)	Group C (Newell sandstone)	Group A	Group B	Group C
Width/depth ratio	9·63	12·46	9·81	10·38	14·02	6·92 (−30)
Sinuosity	1·231	1·191	1·064	1·264	1·221	1·044
Width	4·99 m	4·76	4·40	7·63 (+53)	4·19	4·54
Depth	0·495	0·356	0·454	0·648	0·294	0·618
Meander wavelength	29·63	79·83	98·32	23·28	69·32 (−13)	69·70
Stream channel gradient	0·0337	0·0220	0·0343	0·0341	0·0314 (+43)	0·0505 (+47)

cation of at least one stream channel geometry parameter following a 20 per cent precipitation increase, but whereas width increased for Group A, meander wavelength and stream channel gradient changed significantly on Group B, and width/depth ratio and stream channel gradient changed significantly on Group C. Thus the changes experienced in a particular area will necessarily depend upon the basin characteristics, including the vegetation, rock type and character of sediment in that watershed. Thus most Group A channels occur immediately above bedrock and so width is increased (+53 per cent) faster than depth (+31 per cent), whereas the channels in areas B and C change less because vegetation has an influence on the channel form.

The hydraulic geometry of stream channels necessarily conforms to physical laws but it provides opportunities to compare areas, to analyse the influence of basin characteristics, and to surrogate process measurements from channel form. These opportunities derive from the fact that the form—process relations of a river channel are representative of a system in quasi-equilibrium (Langbein and Leopold, 1964), which will represent the most probable state between two opposing tendencies, namely, that towards minimum total work in the whole fluvial system and that towards uniform distribution of energy throughout the system. Thus in accommodating a change in stream power a channel changes so that each component of power changes as equably as possible (Langbein, 1964).

5.3 The channel reach

Quasi-equilibrium involving an apparent equilibrium and an adjustment between the parameters of form and process of the river channel has also been applied in studies of reaches of rivers or of river long profiles. The concept of quasi-equilibrium was preceded by the idea of the graded profile (p. 385) which depended upon the idea that the slope of the river was delicately adjusted to provide just the velocity required to transport the available sediment load within the framework provided by the existing channel geometry and the available discharge (Sunley, 1969). It has been shown that many long profiles are composed of one or more segments which are concave upwards and from which irregularities have been eliminated. However, this is not always the case and in a study of the rapids and pools of the Grand Canyon, for example, Leopold (1969) demonstrated that over a section 280 miles long a fall of 2200 ft is largely achieved by rapids which account for only 10 per cent of the distance. A dateable lava flow indicated that there has been little deepening of the canyon during the last million years and so although the river has had time and opportunity to eliminate these rapids, they have instead been maintained as an integral part of the quasi-equilibrium. The long profile of the Colorado through the Grand Canyon was shown by Leopold to be composed of an alternation of deeps and shallows and such an alternation is widely reported. Pools alternate with shallower sections called riffles, and the distance between the pools is usually of the order of five to seven times the channel width.

Such an alternation of pools and riffles, paralleled by alternations of deeper and shallower sections along the course of bed rock profiles, have been described from several areas and associated with various patterns which occur along the length of a channel reach. G. K. Gilbert in 1914 noted that 'a free stream does not tolerate a straight channel' and subsequent work has demonstrated that perfectly straight channels are unusual in nature, that channels are seldom straight for distances greater than ten times the channel width, and that in straight channels a wandering thalweg frequently occurs. More common than straight channels are meandering and braided channel patterns (Figure 2.10D) and meandering may be identified where stream sinuosity is greater than 1·5. These patterns are not mutually exclusive in either space or time. For instance, different patterns can occur along the same river course and along the Menderes in Turkey, which provides the name meander, both meandering, straight and braided channel reaches occur (Russell, 1954). Variations in pattern are also found at different times of the year according to the flow regime. The White River, a glacial stream draining from Mount Rainier in Washington, changes from a meandering pattern to a braided one with the onset of high summer discharges and it returns to a meandering pattern with the lower discharges of summer and autumn (Fahnestock, 1963).

5.3a Meandering channels

Studies have been devoted to the problems of meandering channels since the mid-nineteenth century and results over a century have been achieved by field observa-

tion, by laboratory study and by statistical and theoretical analysis. These methods of approach have been focused particularly upon meander geometry, upon the relations between meander geometry and stream channel process, and upon the explanation of the shape, size and development of meanders. The legacy of the early work demonstrated that the stability of the channel depends upon the variation of curvature, that the line of the thalweg is close to the concave bank while sand and silt are deposited on the convex bank, that the depth of the pool on the concave bank increases with the curvature of the banks, and that the deepest point and the greatest width of the beach on the convex bank are downstream of the point of greatest curvature. Over the years meandering patterns have been described in a wide variety of situations and at a variety of scales. Zeller (1967) described four types in Switzerland, namely, alluvial meanders which may be free, restricted, incised, disturbed or under-calibrated; rock meanders cut into bedrock; ice meanders which include small channels with catchment area of 0·002 to 0·050 km² on the Morteratsch glacier, for example; and furrow meanders which are micro forms of furrows 2–20 m long on unvegetated limestone surfaces in the Silbern region. Meandering patterns have also been identified in the waters of the Gulf Stream in the absence of sediment (Leopold and Wolman, 1960)(Plate 16).

Against the background of these observations of character and distribution it is obviously necessary to define the geometry of meanders very precisely, to analyse the inter-relationship of the meander dimensions, and to establish the ways in which these dimensions are related to channel process. Enquiries utilising the field, laboratory and theoretical sources have been focused upon a variety of parameters of meander dimension (Figure 5.4A). From such studies it has been demonstrated that meander wavelength (M_L) is usually six to ten times the channel width, that the width of the meander belt (M_B) is commonly fourteen to twenty times the channel width and is greater than the meander wavelength (Table 5.4), and relations have also been established between average curvature and average depth of meanders (Leliavsky, 1955). Such studies emphasise the regularity of meander geometry and also underline the extent to which numerous parameters are inter-related. The size of meanders has been related to catchment area and this reflects the fact that meander geometry should be related to some parameter of stream discharge (for example, Figure 2.10D; Figure 5.4C). Several indices of streamflow have been employed (Table 5.4) including dominant discharge and bankfull discharge but in a study based upon several rivers of the United States, Carlston (1965) concluded that the discharge controlling meander wavelength is a range of flows, possibly falling stage flows, between the mean annual discharge and the mean discharge of the month of maximum discharge and that meander migration takes place during these stages.

Relations between meander size and stream discharge are necessarily more complex than a simple two variable relationship would suggest and so some relationships have been derived incorporating three or more variables (for example, Ackers and Charlton, 1970; Table 5.4), and channel slope is frequently introduced as an additional variable. From laboratory data Ackers and Charlton (1970) showed that the relationship of channel width to discharge for both straight and meandering channels

Plate 16 Meandering River Minnesota
Well-developed meanders, ox-bow cutoff and flood-plain landscape.

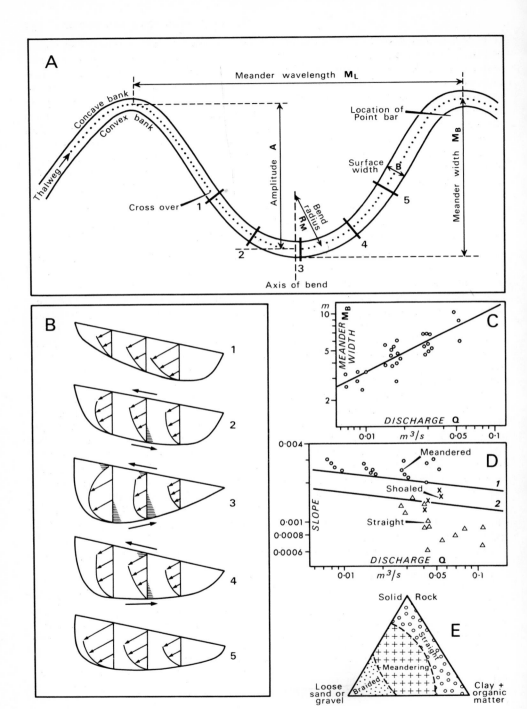

A

Meander wavelength M_L

Concave bank

Convex bank

Location of Point bar

Thalweg

Amplitude A

Surface width B

Meander width M_B

Cross over

Bend radius R_M

Axis of bend

1 2 3 4 5

B

1
2
3
4
5

C

m
10

$MEANDER$ $WIDTH$ M_B

5

2

$DISCHARGE$ Q

0·01   0·05   0·1 m^3/s

D

0·004

Meandered

Shoaled

$SLOPE$

1

2

Straight

0·001
0·0008

0·0006

$DISCHARGE$ Q

0·01   0·05 m^3/s 0·1

E

Solid Rock

Straight

Meandering

Braided

Loose sand or gravel

Clay + organic matter

(Figure 5.5) indicated that the width of meandering channels is twice that of a straight channel. In specific cases meandering channels occur on lower angle slopes than those of braided reaches for the same discharge (Figure 2.10D). When bankfull discharge is plotted against channel slope (Figure 5.5) braided channel patterns usually occur in reaches with higher slopes (Leopold and Wolman, 1957). From a laboratory study Ackers and Charlton (1970) showed that straight, shoaled and meandered channels can be distinguished according to slope (Figure 5.4D). Shahjahan (1970) assembled the results from numerous studies (Table 5.4) and according to laboratory studies, using a conventional small scale sand plain in a wide tilting flume and also using a pressurised system, he concluded that correlations of three parameters are more useful than of two. He suggested that the geometry of a freely meandering stream depends largely upon the relative stream size $\left(\dfrac{Q^{2/5}}{g^{1/5}}/D\right)$, the valley slope (Sv), the sediment charge (Qs/Q) and the relative settling size of the sediment $\left(D/\dfrac{V^{2/3}}{g^{1/3}}\right)$. Although not all workers have considered sediment character as significantly related to meander dimensions the results of these studies indicate (Shahjahan, 1970) that an increase of sediment charge is associated with a decrease of meander wavelength and of channel width, but with increases of meander width, bend radius and depth of flow. The work of Ackers and Charlton (1970) also indicated that there appears to be a threshold rate of sediment charge below which channels meander.

Experiments by Schumm and Khan (1971) employing a flume 30·5 m long, 7·3 m wide and 1 m deep containing poorly sorted sand, demonstrated how the rate at which sand was fed into the channel was related to slope. Two threshold values of slope were identified at which significant changes in the hydraulics of flow and sediment transport occurred. The channel was straight (Sinuosity = P = 1) at slopes less than 0·2 per cent; at slopes above 0·2 per cent the channel pattern changed as alternate bars formed and a sinuous thalweg developed (up to P = 1·25). At slopes greater than 1·3 per cent alternate bars began to erode and for slopes in excess of 1·6 per cent a braided channel pattern developed. Thus Schumm and Khan (1971) envisaged two threshold values of sediment load at a given discharge. They thus proposed that, although a considerable increase of slope and/or sediment load may in some cases have little effect on the pattern of the thalweg and on sediment transport, if a river section has characteristics near a threshold zone then slight sediment or slope changes could have striking repercussions for the channel pattern and for the sedimentary deposits produced by the river.

The significance of sediment has been suggested in a general manner by Tanner

Figure 5.4 Meandering channels
Parameters of meander geometry are denoted in A, generalised flow distributions at sections 1–5 (diagram A) are shown in B based upon Leopold and Wolman (1957). At the cross over (1) the channel shape is not completely symmetrical and the velocity distribution incorporates the effect of the previous bend. Towards the bend the channel is more symmetrical and the velocity distributions more uniform (2); the maximum velocity is found just downstream of the axis of the bend.
An example of the relation between meander width and discharge is shown in C.
The relation of slope and discharge based upon flume studies (Ackers and Charlton, 1970) distinguishes straight channels with slope $< 0.0015Q^{-0.12}$, shoaled when $0.0015Q^{-0.12} <$ Slope $< 0.0021Q^{-0.21}$ and meandered channels when slope $> 0.0021Q^{-0.12}$. The general distribution of channel patterns according to locally available material is shown in E based upon Tanner (1968).

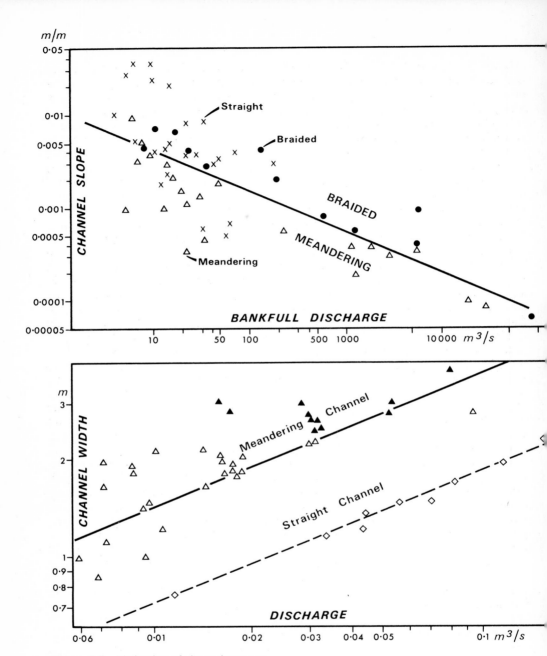

Figure 5.5 Distinction of channel patterns

The relation of slope (S) to bankfull discharge (Q) of the form $S = 0.06Q^{-0.44}$ was suggested by Leopold and Wolman (1957) to distinguish braided (above the line) from meandering (below) patterns.

Relations between channel width (B) and discharge (Q) shown by Ackers and Charlton (1970) to be $B = 7.2Q^{0.42}$ for meandered, and $B = 3.6Q^{0.42}$ for straight flume channels showing that the width of a meandered channel is twice that of straight channel.

(1968) (Figure 5.4E) and a study of 36 stable alluvial river channels in Australia and the U.S.A. demonstrated (Schumm, 1967) that meander wavelength is dependent not only upon the discharge but also on the type of sediment moving through the channel. Therefore the meander wavelength of rivers transporting a high proportion of the total sediment load as sand and gravel will be greater than the wavelengths of channels of similar discharge but which are transporting mainly fine sediment loads.

Meander dimensions therefore appear to be in quasi-equilibrium with the channel slope, the stream discharge, the amount and character of the sediment, and with the viscosity and density of the liquid. Adjustments of meander size and position are achieved by erosion of the concave banks, expressed through slumping in cohesive materials giving irregular channel margins, and through sloughing in non-cohesive materials often giving smoother banks; and by deposition on the convex banks. Deposition within the meander loop is as point bars composed of scroll-shaped ridges approximately reflecting the form of the channel. Along the Klarälven river in Sweden the bar ridges are 9–27 m apart and are up to 1 m above the height of the intervening swales (Sundborg, 1956) whilst along the Mississippi there may be differences of 3·5 m between the ridges and swales of point bars. Sediments are accumulated in the channel chiefly by deposition of bedload materials on advancing dunes or ripples in the form of scroll bars which are aggregated into point bars. Sand predominates whereas gravel, silt and clay are comparatively rare. The flow of water in a meandering stream follows a characteristic, helicoidal, pattern. The water movement therefore includes a lateral component of up to 10 to 20 per cent of the downstream velocity. The maximum velocity and turbulence are found near the concave bank and the lateral movement is towards the concave bank on the surface and towards the point bar at depth (Figure 5.4B). Associated with the movement of water and sediment in meandering channels migration of meanders has been noted at rates of as much as 760 m per year, and migration occurs in a downstream direction on all four types of meanders noted in Switzerland (Zeller, 1967).

Increasingly the incidence, dimensions and shape of meandering channels have been appreciated to be a common response to stream channel processes. This appreciation has been aided by laboratory studies which have shown that water flowing down a slope rapidly develops to give a sinuous channel which is maintained in quasi-equilibrium. Explanations of meander development have to account for the facts that all stable unbraided channels follow a sinuous course, to explain why meander geometry is maintained, and to indicate what causes the sinuous pattern. They have in addition to reconcile the controversy surrounding the significance of sediment, the pattern of helicoidal circulation and the channel slope, all of which have been considered significant by some workers and rejected by others. Although the water circulation in a meandering channel (Figure 5.4B) is undisputed it cannot be responsible for initiating the pattern (Tanner, 1960). The effect of local disturbances (Werner, 1951) including secondary currents and turbulence which could create a transverse oscillation in a straight channel and divert the flow from a straight course does not seem capable of universal application to the problems posed by the regularity of meander geometry. Langbein and Leopold (1966) proposed a theory of minimum variance which perceived the meandering pattern as the most probable form that a river can take and one in which the river does the least work in turning. In a review of previous explanations Yang (1970) suggested that whereas Langbein and Leopold's

Table 5.4 The geometry of meandering streams from model and river data (*from M. Shahjahan, 'Factors controlling the geometry of fluvial meanders', Bull. Int. Ass. Sci. Hydrol., 1970, XV (3), 13–24*)

Area source	Authority	Geometry of meander channel	Average particle size D_{mm}	Initial valley slope S_v	Range of discharge Q ft³/sec
River data U.S.A.	C. C. Inglis (1949)	$M_L = 6{\cdot}6\ B^{0{\cdot}99}$; $M_B = 18{\cdot}65\ B^{0{\cdot}99}$ $M_B = 1{\cdot}7\ M_L^{1{\cdot}06}$ $M_B = 10{\cdot}9\ B^{1{\cdot}04} = 14B$	—	—	—
Orissa rivers (India)		$M_L = 27{\cdot}4\ Q_{max}^{0{\cdot}5} \pm 14{\cdot}33$	—	—	29,000–61,000
India		$B = 4{\cdot}88\ Q_{max}^{0{\cdot}5}$; $M_L = 6{\cdot}46\ B$ $M_B = 57{\cdot}8\ Q_{max}^{0{\cdot}5}$	—	—	—
U.S.A.	C. W. Carlston (1965)	$M_L = 106{\cdot}1\ Q_a^{0{\cdot}46}$; $M_L = 80\ Q_{mm}^{0{\cdot}46}$ $M_L = 8{\cdot}2\ Q_b^{0{\cdot}62}$; $M_B = 65{\cdot}8\ Q^{0{\cdot}47}$	—	—	31–562,800
U.S.A. and Australia	S. A. Schumm (1969)	$B = 2{\cdot}3\ Q_{max}^{0{\cdot}58}\ M^{-0{\cdot}37}$ $M_L = 234\ Q_{max}^{0{\cdot}48}.\ M^{-0{\cdot}74}$	0·11–1·1	—	580–48,000
River and laboratory data	L. B. Leopold and M. G. Wolman (1957, 1960)	$M_L = 36\ Q^{0{\cdot}5}$; $M_L = 10{\cdot}9\ B^{1{\cdot}01}$ $M_L = 4{\cdot}7\ R_m^{0{\cdot}98}$; $S = 0{\cdot}06\ Q_b^{-0{\cdot}44}$	0·2–42·5	—	0·021–1,000,000
Laboratory experiments	J. F. Friedkin (1945)	$M_L = 14{\cdot}0\text{–}34{\cdot}0$ $M_B = 1{\cdot}84\text{–}17{\cdot}8$ $M_L = 36{\cdot}4\ Q_d^{0{\cdot}5}$; $M_B = 16\ Q_d^{0{\cdot}5}$	0·20–0·45	0·006–0·009	0·05–0·30
	C. C. Inglis (1947) N. E. Kondratév (1962)	(a) $M_L = 7{\cdot}78$; $M_B = 3{\cdot}52$ (b) $M_L = 12{\cdot}5\text{–}19{\cdot}7$; $M_B = 3{\cdot}94\text{–}8{\cdot}73$ (c) $M_L = 9{\cdot}75\text{–}13{\cdot}9$; $M_B = 2{\cdot}5\text{–}5{\cdot}35$	0·20 0·22 0·30–0·45 0·27	— 0·005 0·006–0·008 0·005–0·008	0·175–0·40 0·0339 0·05–0·141 0·035–0·071

Table 5.4—*continued*

Area source	Authority	Geometry of meander channel	Average particle size D_{mm}	Initial valley Slope S_v	Range of discharge $Q ft^3/sec$
	F. G. Charlton and R. W. Benson (1966)	$\dfrac{M_L}{D} = 27\cdot2\left[(Q^2/gD^5)^{0\cdot235}\cdot\left(\dfrac{Q_s}{Q}\right)^{-0\cdot033}\right]$	0·15–0·70	—	0·0128–2·0
	H. S. Nagabhushaniah	$\dfrac{M_B}{D} = 0\cdot76\,[(Q-Q_c)D^{-3\cdot0}\cdot S_v t]^{0\cdot5}$	0·5	0·003–0·012	0·019–0·147
	S. N. Gupta *et al.* (1966)	$\dfrac{M_L}{M_B} = 44\cdot78\,S^{0\cdot55} = 3\cdot71\left[\left(\dfrac{Q_s}{Q}\right)\cdot Fr\right]^{0\cdot091}$	0·23	0·0022–0·0040	1·0–3·0
	H. H. Hill (1964)	$M_L = 29\cdot0 - 37\cdot5$; $M_B = 10\cdot0 - 16\cdot4$	0·26	—	0·820–0·875
	P. Ackers and F. G.	$M_L = 37\cdot8\,Q^{0\cdot506}$	0·15	—	0·238–2·74 cf
	Charlton (1970)	$M_L = 38\cdot0\,Q^{0\cdot467}$	0·15	—	0·25–2·0

where B = channel width
M = percentage silt clay in perimeter of channel
Q = flume discharge
Q_a = mean annual discharge
Q_b = bankfull discharge
Q_c = critical discharge at which critical shear velocity corresponding to D exists
Q_d = dominant discharge
Q_{ma} = mean annual flood discharge
Q_{max} = maximum discharge
Q_{mm} = mean maximum monthly discharge
$\dfrac{Q_s}{Q}$ = sediment charge in parts per million
S = channel surface slope
S_v = initial valley slope
t = time of run in hours
F_r = froude number
D = mean sediment size
g = acceleration due to gravity

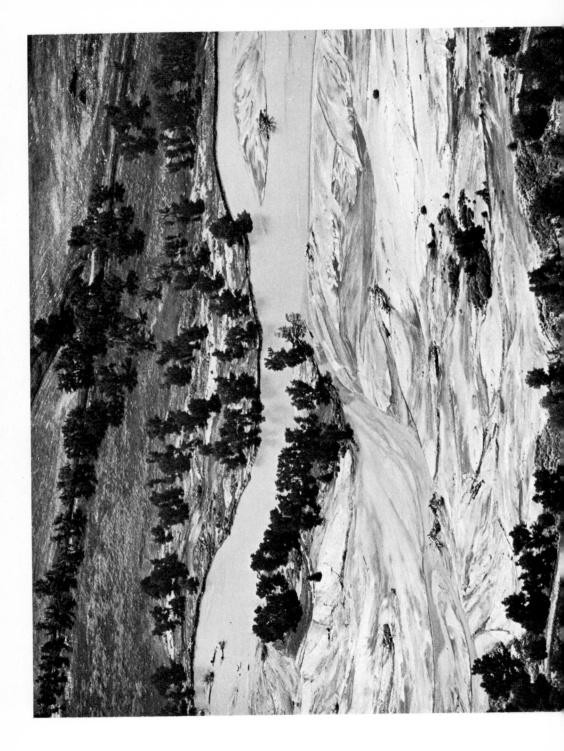

explanation involves minimising the time rate of potential energy expenditure per unit mass of water, it is possible to minimise the total energy expenditure. The natural stream channel will therefore adjust channel slope and geometry to minimise its overall time rate of potential energy expenditure per unit mass of water. These theoretically-supported conclusions therefore emphasise that meandering affords a mechanism whereby channel slope may be adjusted and that within the conditions which are locally prevalent the adjustments made to maintain quasi-equilibrium will be those which minimise energy expenditure—this is the source of the meandering channel.

5.3b Braided channels

Although they have received less attention than meandering channels, braided patterns have been widely described from areas which include semi-arid regions of low relief which receive discharge from mountain areas, from highland areas in a variety of climates, from glacial outwash plains, and from periglacial areas especially over permafrost, but the quantitative description of braiding is not easily accomplished. A variety of braided forms exists ranging from a channel with islands to one with numerous anastomosing channels as illustrated by the Kuskokwim river in Alaska (Figure 5.6A). A braiding index was devised by Brice (1960) (pp. 56–7) but an excess segment index (Howard, Keetch and Linwood Vincent, 1970) and a bed relief index (Smith, 1970) have also been employed. The distinctive features of braided channels include (Chien Ning, 1961) a wide, shallow bed choked with sand bars, together with rapid shifts of sand banks and channels. These shifts can be as much as 90–120 m per day on the lower Yellow River of China. A braided river course also moves rapidly, so that movement can be as much as 130 m per day laterally in the upper course and 50 m per day in the lower course of the Lower Yellow River. Braided rivers are also identified by a distinct lack of well-developed river bends. Wandering of the thalweg occurs at varying rates and in the case of the Lower Yellow River the thalweg moves spasmodically; by large amounts during the flood season of September and October but much more gradually during the lower flows of July and August (Figure 5.6C1).

Braids appear to develop in deposits coarser than those supporting meanders (Figure 5.4E), they are initiated as short submerged bars which are pointed downstream, once initiated they accrete rapidly as the finer material is trapped allowing downstream extension, and subsequently the growth of a braid leads to reduced channel width which can encourage bank erosion. Krigstrom (1962) considered this as one method whereby braids could develop in straight river courses, at the junction of two channels or in curved channels over point bars. In addition braids may also succeed strong aggradation when the bed of the river is raised: when the water level is high enough flow occurs over the bank top in places and if several of these channels are permanent features they can lead to a permanent forking. Along the South Platte River in Nebraska and Colorado, two types of channel bars (Figure 5.6B) were identified (Smith, 1970). Aggradation of longitudinal bars by the construction of channel ridges in poorly-sorted sediment, and by the dissection of transverse bars

Plate 17 Braided stream deposition environment
South Platte River (see pp. 257–8).

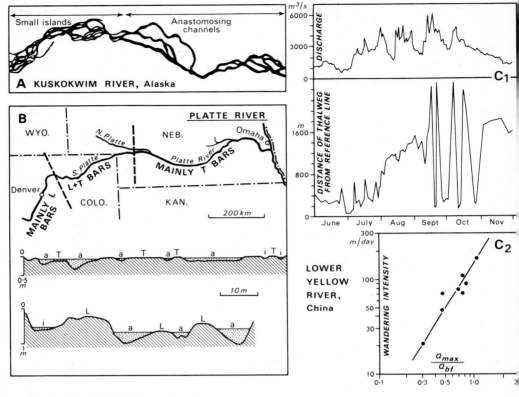

Figure 5.6 Braided channels
A shows the gradation from distributary channels to one channel with numerous small islands (Howard, Keetch and Linwood Vincent, 1970), B illustrates generalised sections across areas of Transverse (T) and Longitudinal (L) bars (Smith, 1970), C1 indicates the correspondence between peak discharge during the year and movement of braided channels along the Lower Yellow River and C2 expresses wandering intensity of thalweg as a function of the maximum discharge (Q_{max}) to the bankfull discharge (Q_{bf}) (Chien Ning, 1961).

and to a lesser extent of longitudinal bars were shown to be the basis of braid development. Longitudinal bars extend upstream whereas transverse bars, composed of finer material, grow by downstream migration of foresets more or less perpendicular to current direction. These transverse bars are more frequent in downstream locations (Figure 5.6B) associated with decreased grain size and an increase in sorting of channel sediment (Plate 17).

Conditions which favour braiding are suggested by the types of area in which the pattern occurs. Bedload is an essential pre-requisite and this is available in large quantities in areas of glacier outwash as across the sandur of Iceland, or where channel banks are easily eroded especially during the falling stage when banks are saturated. Availability of bedload does not provide a complete explanation. On the Adur river of France for example, meandering and braiding have alternated over the last two centuries and accordingly Tricart and Vogt (1967) suggested that size of material and irregularity of streamflow were both responsible. During the rising flood hydrograph bank-sapping occurs and the rapid rise of the hydrograph, diagnostic of regimes with irregular discharges or with sharp seasonal peaks, has been proposed more widely as a reason for the development of braiding. Such high discharges can be occasioned by melting snows, by melting ice, by occasional glacier bursts (Jökulhaups of Iceland), or in semi-arid areas by high intensity storms. Intense precipitation can also, through its influence on peak discharges, give rise to braiding and particularly in monsoonal areas high seasonal discharges are a factor contributing to the development of braiding. Therefore in the Lower Yellow River the rate of wandering of the braided channel is quantitatively related to the peak discharge (Figure 5.6C2). A final factor which influences the occurrence of braided channel patterns is the presence of high regional slopes as in mountain areas and particularly at the junction of high relief and low relief areas. The significance of slope is underlined in Figures 2.10D, 5.4, 5.5. Although these four factors of bedload availability, irregular discharge, precipitation, and slope, have been proposed at different times to be solely responsible for the incidence of braiding, in many cases it is the way in which the system is changed that may result in braiding as a response to restore the quasi-equilibrium. Accordingly Chien (1961) used data from 31 gauging stations to express the wandering index as a function of stream discharge, channel slope, and sediment characteristics. See Plate 5.

5.3c Alluvial fans and deltas

Whereas braided stream channels may occur along reaches with a comparatively high slope, where the slope changes abruptly an alluvial fan may develop. They are frequent at mountain fronts in the European Alps, the Himalayas, in Japan and also in California where about 20 per cent of the State is covered by alluvial fan deposits (Bull, 1964). In addition alluvial fans occur in periglacial areas, in areas with sparse vegetation and in regions with low, or intermittent, rainfall which is often intense as in the south western parts of the U.S.A. and in Australia. In size alluvial fans vary in radius from as much as 60 km to as little as 10 m; in profile they are concave with a slope which decreases to the margin but is usually less than 10 degrees; in composition they are made of coarse material usually not sorted; and in character they possess a radial distributary system of wide, braided channels. In addition to stream flow in the major channels, Blissenbach (1954) recognised streamfloods composed of large amounts of water and sediment transported in the channels over the fan, and sheetfloods which involve waterflow over the fan surface. In the White mountains of California and Nevada the most significant processes are streamfloods and debris flows following thunderstorms (Beaty, 1963) and dating of one fan indicated an average rate of fan accumulation of 7·6 to 15·2 cm per thousand years (Beaty, 1970). The factors leading to fan development appear to be principally a change of slope which encourages deposition, accompanied by periodically high discharges which facilitate supply of material. Alluvial fans are in some ways the most sensitive landform indicators of the sum total of catchment characteristics, and of climatic inputs into the drainage basin. Thus in Western Fresno County California alluvial fan characteristics have been related to catchment characteristics (for example, Figure

5.13C) and two periods of fanhead trenching since 1854 have been associated with two periods of high rainfall from 1875–95 and from 1935–45 (Bull, 1964).

Overall catchment characteristics, especially the sediment supply from the drainage basin are also significant in contributing to the development of deltas but in addition shore processes on the delta margin exercise a substantial influence and accordingly much of the research on deltas has been the province of the coastal, rather than of the drainage basin, geomorphologist (for example, Russell, 1967; van Straaten, 1964; Morgan, 1970).

5.3d Flood plains

The relatively smooth strip of land bordering the river channel, embracing the river pattern, and inundated at times of high stage is described as the flood plain. The low-lying plain bordering the river channel is not always a contemporary feature and the meander belt width may not necessarily be equivalent to the width of the flood plain. Melton (1936) distinguished meander plains composed largely of meander scrolls, covered plains consisting largely of overbank deposits, and bar plains around braided river channels with no levées or meander scrolls. The dependence of the flood plain upon the stream channel pattern is apparent from this classification and the features of flood plains are of two main types, firstly those due to lateral accretion and secondly those formed by overbank deposits respectively. These two types are classified in Table 5.5 together with intermediate forms and associated ancillary features (Plates 16, 18, 25).

In the stream channel vertical erosion and aggradation, or scour and fill, combined with lateral erosion and deposition associated with the shift of meanders, are collectively responsible for the accumulation and redistribution of sediment on flood plains. Point bar accretion is perhaps the main agent in some areas. Point bars are built up from layers of coarse sands and perhaps with some gravel, they have layers of sand inclined and pointing downstream, they classically occur on the convex side of the river bend and deposition is greatest downstream from the axis of curvature. Point bar islands (Carey, 1969) may be formed when a rapidly caving bank recedes faster than the opposite bar builds out and in mid-stream a point bar emerges as an island, which after building up and being stabilised by vegetation, may eventually coalesce with the original point bar (Figure 5.7). At the margins of the river channels levées may form from the coarser materials carried by turbulent water, but during a stage greater than bankfull the flood water extends to areas beyond the levées, the water becomes less turbulent and flows as sheets. Bedload is therefore deposited in and close to the channel but the suspended load, with silts and clays, accumulate to form the backswamp deposits. The rate at which sediment accumulates depends upon the frequency of flooding and upon the quantity and size of sediment available. Although coarse material may settle quite rapidly, finer materials take longer (perhaps

Plate 18 Flood-plain landscapes
Above, the flood plain of the Narew River, Poland (May 1971) showing flood-plain features and inundation after spring peak discharges.
Below, aerial photograph of lower Mississippi in Mississippi showing point bars, cut-offs (*U.S. Geol. Survey*).

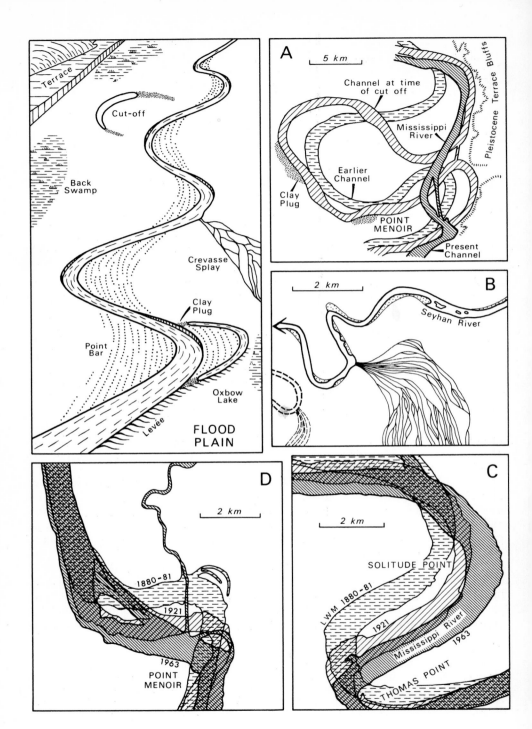

FLOOD PLAIN

Terrace
Cut-off
Back Swamp
Crevasse Splay
Clay Plug
Point Bar
Oxbow Lake
Levee

A
5 km
Channel at time of cut off
Mississippi River
Earlier Channel
Clay Plug
POINT MENOIR
Present Channel
Pleistocene Terrace Bluffs

B
2 km
Seyhan River

C
2 km
SOLITUDE POINT
LWM 1880-81
1921
Mississippi River
1963
THOMAS POINT

D
2 km
1880-81
1921
1963
POINT MENOIR

Table 5.5 Flood-plain features

Lateral accretion:	Point Bars	a succession of point bars with intervening swales comprises a meander scroll
	Channel Bars	
	Alluvial Islands	
Intermediate types:	Cutoff channels	*chute cutoff* where a new channel is developed along a swale in a bar
		neck cutoff where the entire meander loop is abandoned
	Channel Fills	*clay plugs* develop by slow accumulation of fine material and organic matter in cutoffs
Overbank features:	Levées	wedge-shaped ridges of sediment bordering stream channels, best developed on concave banks and may be 4·5–5 m high on banks of Mississippi
	Crevasse Splays	a system of distributary channels on the levée slope when water escapes through low sections or breaks in natural levées
	Flood basins	backswamps which are poorly drained, flat, relatively featureless with little or no relief
Ancillary features:	Lakes	in cutoffs, abandoned channels, meander scrolls, or where a tributary is blocked
	Deferred tributaries	where a tributary flows parallel to the main river for some distance because of an aggraded alluvial ridge
	Alluvial ridge	an aggraded meander belt above the general flood-plain level. Avulsion may occur if the river suddenly abandons its course for a new course at a lower level on the flood plain

20 cm per day for clay but 3 m per minute for coarse sand) and back swamp deposits are slowly formed of colloidal clay, and of fine silt particles together with organic and in-solution materials.

According to their studies in the United States Leopold and Wolman (1957) concluded that lateral accretion and channel deposition are most significant in the formation of flood plains, and that this mode of deposition may account for as much as 80 to 90 per cent of flood plain deposits whereas the remaining 10 to 20 per cent are composed of overbank deposits. The supremacy of channel deposition may reflect the infrequent discharges inundating the flood plain; the fact that the highest discharges may be associated with lower concentrations of suspended sediment than are slightly lower discharges because of overbank flow; the removal of some material from the flood plain at different stages of the flood; and perhaps also the lack of direct comparability between streamflow and sediment hydrographs (pp. 215–25). This dominance of channel deposits may be a regional phenomenon however because where

Figure 5.7 Flood-plain characteristics
Indicated diagrammatically in the top left and specific examples of cutoff development in A, False River longest cutoff of Lowest Mississippi (Russell, 1967), of crevasse splay development on Seyhan River, Turkey, in B (Russell, 1967), of point bar development along the Mississippi in C with average annual rate of bank recession of 23 m (Carey, 1970), of point bar island evolution showing downstream migration of Fancy Point Island over 80 years (Carey, 1970).

flooding is more frequent, and perhaps seasonal, and where fine-grained material is dominant in sediment transport, overbank deposits may assume greater significance. Such situations may occur in the humid tropics, and in Papua for example Blake and Ollier (1971) have shown that, along the Fly and Strickland rivers, channel and lateral accretion deposits predominate along the meander belts but that elsewhere, including backswamps and alluvial plains of minor rivers, overbank deposits are dominant.

5.4 The drainage basin

There are many ways in which the physical characteristics of drainage basins may be expressed and an equal variety of methods whereby the drainage processes may be represented. Relationships between these two are required to assist in the explanation of processes, to facilitate the prediction of processes, and to indicate changes over time past and future. Relationships have been obtained parametrically, stochastically and experimentally. The parametric approach involves a knowledge of physical hydrology and aims to relate actual measurements of drainage basin process to specific quantitative attributes of basin character. Such relations can be achieved by several types of model (see pp. 225–32) and these range from the simple relation of two variables, through three-variable relationships including coaxial regression, to the case of a number of variables analysed by multivariate techniques. The experimental approach, analogous to that used in the study of basin characteristics and channel patterns (pp. 79–81), may be used because hardware models of drainage basins having defined characteristics can be constructed in the laboratory, they can be subjected to clearly specified inputs and hence can afford the basis for a study of the relations between basin form and process. Such methods need to overcome the scale problem but they can be employed to model either an entire catchment or particular processes or subsystems of the drainage basin (Amorocho and Hart, 1965). Hardware models of basins have been used by Cherry (1966) and by Roberts and Klingeman (1970) who used layers of plywood and carved styrofoam sheets to construct a basin on which a drainage network was superimposed (Figure 5.8). The application of specified amounts of rainfall to the modelled basins illustrated the way in which particular basin characteristics produce streamflow in response to specific inputs (Figure 5.8).

Many studies have been achieved relating drainage basin form to drainage basin processes but the use of various methods (pp. 225–32), and the direction of the analyses to varied purposes, hinder the easy reconciliation of results from different areas. This difficulty is emphasised by problems of comparability, in that several workers have chosen different indices from the range available; of availability, in that data is not always comparable from one area to another; and of applicability, in that some statistical techniques are appropriate to some problems more than to others. Many analyses have been conducted to describe the relations between basin characteristics and basin process. When such studies are used as an aid to understanding and illustrating the way in which a particular basin characteristic influences streamflow and sediment yield, it is imperative to recall the facts that the watershed is a dynamic unit (for example, Figure 2.13), that records from short time periods cannot easily be extended to longer periods because of time series problems (pp. 184–6), and that different relationships may be obtained for each of the varied output indexes of stream-

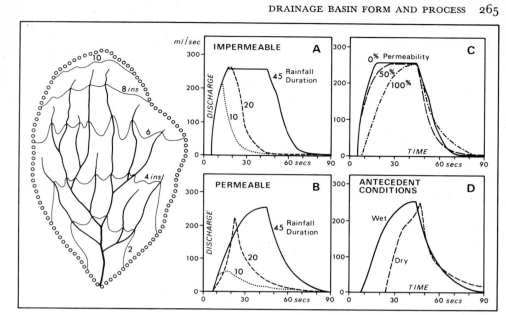

Figure 5.8 A laboratory watershed

Based upon Roberts and Klingeman (1970). Illustrating the watershed constructed from layers of plywood and carved styrofoam sheets, coarse painted to simulate land roughness and irregularities. The drainage network is indicated, degrees of permeability were simulated by laying thicknesses of absorbent material over different proportions of the basin surface, and input was applied using a rainfall simulator. In A and B with moderate rainfall intensity, wet antecedent conditions and a stationary storm the effects on outflow of different rainfall durations are illustrated for a completely impermeable (A) and a permeable (B) watershed surface. Three degrees of permeability are represented in C where, under the same conditions as A and B, an absorbent covering was placed over the whole (100 per cent), or half (50 per cent), of the basin or was not used (0 per cent). In D the basin was completely covered by an absorbent layer (100 per cent permeability) but in one case was completely dry and in the other was wet to simulate storage which was filled.

flow and sediment yield which have been proposed (pp. 191–210) as expressions of drainage basin process.

5.4a Topographic characteristics

Area and size, shape and pattern, relief and slope, and drainage density were denoted as the salient topographic characteristics (2.1) and these individually and collectively influence watershed processes.

Drainage basin area is in many respects the easiest basin characteristic to relate to drainage basin processes but because it is in turn correlated with other characteristics (for example, Figure 3.1) its significance may not be easy to interpret. It is for this reason that Anderson (1957) designated it as the 'devil's own variable'. The significance of catchment area obviously lies in the fact that an area with homogenous rock type, soil mantle, vegetation character and topographic properties, if subjected to uniform inputs of precipitation, should provide a streamflow response varying in

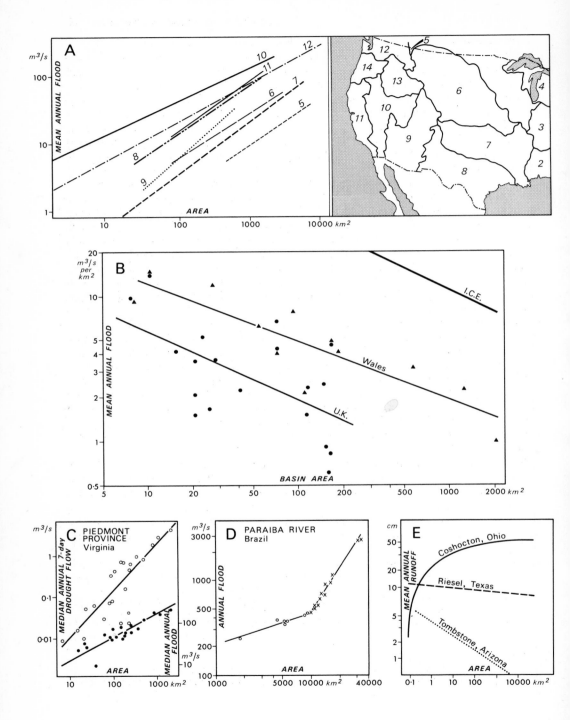

magnitude according to catchment area alone. However such uniformity of basin character is seldom encountered. In addition, because area is related to other basin characteristics such as relief, in that the highest relief ratios tend to occur in the smaller basins; because of restricted distribution of inputs; and because a flood wave decreases in intensity as it is routed through the basin, the significance of area is not easily ascertained. For many years however a simple relation between an index of streamflow (Q) and catchment area (A) has been used as a guide to discharges of a particular value. Such relations have usually taken the form $Q = aA^b$ and have been used particularly for predicting flood events. Relations of this type are compared for eight areas of the United States in Figure 5.9A.

Due to the nature of basin characteristics and to the restricted areal occurrence of intense precipitation of large amount, storm discharge per unit area, expressed as mean annual flood for example, may be inversely proportional to catchment area. Thus when mean annual flood values calculated for Wales and for the U.K. respectively are plotted per unit catchment area and related to basin area (Figure 5.9B), the highest floods per unit area are found to be characteristic of the smallest catchment areas. This feature has been utilised as a basis for tentatively indicating the maximum flood for certain regions, employing formulae of the kind $Q = aA^b$. The Institution of Civil Engineers regarded the normal maximum flood for the British Isles (Figure 5.9B) as being given by this equation where $a = 3000$ and $b = -0.5$. Comparable formulae developed in other countries have utilised different values of the exponent b; $b = 0.75$ for India and $b = 0.33$ for Italy. This inverse relation arises partly from the storage potential of the basin, including storage in channels as well as in lakes, and this storage facility is more obvious in the case of total sediment yield because small catchments may be dominated by erosion whereas deposition may figure more prominently in larger catchments. Thus in many areas sediment yield per unit area has been observed to decrease as catchment area increases and according to measurements in the United States it appears that, on average, watersheds less than 25 km² produce seven times as much sediment per unit area as watersheds greater than 2500 km² (Table 5.6).

Not only does the significance of catchment area vary according to scale of basin but also with catchment characteristics, and area has a different significance according to the index of streamflow related to basin area. Thus when mean annual runoff is related to drainage area, Glymph and Holtan (1969) illustrated different types of relationship which can be obtained (Figure 5.9E). In some areas, such as the humid Appalachian region instanced by Coshocton, Ohio (Figure 5.9E), upland infiltration

Figure 5.9   Some implications of basin area
A   The relations of mean annual flood and catchment area for eight areas of the United States shown on the map are based upon *U.S. Geol. Surv. Water Supply Papers.*
B   Mean annual flood per unit area is plotted against catchment area using data contained in Rodda (1969) for the U.K. and in Howe, Slaymaker and Harding (1967) for Wales. The I.C.E. line indicates the position of the normal maximum flood according to the Civil Engineers (see p. 267).
C   Illustrates the way in which indices of peak flow and of low flow differ in their relationship with catchment area (Guisti, 1962).
D   Shows the variation with catchment area within a single catchment (Santos, 1966).
E   Demonstrates three different relations between mean annual runoff and catchment area (Glymph and Holtan, 1969) (see pp. 267–8).

Table 5.6   Average sediment production for groups of drainage areas in the United States (*after Gottschalk, 1964*)

Watershed size range	Arithmetic average annual sediment production	Number of measurements
Less than 26 km^2	1810 m^3/km^2	650
26–260 km^2	762 m^3/km^2	205
260–2600 km^2	481 m^3/km^2	123
More than 2600 km^2	238 m^3/km^2	118

returns in part to channels downstream causing a gain in streamflow per unit area of basin as basin area increases. In regions where channels absorb streamflow, runoff per unit area conversely decreases with increasing basin size. Thus in Arizona the effects of land management practises retaining water on the land, the result of influent seepage beneath ephemeral channels, the consumptive use by phreatophytes, and the loss by evaporation can produce a downstream decrease in runoff per unit area (Figure 5.9E). Whereas in humid and arid regions the significance of area is contrasted, in other areas where interflow is small and where channel gains and losses nearly balance, the volume of runoff may be comparable for small and larger watersheds; this situation is exemplified by the Riesel draining the Texas Blacklands (Figure 5.9E).

The effect of basin characteristics may also be apparent in the way in which different indices of flow vary according to area. In the Piedmont province, Virginia, Giusti (1962) showed that the relation between flood flow and area differs, according to the slope of the regression line, from that between drought flow and area (Figure 5.9C). Furthermore in some cases streams with the lowest base flows may have the highest flood flows. These qualifications for the significance of catchment area arise when streamflow indices are compared for several basins, but it is sometimes apparent that variation in annual flood occurs according to area drained within a single catchment. Thus Santos (1966) showed that the Paraiba river of Brazil behaves as if it were two rivers because the change of flood flow relative to drainage area is at a lower rate in the upper than in the middle and lower reaches of the basin (Figure 5.9D). This was interpreted by reference to the climatic and topographic characteristics of the two contrasted parts of the area drained and is reminiscent of the dog-leg phenomenon identified in the analysis of flood frequency over time (p. 195).

Other indices of size of basin have been employed in relationships with drainage basin process. In the case of water yield, length of longest stream (L) was utilised by Morisawa (1967) in a relation with mean annual discharge (Q) for 96 watersheds in six different physiographic areas of the eastern U.S.A. and five of the relations, of the form $Q = aL^b$, were different at the 0·005 probability level. Use of length of longest stream, or of total channel length, is sometimes quicker, and sometimes more relevant than use of basin area, but both are closely related to area. Basin order has been employed in relation to measures of stream discharge but this is of restricted value unless a method of ordering is used which is directly relevant to runoff production. Some progress has been made and Stall and Yu-Si-Fok (1967) correlated order with 10, 50 and 90 per cent indices of flow duration curves for eight gauging stations in the basin of the Embarras river in Illinois.

Relief, as a drainage basin characteristic, has been expressed in several ways (pp. 58–60) and it undoubtedly exercises an influence over runoff and sediment production in the basin. This influence derives from the fact that higher relief or steeper slopes potentially provide more available energy than do more subdued basins. Thus in the Manning equation (pp. 236–7), if channel size and roughness were constant from one basin to another then stream velocity would be directly proportional to slope. Parameters of the streamflow hydrograph will vary according to basin relief and to basin slope, and in particular, time of hydrograph rise and lag time will be shorter, and peak discharge rate may be higher in the basins with the highest relief ratio (Figure 5.10). In a study of United Kingdom catchments Nash (1966) expressed mean delay between precipitation and outflow (m_1) in terms of basin area (A) and basin slope (S, parts per 1000) in the form $m_1 = 27 \cdot 6 \ A^{0 \cdot 3} S^{-0 \cdot 3}$. The influence of relief is most relevant to indices of peak streamflow and, as sediment is transported during these peak events, relationships between sediment yield and relief measures

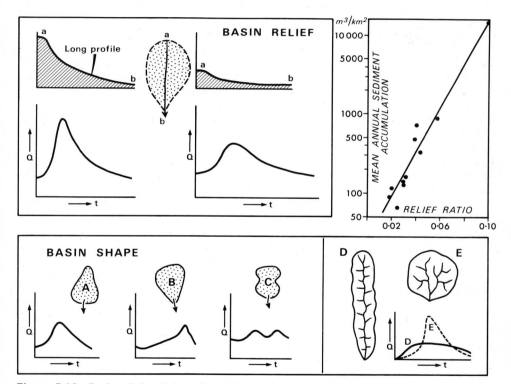

Figure 5.10 Basin relief and shape in relation to drainage basin process
The influence of basin relief upon stream hydrograph form is contrasted generally for two basins with different long profiles, and the relation of mean annual sediment accumulation to relief ratio is based upon Schumm (1954).
    The significance of drainage basin shape (A, B, C) for hydrograph form is based upon DeWiest (1965) and the effect of drainage networks with contrasted bifurcation ratios follows Strahler (1964).

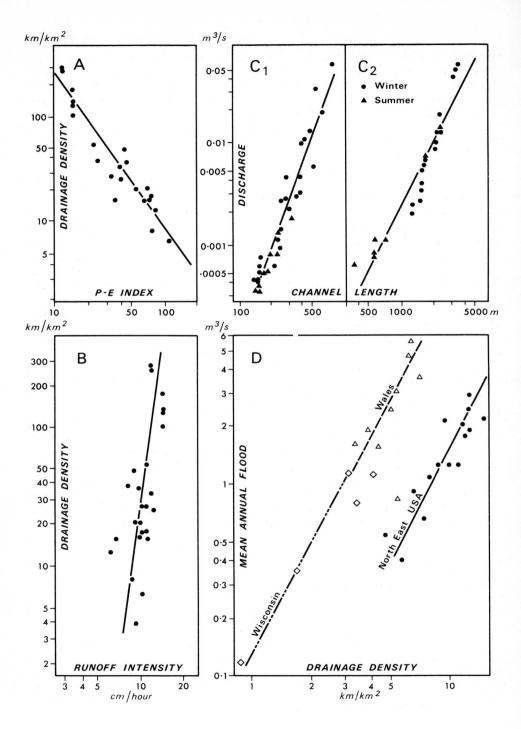

have been obtained. Using data on sediment production from small basins of the Colorado plateau province Schumm (1954) derived a relationship between relief ratio and mean annual sediment accumulation in reservoirs (Figure 5.10). In a study of 25 basins in the Red Hills area of southern Kansas, western Oklahoma and western Texas, Maner (1958) demonstrated that sediment delivery rate is a function of several basin characteristics which are related to, and adequately expressed by, relief ratio. At the scale of large mid-latitude drainage basins Ahnert (1970) showed that mean denudation rate is directly proportional to mean basin relief.

The influence of relief is inextricably bound up with other basin characteristics and is of greater significance to some indices of basin response, particularly peak runoff rates and sediment delivery, than to others. These same points epitomise the significance of basin shape. In addition the effect of shape has proved more elusive to express (Table 2.3) because its significance for basin processes depends partly upon the shape of the basin and partly upon the dependent network shape. In general terms the significance of basin shape has been appreciated because it determines lag time, the time of rise and other hydrograph parameters. The shape of the basin thus exerts an effect on the shape of the hydrograph (Figure 5.10A, B) and where a particular basin embraces two partly separated areas, the hydrograph may have two peaks (Figure 5.10C). The significance of basin shape is often expressed through the pattern of the drainage network. In Figure 5.10D and E the more efficient network (E) gives a slower rise but a higher peak, whereas an attenuated network (D) produces a quick rise but a lower and more protracted hydrograph peak.

Some studies have revealed clear relationships between topographic characteristics of drainage basins and parameters of water and sediment yield, but the exact significance of a particular topographic variable of area, relief or shape is not universally agreed. This may arise because a particular topographic parameter may exercise an influence over flood flow but not over mean flow or over flow duration indices, and also because the catchment is dynamic whereas many of the indices proposed express the overall size, shape or relief of the basin. In fact it is now increasingly appreciated that only part of the basin actually produces runoff and sediment at a particular time. Therefore of all topographic characteristics perhaps drainage density is the most valuable, and potentially the most useful single index in relation to drainage basin processes. The significance of drainage density stems from the facts that water and sediment yield are very much influenced by the length of water courses per unit

Figure 5.11 Drainage density and drainage basin process
In A Drainage density is inversely related to (P–E) index (r = −0·943 n = 23) for basins in Colorado, Utah, Arizona and New Mexico.

$$\text{Where (P–E) index} = 115 \sum_{12} \left(\frac{P}{T-10}\right)^{1\cdot11}$$

and   P = average precipitation for each month
        T = monthly average daily temperature   (Melton, 1957)

and in B runoff intensity is directly related to drainage density for the same 23 basins (Melton, 1957). The dynamic network is illustrated by relating stream discharge to actual channel length in two basins (C1 and C2) (Gregory and Walling, 1968) and the relation of mean annual flood to drainage density is based upon Orsborn, 1970, (Wisconsin), Howe Slaymaker and Harding 1966 (Wales) and Carlston 1963, 1966 (northeast U.S.A.).

area, that drainage density is closely related to and expressive of other basin character-istics, and is also related to the input to the basin. Drainage density therefore occupies a central position because it can be regarded as both an output or response to input, or to the sum total of drainage basin characteristics, and it can also be viewed as a characteristic which affects amount and rate of output from the drainage basin. Accordingly, attempts have been made to relate drainage density to climatic inputs and also to drainage basin outputs.

To relate density of drainage to climate Peltier (1962) derived four relations between average slope and mean number of drainageways per unit distance for four world climates, namely moderate or mesothermal, semi-desert, desert, and tropical. More detailed measurements are required to relate drainage density to climate but it is not sufficient to utilise rainfall alone because the additional factors of rainfall intensity, temperature and evapotranspiration reflecting the character of the vegeta-tion cover are also important. Specifically it has been proposed that drainage density increases as mean annual precipitation (Williams and Fowler, 1969) and as rainfall intensity increase (Chorley, 1957; Chorley and Morgan, 1962), and as runoff intensity increases (Figure 5.11B). In the Sungai Klang drainage basin in west Malaysia magnitude of 10-year daily rainfall and seasonality of the rainfall regime were shown to be the most significant climatic controls of drainage texture (Morgan, 1970). Climatic indices have also been used by Melton (1957) who showed that drainage density was inversely related to the Thornthwaite precipitation effectiveness index (P-E) (Figure 5.11A) but that it also depended upon runoff intensity, per cent bare area, infiltration capacity, and soil strength; and by Chorley (1957) who related drainage density to a climate-vegetation factor (P-E index/Mean monthly precipita-tion × Intensity of precipitation). Although Carlston (1963) interpreted variations of drainage density according to terrain transmissibility, Cotton (1964) concluded that differences in drainage density between western Europe and the eastern U.S.A. are controlled by climatic factors.

The general consensus of opinion is that drainage density reflects precipitation intensity and that local variations can be inspired by other basin characteristics such as rock type, soil and land use. However it is not easy to express the relationship of drainage density to climate in different areas and instead of using the mean density for a number of sampled basins, variation can be illustrated by using the relation between total stream length and basin area and by considering the way in which the constant of the power relation varies according to climate (Figure 5.12). Despite the differences in sources and methods (pp. 47–8) this demonstrates that the highest drainage densities are encountered in semi-arid areas, illustrated by Australia and by the western U.S.A., and that lower densities are common in humid temperate land-scapes.

Perhaps more significant is the relation of drainage density to basin output; Sokolov (1969) has argued that 'drainage density is certain to be the most important factor characterising the conditions of flood flow formation. Under other identical conditions it indirectly characterises the infiltration capacity of soils forming the basin surface. Nature itself creates a drainage network of a density necessary for outflow of water excess from a watershed surface.' If channel characteristics and pattern and other basin characteristics were constant then discharge should be directly related to density of channel because this provides an index of the availability of channel flow

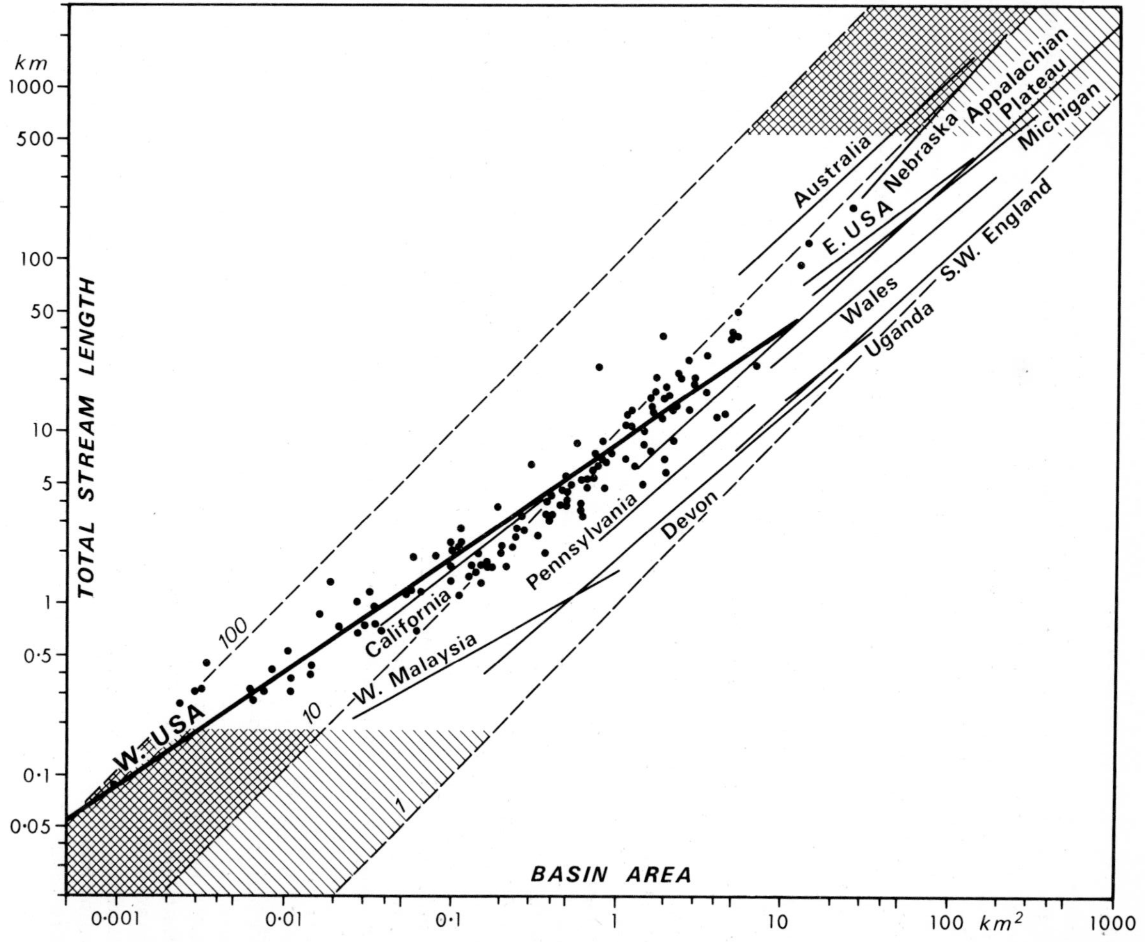

Figure 5.12
Drainage density variation

Expressed by relating total channel length to basin area for data presented for Australia (Woodyer and Brookfield, 1966), Western U.S.A. (Melton, 1957), Eastern U.S.A. (Carlston, 1963), Appalachian Plateau (Morisawa, 1962), California (Smith, 1950), Pennsylvania (Smith, 1950), western Malaysia (Eyles, 1968), Michigan (Hack, 1965), Wales (Howe, Slaymaker and Harding, 1967), Uganda (Doornkamp and King, 1971), south-west England and Devon (Gregory, 1971), Nebraska (Brice, 1966). Dashed lines represent drainage densities of 0·1, 1·0, 10 km/km² respectively and the distribution of points is shown for data from Western U.S.A. (Melton, 1957).

which is faster than the alternative forms, of overland flow or throughflow. Character of the channels can also exercise a significant influence upon the supply of sediment. In a shale area of South Dakota sediment production was found to be related to density of incised channels in the watershed (Gottschalk, 1946) and, in Illinois, Stall and Bartelli (1959) found that the reciprocal of the density of non-incised channels was a significant factor related to sediment yields from small watersheds. It has been shown that runoff intensity is dependent upon drainage density (Melton, 1957) (Figure 5.11C), that mean annual flood ($Q_{2.33}$) is usually related to drainage density (D_d) in the form $Q_{2.33} \propto D_d^2$, (Figure 5.11D) and that indices of base flow can be related to drainage density (Orsborn, 1970; Trainer 1969). Such relations, although indicating the general nature of the relationships, are static interpretations and can give ambiguous results if the same values of drainage density are related to different flow indices (Gregory and Walling, 1968). An alternative method of assessing the significance of drainage density is to consider its dynamic nature by analysing the changes of drainage density in a single catchment. In two small catchments in southeast Devon it was demonstrated (Figure 5.11C) that total channel length (ΣL), which is equivalent to drainage density, increases in relation to actual discharge values (Q) according to a relation of the form $Q \propto \Sigma L^2$ (Gregory and Walling, 1968). The total discharge is used in this case but if surface runoff (Q_s) is employed the relation may be of the form $Q_s \propto \Sigma L^1$ (Weyman, 1970).

Drainage density therefore expresses the significance of effective climate on the basin characteristics, it influences output from the basin, and it therefore emphasises the complex way in which topographic variables are inter-related. The dynamic nature of these variables must always be considered because the stream network is really a composite net which derives from perennial sources and also is supplemented by expansion according to particular conditions (see pp. 85–6).

5.4b Rock type

Geomorphologists have perhaps tended to ignore the opportunities for quantitative study of the nature and magnitude of the significance of rock type. This omission has been noted by Yatsu (1966) who reviewed the material available and commented that 'geomorphology should be constructed on an exact scientific basis, especially exact dating, correct knowledge of processes, and physico-chemical and mechanical understanding of rocks'. The omission is not applicable in the study of limestone terrains however and these have fostered the development of limestone geomorphology (Jennings, 1971; Sweeting, 1972) and the study of the hydrology of carbonate terrains (Stringfield and Le Grand, 1968).

Rock type is significant in the drainage basin statically and dynamically although both are necessarily inter-related. The static significance may be regarded as the capacity of a rock to absorb and to retain water so that, due to permeability and porosity, different rock types vary in their ability to store water. The significance of this storage is felt dynamically in the basin because it controls the rate of water outflow; the rock type dictates the character and rate of weathering, the weathering products obtained and thence the nature of the sediment and solutes supplied to the stream. For these static and dynamic reasons rock type is a significant factor influencing drainage basin dynamics. The significance of rock type affecting other parts of the hydrological cycle is properly the field of groundwater hydrology (Davis and De Wiest, 1966; Todd, 1959) and the field of groundwater hydraulics (Lohmann, 1972).

Spaces within a rock can be either intergranular, which occur between rock particles, or massive which refers to those along joints, faults or structural lines. Capillary interstices are those capable of retaining water by surface tension, sub-capillary interstices are so small that water is retained in them by molecular forces, and supercapillary interstices are larger than those in which capillary action is possible and can be as large as cave-size. Water in these spaces may be either original, in which case it was included during formation of the rock, or secondary which occurs due to movement of water from the basin surface through the rock. Rate of movement will depend upon the permeability (pp. 61–2), amount of storage will depend upon the porosity, but the vertical distribution of water usually conforms to a general sequence. Below the soil water is the aeration zone which includes intermediate vadose water which is moving downwards under the influence of gravity and can vary in thickness from nothing to several hundred metres in arid regions and this zone is succeeded by the capillary zone or capillary fringe. Below the capillary fringe is the water table which is maintained by atmospheric pressure as a theoretical level between the aeration zone above and the saturated zone below. In the saturated zone, or zone of phreatic water, water obeys the laws of hydrodynamics and is subject to pressure differences. This zone merges at depth with a zone of dense rock which contains some water in pores which are not connected. The depth to dense rock varies with rock type but can be more than 10000 m in deep sedimentary basins.

This model of water distribution is diversified in two principal respects. Firstly, different rock types may occur in juxtaposition and so water in the aeration or saturated zones may be unconfined, if it is accessible to the atmosphere by open spaces, or it may be confined where this is not the case. In the confined situation, often where an impermeable rock overlies an aquifer, several water tables may exist dictated by the vertical sequence of rock types, and water tables produced by the presence of impermeable strata or layers are designated perched water tables. Secondly, the position of the saturated and aeration zones may fluctuate seasonally and from year to year in accord with the supply of water received below the soil which can recharge the saturated zone, and with the loss of water from springs and seepages. This variation is illustrated by measurements from two ground water observation boreholes in Devon which indicate the seasonal fluctuation in response to rainfall regime and which also demonstrate the variation over the 3·5 year period according to rainfall amount (Figure 5.13). The fluctuation of ground water levels will depend upon a number of factors including the nature of water movement, the storage and release of water in the several media involved, in relation to the sorting and packing of grains, the distribution and interconnection of joints and fractures, the degree and depth of weathering and, where pertinent the development of solution channels and caves. Thus Maxey (1964) has proposed that 'hydrostratigraphic units' should be recognised as 'bodies of rock with considerable lateral extent that compose a geologic framework for a reasonably distinct hydrologic system'.

The dynamic significance of rock type is first apparent by consideration of supply from ground water to the surface. The baseflow component of streamflow may be

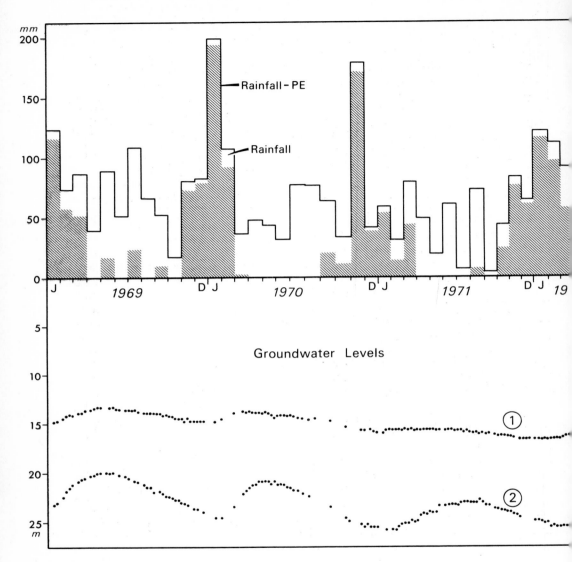

Figure 5.13 Ground-water level variation
Variation in water levels for two ground-water observation boreholes are show over a period of
$3\frac{1}{2}$ years in relation to monthly rainfall and to monthly rainfall minus estimated monthly potential
evapotranspiration (shaded). The two boreholes are located on the dip slope of the Pebble Beds
aquifer on Woodbury Common in south-east Devon and the borehole (2) near the crest of the
dip slope records more substantial annual variations than the borehole located in the middle of
the dip slope (1) although both exhibit a variation in response to precipitation fluctuations over
the $3\frac{1}{2}$-year period.

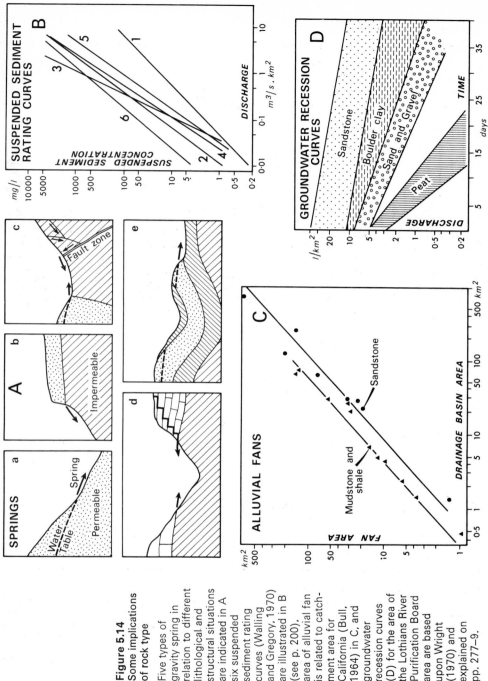

Figure 5.14
Some implications of rock type

Five types of gravity spring in relation to different lithological and structural situations are indicated in A six suspended sediment rating curves (Walling and Gregory, 1970) are illustrated in B (see p. 200), area of alluvial fan is related to catchment area for California (Bull, 1964) in C, and groundwater recession curves (D) for the area of the Lothians River Purification Board area are based upon Wright (1970) and explained on pp. 277–9.

derived by several mechanisms. Although water may percolate through a stream bed as influent seepage (for example, Figure 5.9), where the stream flows at the saturation level water may reach the stream or a lake by effluent seepage from an aquifer and this kind of seepage may also occur as diffuse percolation over an open slope. Capillary seepage obtains where water from the water table is brought by capillary action to be discharged on the surface. Any natural discharge of water large enough to produce streamflow is usually termed a spring and the occurrence of springs will be dictated by the position of the water table intersecting the surface (water table springs), by the presence of vertical or horizontal variations of permeability (contact springs), by the occurrence of structures such as faults or jointing, or due to the presence of folded strata (Figure 5.14A). Whereas these factors account for the distribution of gravity springs there are non-gravity springs, explained by volcanic activity for example where thermal springs occur. The discharge of springs varies appreciably and while the world's largest may be Trebisnjica in Yugoslavia with a mean annual discharge of 41 m^3/s (Burdon, 1966), the discharge will vary according to aquifer permeability, area contributing to recharge of the aquifer, and the quantity of recharge. Meinzer (1923) proposed a quantitative classification of eight magnitudes of springs classified according to their discharges and he recorded 65 first order magnitude springs (greater than 2·83 m^3/s discharge) in the U.S.A. Of these 38 occurred in volcanic rocks, 24 in limestones and three in sandstones.

Reminiscent of water table fluctuations, springs also fluctuate in discharge and vary in position. Seasonal variations in discharge can reflect precipitation distribution and régime, and diurnal variations have also been noted in response to varying use of water by vegetation. Seasonal movements of the water table provide the basis for a classification of springs as perennial which flow all the year, as intermittent which flow for part of the year, and as periodic which flow at intervals and are not necessarily related to the occurrence of precipitation but to other factors including variation of barometric pressure or dormant vulcanicity where geysers occur.

Variation in the position of springs and also in their discharge promotes variation in streamflow. Classically the base flow component of stream discharge is provided by the saturated zone through springs or seepages, and low flow and base flow characteristics have been expressed in a number of ways. Both the absolute value of low flow and the character of the rate of change of low flows, indexed by the form of the recession curve (pp. 200–201), will reflect rock type and its storage properties. Thus the baseflow recession curve may be employed to compare drainage basins according to rock type and to express ground water discharge from a basin. Frequently the relation $Q = Q_o e^{-ad}$ has been used to relate the discharge on a particular day (Q), the discharge d days previously (Q_o), the number of recession days (d) and a constant (a) governed by the drainage basin characteristics and particularly by rock type. Thus Wright (1970) in a study of the area of the Lothians River Purification Board in Scotland plotted ground water recession curves (Figure 5.14D), calculated a geology index (G) appropriate to the geological formations beneath the catchments studied and then developed an equation relating mean lowest annual daily mean flow (Q_L l/s), mean catchment slope (S, per cent$^{0.5}$), mean annual runoff (R, mm) and catchment area (A, km^2). The relation of the form $Q_L = 0.00175G^{0.44}S^{0.53}R^{0.63}A^{1.10}$; (r = 0·98) could be employed with geological indices quoted as follows:

Geological Index (after Wright, 1970)

Peat 0·8   Limestone 2·5   Boulder Clay 4·2
Igneous rock 1·9   Shale 2·5   Sandstone 4·7
Sand and Gravel 1·9 Alluvium 3·0   Chalk 5·0

The baseflow duration curve may also be employed to demonstrate the groundwater discharge from a basin and this technique is similar to the flow duration curve (pp. 192–4) but shows the percentage of specified time which a daily base flow is equalled or exceeded. Kunkle (1962) used the baseflow duration curve in a study of six locations of the Huron river above Ann Arbor, Michigan and showed that the curve contains two components representing discharge from bank storage and from basin storage respectively, and that the curve assumes a particular form according to the character of the deposits drained. In this study of an area covered by thick glacial deposits, the discharge ratio was shown to be 1 from relatively impermeable till, 5 from moderately permeable till, and 9 from permeable outwash.

Although the significance of rock type is expressed largely in base flow components of streamflow, type of rock will also influence directly, and especially indirectly, the rate at which other forms of flow contribute to the streamflow hydrograph. The direct influence is exercised where rocks have low permeability and thence compel surface flow and flow in the unsaturated zones. These indirect influences are wide-ranging firstly because the rock type is one factor which may determine the character of weathering processes and therefore of the overlying weathered material (Ollier, 1969); and secondly because the rock type influences other basin characteristics including topographic ones. Thus in alluvial fans in California the relation between fan area and drainage basin area for mudstone and shale basins differs from the same relation plotted for sandstone basins (Bull, 1963) and this contrast (Figure 5.13C) may be ascribed to the differences in runoff rate conditioned by the two groups of rock type.

More significant is the effect of rock type in determining the amount and nature of the sediment and solutes available to the stream for transport in the drainage basin. The chemical composition of a rock will, in conjunction with the flow properties of the strata and the antecedent conditions, determine the chemical composition of groundwater and thence of streamflow (see pp. 319–22) which contains solutes transported as ions. The ease of removal of elements as ions is indicated in Table 5.7A, groups of compounds classified according to their potential for transport are shown in Table 5.7B, and in Table 5.7C the chemical composition of water in three contrasted rock

Table 5.7
A   Migrational series of elements (*after Strakhov, 1967*)

		Index of order of magnitude of migrational capacity
Actively removed	Cl (Br, I) S	$2n \times 10$
Easily removed	Ca, Na, Mg, K, F,	n
	Sio_2, silicates	$n \times 10^1$
Mobile	P, Mn, Co, Ni, Cu	
Inert (weakly mobile)	Fe, Al, Ti	$n \times 10^{-2}$
Practically immobile	Sio_2 (quartz)	$n \times 10^{-\infty}$

Table 5.7—*continued*

B   Compounds classified according to potential for transport (*after Strakhov, 1967*)

1   Easily dissolved salts—NaCl, KCl, MgSo$_4$, MgCl$_2$, CaSo$_4$, CaCl$_2$—true ionic solutions
2   Carbonates of alkali and alkaline earths—CaCo$_3$, MgCo$_3$, Na$_2$Co$_3$, and silica
3   Fn, Mn, P, and minor elements (V, Cr, Ni, Co, Cu)—very low solubilities in water. Tend to form colloidal solutions in addition to true solutions. Only phosphorus compounds form true ionic solutions
4   Quartz and various silicate and aluminium silicate minerals

C   Composition of water in rock types (*from Downing, Allender and Bridge, 1970*)

	Lower Keuper sandstone Brindley Bank near Rugeley	Coal measures Goldthorn Hill, Wolverhampton	Buxton Spring
Calcium as Ca	71·0 mg/l	141	58·5
Magnesium as Mg	5·0	9	19·2
Sodium as Na	20·4	90	22·8
Carbonate as Co$_3$	70·8	Free Co$_2$ 13	255·9
Sulphate as So$_4$	41·1	139	13·6
Chloride as Cl	41·3	22	40·0
Nitrate as No$_3$	14·2	—	1·0
Iron as Fe	—	0·03	Silica 13·7
			Potassium K 3·8
Total dissolved solids	302·0	537	424·0
Carbonate hardness	124·0	187	210·0
Non-carbonate hardness	70·0	151	15·0

types is indicated. The suspended sediment available for transport will also depend upon the characteristics of the rock available although in many cases this will have been previously weathered, but suspended sediment rating curves (Figure 5.14B) may be of different forms according to the rock type in the catchment. In this case (Figure 5.14B) the rating curve from shales (6) is higher than those from catchments underlain by marls, greensand and clay-with-flints (2, 3, 4, 5) whereas a catchment developed on clay-with-flints alone produces the smallest amounts of suspended sediment (1). Prior weathering may also provide material for bedload transport but this may sometimes be obtained directly by erosion of the bedrock if it is exposed in the stream channel. Such erosion may be accomplished by corrasion which entails the mechanical action of transported particles on exposed bedrock, by corrosion which is the chemical process whereby water erodes available rock, and occasionally by the force of the water (evorsion) and by the process of cavitation at high water velocities. The extent of removal and transport of material as bedload will depend to some extent upon the physical and chemical properties of the rock. Rocks which are well-foliated and easily fractured are especially prone. The removal of material and the erosion accomplished during transport produce a number of features upon rock beds, and potholes (Plate 4) are frequent minor features, and irregular cross profiles and accidented long profiles broken by rapids, or waterfalls (Plate 4) may be a response to variations in rock character.

The consequence of transport of rock particles of any size is that changes will occur in the amount and relative composition of constituents especially as transport proceeds. These changes are related to composition, size and shape of particles. In general rivers in uplands and mountain courses will have a greater amount of felspathic fragments, will be poorly sorted, and will show rapid changes of size and a prevalence of angular and subangular grains. A lowland stream however will have deposits in which quartz grains are prevalent, will be better sorted with a dominant sand fraction, and have sand grains and larger particles which are rounded or well-rounded. The changes in size, sorting, composition and shape will reflect the contributing rock type, although these parameters are closely inter-related so that roundness and size are related (Figure 2.11) because size and density will determine whether a particle will travel by saltation rather than traction. During transport particle size decreases downstream to a weight W according to the original weight (W_0) and distance travelled (x) in the relation $W = W_0 e^{kx}$ where k is a constant in this Sternberg Law, depending upon the nature of the rock type. Particles also change in roundness, and using the various indices proposed (pp. 71–3) studies have been conducted to ascertain the rate and extent at which changes take place. Thus Ouma (1967) in a study of sediments along the Hacking River in New South Wales showed that unweighted mean roundness does not approach 1·00 asymptotally downstream but first increases to a maximum, depending on rock type and local conditions, and then declines in a downstream direction. He also showed that changes take place at different rates in several size ranges. Changes of particle shape, size and composition have also been studied by laboratory experiments (for example, Krumbein, 1941) and by field analyses under closely controlled conditions. Thus Pittman and Ovenshine (1968) analysed the changes in roundness and other properties during transport along the Merced river California and demonstrated that the regimen and type of channel affect roundness as well as the physical properties of the rock, that in some cases pebbles may decrease in roundness downstream as a result of high energy transport and they noted a decrease in roundness which occurred immediately downstream from the high energy environment of a cataract.

The results of these studies has been to evaluate the comparative significance of lithologies for fluvial transport. A general indication of the resistance of certain rock types is given in Table 5.8. In general it is thought that in the case of the more

Table 5.8 Resistance of pebbles during river transport (*based upon Kutal, 1971*)

Rock type	Length of transport
Sandstone	Not more than 15–20 km unless quartz sandstone
Shale	Completely disaggregated after 40 km
Gneiss	Usually not greater than 20 km
Granite	Hundreds of kilometres if original weight more than 20 gm
Limestone	About 100 times less resistant than granite
Greywacke	Weight loss after 450 km was 21 per cent

resistant lithologies a particle soon attains a roundness value typical of the energy environment. It then maintains this value as a steady state or equilibrium form, although the initial rate of rounding will depend particularly upon the rock type

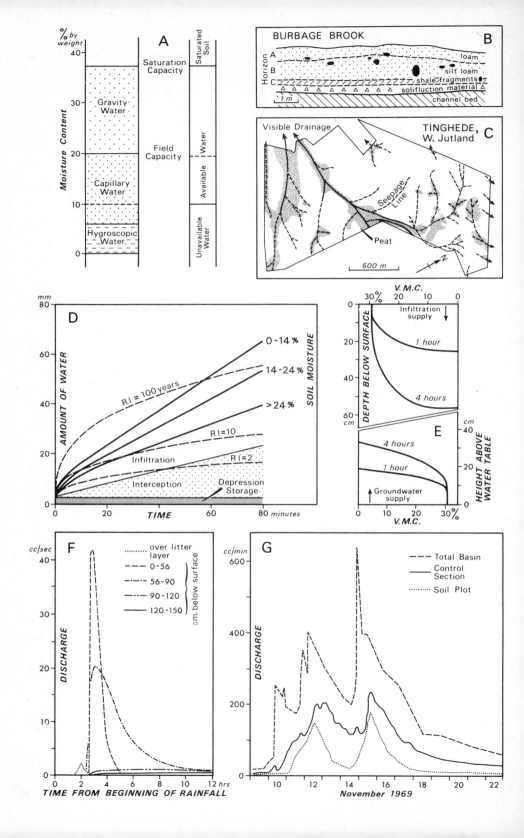

available, and the subsequent maintenance of this shape will depend upon the energy environments and channel characteristics available along the river course.

5.4c Soil

Rock type is sometimes indirectly a factor influencing drainage basin processes. One of the most obvious indirect expressions is found in the character of the soil and the superficial deposits which owe their presence and character to the soil-forming factors of local rock type combined with topographic, climatic, biotic and temporal factors. Although in some areas the soil or superficial material may have been derived by development in situ, in other areas erosional history may have provided a legacy of glacial or slope deposits which are not directly related to the rock type beneath. Whatever their history soils, like rock type, may be perceived statically and dynamically. The static influence derives from the water-holding, whereas the dynamic influence reflects the water-transmitting and sediment-providing properties of the soil.

The static significance relates to the type and amount of water which can be contained by the soil and this is modelled for a fine sandy loam soil in Figure 5.15A. Some small percentage of water is held in combination with soil particles and approximately 6 per cent may be hygroscopic water in combination with particular salts. Capillary water is held to individual soil particles by surface tension, and gravity water which is usually the largest representative is that which is free to move through the profile under the influence of gravity. These classes of water in the soil can be the basis for certain definitions of water content (Figure 5.15A) so that saturation capacity occurs when all available spaces are utilised by water, field capacity obtains when all the gravity water has drained from the profile and this may occur several days after rainfall above a freely drained soil. If subsequently, the reserves of capillary water are depleted, due to evapotranspiration, the soil moisture content may decrease to the level at which plants cannot obtain sufficient water for transpiration and this level from which plants cannot recover is designated the permanent wilting point. The permanent wilting point will vary according to the 'habit' of a particular plant and also according to the soil characteristics. These characteristics, particularly texture, structure, organic matter content, horizon sequence and chemical composition will also influence the saturation capacity and the field capacity, and they will dictate the potential water capacity at a particular time. The amount of water in the soil at any

Figure 5.15 Soil characteristics in relation to drainage basin process
The general distribution of water in the soil (A) is based upon Todd (1970), a river bank section including pipe outlets (B) shown in black was documented by Jones (1971), and the pattern of seepage lines (C) in relation to the drainage network and to areas of peat (stippled) was mapped by Bunting (1961). D compares expected U.K. rainfall intensities (dashed lines) with infiltration rates likely for clay-pan soils with good grass cover (solid lines) to demonstrate the relative frequency of overland flow in Great Britain (Kirkby, 1969) and E illustrates the distribution of water in a sandy soil mass during infiltration (above) and during capillary rise (below) according to volumetric moisture content, VMC, (after Liakopoulos, 1965). In F discharge hydrographs from a simulated storm of 5·1 cm/hour on a 16° slope succeeding drainage for more than 4 days are shown for different levels below the surface (Whipkey, 1965) and in G water discharge over a 14-day period in the East Twin Brook (Mendips) catchment is shown for the total basin, for a central section between two gauging stations, and for a soil plot (Weyman, 1970).

one instant will obviously depend upon the antecedent conditions especially the supply and loss of water to a particular profile.

The content of water at a particular instant will be responsible for the dynamic characteristics of the soil because these will dictate how much water can enter the profile and at what rate, how much can be retained, and how much can be transmitted. A simple analogy is often drawn between the soil and a sponge: if water is supplied to a dry sponge it may be contained but if the sponge is very wet water will be discharged from the sponge. In this way the soil layer is fundamentally significant in the functioning of the drainage basin. Water received from precipitation on the surface of the basin may initially collect in depressions as surface detention but as the amount of such detention increases it will either flow over the surface or infiltrate into the soil. The proportion of each depends not only upon the soil character at a particular time but also upon the land use, the topographic character of the ground especially slope and of course upon the characteristics of the precipitation received, especially amount and intensity. Infiltration is greater for dry than for moist soils and the rate of infiltration, defined as the maximum rate at which water can enter the soil, decreases as the storm proceeds. Horton (1945) considered that the infiltration capacity of a soil was the maximum value of infiltration which could be sustained by a soil under specified conditions and he considered that when this capacity was exceeded then surface detention and runoff would occur. Experience has shown that infiltration rates are seldom less than rainfall intensities. Some typical infiltration capacities are given in Table 5.9. Figure 5.15D was used by Kirkby (1969) to com-

Table 5.9 A Infiltration capacities (*after Kohnke and Bertrand, 1959*)

Clay loam	2·5– 5·0 mm/hour
Silt loam	7·5–15·0
Loam	12·5–25·0
Loamy sand	25·0–50·0

B Relative infiltration correlated with soil properties (*based upon results of correlation coefficients between relative infiltration rates and soil properties for 68 soil profiles studied by Free, Browning and Musgrave, 1940*)

Soil property	Correlation coefficient	Multiple correlation coefficient
Non-capillary porosity, subsoil	0·54	r = 0·71
Organic matter, surface	0·50	
Clay content, subsoil	−0·42	
Organic matter, subsoil	0·40	
Non-capillary porosity, surface	0·36	
Total porosity, subsoil	0·36	
Volume weight, subsoil	−0·33	
Aggregation, surface	0·30	
Moisture equivalent, subsoil	−0·30	
Suspension, surface	−0·29	
Total porosity, surface	0·24	
Silt clay, subsoil	−0·24	
Volume weight, surface	−0·24	

pare typical infiltration rates with expected rainfall rates for Great Britain to indicate the possible frequency of overland flow. Infiltrating water has two components, one representing a steady water movement through the soil (a transmission component determined by character of soil and a transmission constant A) and one involving the slow filling up of spaces formerly occupied by soil air (a diffusion component controlled by a diffusion constant B of the soil). Philip (1957) proposed a relation between these two constants and the time since beginning of rainfall (t) to give the instantaneous rate of infiltration (f) in

$$f = A + B \cdot t^{-\frac{1}{2}}$$

Although these two types of movement are intricately related it is possible to visualise the progression of the wetting front at the contact of wet and dry soil. Thus Liakopoulos (1965) modelled the distribution of water in a sandy soil during surface infiltration according to the moisture content (Figure 5.15E) and he also showed that the upwards movement of water into the soil from ground water was essentially similar in character (Figure 5.15E). These two situations differ in that the infiltration source is eliminated when the rainfall ceases and subsequently some of the water above the wetting front may drain vertically under the influence of gravity. Factors influencing infiltration for certain soils are given in Table 5.9B.

Horton originally envisaged two forms of water flow which contribute to the stream hydrograph, namely overland flow and base flow, the latter derived from ground water recharged by water infiltrating the soil and proceeding to the saturated zone. More recently it has been appreciated that flow can occur laterally within the soil. Such flow may be diffused or may occur along concentrated lines in pipes. Particularly on slopes, water moving vertically through the soil may be deflected laterally as throughflow and this can be encouraged by the existence of less permeable horizons. Frequently permeability is reduced at the base of the A horizon and this may encourage throughflow at rates of perhaps 20–30 cm per hour. Throughflow is not necessarily concentrated at a particular level, and vertically draining water may be gradually deflected by the changes in permeability or structure. A simulated storm lasting for 2 hours with an intensity of 5·1 cm per hour on a 16 degree slope allowed Whipkey (1965) to demonstrate the pattern of discharge from the slope according to the depth of soil (Figure 5.15F). In this experiment water infiltrated to a depth of 90 cm at which permeability decreased, throughflow then occurred and once this was established throughflow subsequently occurred at higher levels in the profile. The significance of throughflow for the stream hydrograph has been demonstrated by Weyman (1970) in a study of the East Twin Brook, a small 0·21 km² basin of the Mendips. This study entailed assessment of the contribution made to streamflow by throughflow from a slope between two streamflow measurement stations. This allowed comparison of total discharge from the basin with discharge from the control section length of stream and with that from a soil plot (Figure 5.15G). The results demonstrated that stormflow derives from the 10–45 cm soil horizon according to the upslope extent of saturated conditions, whereas slow unsaturated flow from the whole soil mass to a small constant zone of saturation supplies base flow from the 45–75 cm horizon.

Flow may thus occur under unsaturated conditions within the soil horizons and a whole watershed will not necessarily contribute water to the streamflow hydrograph which succeeds a particular storm (pp. 28–30). Whereas the dynamic saturated zone, or contributing area, and throughflow may depend upon flow diffused in several layers of soil; concentrated flow has been identified in many areas as taking place in pipes or tunnels. Such piping has been recorded in south west U.S.A., in North Island, New Zealand, in several localities in the United Kingdom (Jones, 1971) and elsewhere. Pipes are usually several centimetres in diameter, they may occur at several levels in the soil, and their location may be dictated by a horizon of low relative stability and low aggregate stability, although animal and rodent burrows may initiate piping in some areas. In distribution, piping tends to occur in soils which have a high silt content, a high percentage of swelling clays, and in areas with periodic high intensity rainfall and a record of devegetation. The location of pipes in a stream bank studied by Jones (1971) is illustrated in Figure 5.15B and although such pipes do not function frequently, they are maintained despite soil creep. The pipes must contribute significantly to water flow in the soil and hence to the discharge of streamflow from the basin in which pipes occur.

Whereas Horton envisaged two forms of water flow, flow in the soil provides an additional component or components of streamflow. The implication of these results is that soil moisture will concentrate in certain areas, termed seepage lines by Bunting (1961) and these may be areas at the head of the stream network, in depressions, and in areas adjacent to stream channels. In some of these areas water concentration may be effected by flow along pipes. Thus the dynamic drainage network (pp. 86–7) receives an added dimension from the soil character and this is illustrated by the map of extensions to the drainage net in an area of Jutland (Figure 5.15C) studied by Bunting (1961). Seepage lines or percolines are the paths along which moisture becomes concentrated, particularly where soils are relatively deep and represent a distinct phase, and the seepage lines form a dendritic pattern related to the pattern of surface stream courses (Figure 5.15C). Jones (1971) has therefore argued that the conceptual model of the drainage basin system should be revised to incorporate channel net, pipe net, pseudopipe net (for example, soil cracks and root channels in the soil), zone of diffuse seepage, zone of diffuse groundwater flow, and groundwater spring net.

The way in which the soils of an area receive, store and transmit water will also influence the production of sediment from the soil and its transport to the permanent stream channels. Subsurface flow, although not fully evaluated for sediment and solute production, is capable of transporting solutes and in this way agricultural fertilisers as well as the inherent composition of the soil may affect the composition of soil drainage. Particularly on slopes, lateral movements of solutes will occur and especially where there are pipes, mechanical eluviation of soils may find its counterpart in the movement of mineral matter suspended in the water draining laterally through the soil. Sheet and channel erosion have long been recognised as two important ways in which sediment is produced from the soil. Sheet erosion embraces the disturbance of particles by raindrop impact and the removal of fine particles by overland flow, and its extent will depend upon the rainfall characteristics (p. 184), the nature of the soil and particularly its erodibility (pp. 65–7) together with local topographic, vegetation and land use characteristics. These factors are basic to the development of the Universal Soil Loss equation (p. 188). Channel erosion connotes the way in

which sediment can be released and removed from banks and beds of existing channels and from soil surfaces as the drainage network extends areally. Channel erosion
has therefore been classified areally into rill, gully and streambank erosion. The
energy of the water in relation to the physical properties of channel materials will
dictate the extent to which erosion occurs, and erodibility of the soils and their
cohesiveness reflecting mechanical, chemical and organic matter content will exercise
a significant influence. The significance of several soil properties such as ease of
dispersion (index D), infiltration capacity of the soil surface (A), permeability of the
soil profile (P), and size of soil particles (p) in relation to erosion (E) has been recognised in various established equations such as that of Baver (1956) who expressed

$$E = K \cdot \frac{D}{APp}$$ where K is a proportionality constant.

5.4d Vegetation and land use

The production of sediment by erosion will necessarily depend upon the vegetation
cover as this determines the exposure of the soil surface, the extent to which the upper
soil horizons are bound by roots, and the magnitude of the modification of precipitation characteristics received at the surface. The significance of vegetation in the drainage basin can thus be visualised in its influence over modifying net input to the basin,
through interception and evapotranspiration; as affecting storage in the basin; and as
influencing the rate at which water and sediment are produced and transmitted
through the basin system. These influences of vegetation need to be understood by the
geomorphologist as a background to the basin. They are particularly pertinent where
land use changes have occurred, often producing significant effects upon drainage
basin dynamics (pp. 345–50), and more detailed treatments can be found in Penman
(1963) and in Sopper and Lull (1967).

The nature of the vegetation cover determines the proportion of gross precipitation
that can reach the ground surface as net precipitation. This is dependent upon interception which embraces several components. Initially water may be retained on
vegetative surfaces as *interception loss* and this water is either absorbed or evaporated.
Subsequently during a particular rainstorm, water may drip from leaves and stems
and this together with the water that passes through the available spaces is characterised as *throughfall*. Some water however may pass over the leaves or stems and flow
down tree trunks to reach the ground surface as *stemflow*. The amount of water
retained by each of these components varies with the species content, age and density
and thus with seasons of the year, as well as with precipitation characteristics and
preceding conditions. Variation in a particular area will also occur during an individual storm event because whereas initially interception losses may be quite high, as
the storm continues the cover will subsequently effect less interception of the rain
falling. These characteristic features must be considered when point values of interception are considered but the general balance is illustrated by average rainfall interception in the rainforest of southern Brazil (Table 5.10A) and some typical values for
interception under different cover types are sketched in Table 5.10B. Analysis of
interception by particular covers have been made and variations in throughfall
according to forest community in Obergurgl, Tyrol (Aulitzky, 1967) are illustrated in
Figure 5.16A.

Table 5.10

A Approximate average rainfall interception, Evergreen rainforest of Brazil (*Freise, 1936*)

Penetrating to rain gauge at 1·5 m	33%	
Evaporated directly from tree crowns	20%	
Running down trunks 46% → evaporated from surface	9·2%	
absorbed by bark	9·2%	
reaching base of trees	27·6% → absorbed by roots	20·7%
	reaching water table	6·9%

B Examples of rainfall interception according to different covers (*from values quoted by Lull, 1964; Todd, 1970; Penman, 1963*)

Type	Rainfall	Interception (%)
Grass—Little bluestem	25·9 mm/hour	50–60
Big bluestem	25·9 mm/hour	57
Buffalo grass	25·9 mm/hour	31

	Gross interception (%)	Stemflow (%)	Net interception (%)
Trees—Northern hardwood	20 (17 without leaves)	5(10)	15(7)
Aspen—birch	15 (12)	5(8)	10(4)
Spruce—spruce-fir	35	3	32
White pine	30	4	26
Hemlock	30	2	28
Red pine	32	3	29

	Interception (%)	
	During growing season	During low vegetation development
Crops—Corn	15·5	3·4
Oats	6·9	3·1
Clover	40	—

   The presence of vegetation cover can account for a difference between gross and net precipitation of as much as one third, and in addition differences have been detected between the solute content of rainfall in the open and that under forest. However the precipitation which reaches the ground surface is further influenced by vegetation through its effect upon infiltration and upon evapotranspiration. Infiltration is encouraged by the presence of vegetation which resists movement of water over the surface and observed infiltration rates may be lowest in fallow land, may increase progressively through crops and pasture to woodland, and under undisturbed forest cover infiltration rate can be maintained at a maximum. The extent of infiltration is facilitated not only by the resistance afforded but also by the drainage lines which vegetation provides in the soil, and along which infiltration can occur. Water is lost to the atmosphere by transpiration of water from plants and by evaporation from plant surfaces and these collectively are termed evapotranspiration. The extent depends upon time of year and of day, character of the vegetation cover, availability of soil moisture, and nature of the air—especially air temperature, humidity, wind

vegetation and land use

Variations in interception indicated by throughfall according to age and character of forest cover at 2000 m in the Austrian Tyrol (Obergurgl) are indicated in A (Aulitzky, 1967), and the significance of drainage (B1) and forest removal (B2) upon water levels in an experimental area of Denmark is shown (Holstenet-Jorgensen, 1967). The effect of different land use covers upon peak discharge is shown (C) by plotting peak discharges for the Severn catchment (870 ha, 2/3 forest) against those for the Wye catchment (1055 ha, sheep grazing) in the Plynlimon Study (Institute of Hydrology, 1971–2) (see pp. 177–8). D shows the increase in runoff according to years since treatment in the Coweeta Watersheds (Hewlett and Hibbert, 1961). E indicates possible differences in hydrograph form according to land-use cover, and F indicates sediment yields for basins in the Serra do Mar Brazil (Haggett, 1961).

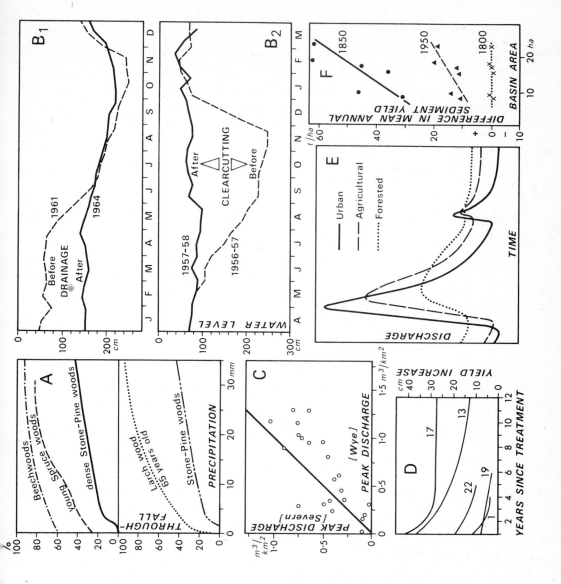

speed, and amount of light. Evapotranspiration can range over a number of values for different plants but for some trees can be as much as some 5 mm per day.

Vegetation cover thus produces losses to, and restraints upon, water movements through the basin. Losses are sustained by interception and by evapotranspiration which reduce the amount of water available, and restraints are placed upon the rate at which water moves in the drainage basin so that stemflow reduces the rate at which water reaches the ground surface. In addition the rate of infiltration will vary according to different types of vegetation cover. The combined effect of vegetation is most readily observed after changes have taken place and changes in ground water levels in an experimental area of Denmark (Figure 5.16B2) illustrate the influence which removal of forest cover exercised upon storage. Before clear cutting water levels varied annually over a substantial range and were low during the latter part of the year reflecting evapotranspiration by the beech stand approximately 75 years old. After clear-cutting however this annual variation did not occur and water levels maintained a fairly constant position (Figure 5.16B2). Change of vegetation cover is also often associated with a change, direct or indirect, in the nature of drainage. In the Spring of 1962 the Danish Forest Experimental Station established a drainage experiment in a 90-year old beech stand (Holstener-Jorgensen, 1967). Drain pipes were installed at intervals of 10 m and laid at a depth of 150 cm. The effect of drainage on the water table was shown by the mean curve for five wells which indicates a marked lowering of the winter levels but little difference was found during the growing season (Figure 5.16B1). This reflects the tendency of the drainpipes to convey moisture which was formerly retained by vegetation and soil and conveyed to groundwater storage.

The effect of vegetation cover is thus especially notable when cover changes have occurred and such changes have produced even more dramatic consequences for sediment production. The solutes produced from forested contrast with those derived from non-forested areas (p. 321) and mechanical erosion has generally been thought to be slower under natural forest than under grassland which is in turn slower than that in areas with sparse vegetative cover. This sequence obtains through the effect of raindrop erosion whereby the impact of rain falling on bare soil breaks down the aggregates, and detaches particles which may be moved downslope in suspension. Subsequently the flow of rainwater as sheets or in rills can remove vegetative residues and material from the litter layer. In this way the vegetation cover affects the stability of slopes in the basin and it can equally exert a binding influence on channel banks. The beginnings of soil conservation movements in the twentieth century were concerned with the significance of land use changes and therefore used slogans such as 'stop the little raindrops where they fall' and 'erosion begins at the top of a hill'. Several studies have demonstrated how land use character, especially cover density, influences sediment production. In a study of Tobacco watershed, Michigan, Striffler (1965) found that the ratio of sediment production varied according to land use type in the ratio 30 from cultivated land, 24 from pasture land, 5 from forest land, and 1 from wild land which included swamps and bogs. Although sediment yields are especially related to eroding banks, certain vegetation covers promote active extension and erosion of channels (pp. 345–8) and the maximum values of sediment production have been detected where all vegetation cover has been removed during building construction. Changes necessarily occur over time, and in the Serro do Mar of Brazil Haggett (1961) suggested the implications of a land use cycle from primary forest (*c.* 1800), through crop and pasture (*c.* 1850) to secondary forest (*c.* 1900) for the difference in sediment yield between a modified and a control watershed (Figure 5.16F).

Similar interest has necessarily been developed in the differences in water yield between catchments because these can give an indication of the significance of vegetation cover. Many methods have been evolved but a comparison of peak flows for two watersheds can be revealing. In the Plynlimon catchment (pp. 177–8) comparison of the peak discharges from the Severn with two thirds forest are generally lower than those from the Wye under mountain grassland (Figure 5.16C). In general if other characteristics are constant the pattern of streamflow from forested areas should show less difference between peak and low flows than is the case from a water-shed covered by other forms of vegetation. Grassland and agricultural land may exhibit a greater range of flows and an earlier and more rapid hydrograph rise (Figure 5.16E). As the extent of vegetation cover is reduced the time of rise and lag time will be reduced, the peak discharge rate increased, and the recession will become more rapid. The lowest low flow values are found from urban areas (Figure 5.16E). A similar feature characterises sediment production which increases to a maximum as vegetation cover is reduced but urban areas can reduce sediment delivery to values lower than those of forest (pp. 364–5).

Temporal variations of runoff will also occur in relation to changes of vegetation cover and this is illustrated in Figure 5.16D by controlled experiments conducted at the Coweeta Hydrologic Laboratory. After cutting the mature hardwood forest, increases in water yield were recorded but these decreased in subsequent years as regrowth occurred (Figure 5.16D). Many other experiments have been conducted to show what effects land cover changes or watershed management can produce. Although the geomorphologist is not primarily concerned with the hydrological implications of such changes, he must be aware of them because of the attendant changes in runoff patterns which can influence sediment production and in this way have a feedback effect upon basin characteristics.

5.5 Drainage basin mechanics

That drainage basin characteristics influence water and sediment yield from a drainage basin is long-established, it was explicitly formulated by Horton (1932, 1945), and it has been the subject of a considerable number of analyses. A variety of indices expressing water and sediment yield from a drainage basin have been related to climatic and to basin characteristics, these relations have often been expressed by multivariate techniques, and some examples are included in Table 5.11. The need to utilise several parameters of climate and basin character arises because they are all inter-related and unless all other variables are constant no single variable will account for a large percentage of the variation in water and sediment yield, although area and drainage density are amongst the most frequently included independent basin variables, in multiple regression studies. Indices of streamflow or of sediment yield can be predicted from such relationships but in addition a knowledge of the hydrograph form is required as well as knowledge of the parameters of the hydrograph expected from ungauged areas. Such knowledge can be provided by the Synthetic Unit Hydrograph which is a dimensionless unit hydrograph, derived using the data from a

Table 5.11 Examples of relationships between basin characteristics and basin response

Index of response	Related to	Area	Source
Streamflow characteristic (Y) 71 indices of streamflow used for each station including 2 for low flow, 3 for duration of daily flows, 6 for flood peaks, 8 for flood volumes, 13 for annual and monthly means $Y = aA^{b1}S^{b2}L^{b3}St^{b4}E^{b5}I_{24.2}^{b6}P^{b7}S_n^{b8}F^{b9}Si^{b10}$	Drainage area (A) Average slope of mean channel between points 10 and 85% of distance upstream from gauging site (S) Main channel length (L) Percentage of total area in lakes, ponds, swamps (St) Elevation (E) Percentage of area under forest (F) Values of potential maximum infiltration in inches during annual flood under average soil conditions (Si) Mean annual precipitation (P) Snow index (S_n) Maximum 24-hour precipitation expected to be exceeded once every 2 years ($I_{24.2}$)	U.S.A. Four areas: Eastern (Potomac) Central (Kansas) Southern (Louisiana) Western (California)	Thomas and Benson (1970) (see Table 6.4 for examples)
Runoff per unit area (M)	Drainage density (Dd) Percentage of swamped area (P) $M = \dfrac{0.24}{P^{0.23}} Dd$	U.S.S.R.	Chebotarev (1966)
Runoff/rainfall ratio (R)	Area (A) Basin length (LB) Relief (H) Perimeter (P) $R = -95.604 + 29.238A - 15.36LB + 0.184H - 9.377P$	New Zealand	Taylor (1967)
Hydrograph time (Base)	Main stream length Distance from outlet to centre Channel roughness Main stream slope	Australia New South Wales	Cordery (1967)

Table 5.11—*continued*

Index of response	Related to	Area	Source
Lag time (L_t)	Length from outlet to centre of gravity of source area L_{sa} Average width of source area W_{sa} Average slope of source area S_{sa} Drainage density (Dd) Where source area = half of watershed with highest land slope $$L_t = 23\left(\frac{\sqrt{L_{sa}+W_{sa}}}{S_{sa}\sqrt{Dd^{0.65}}}\right)$$	U.S.A. Southwest	Hickok, Keppel, Rafferty (1959)
Low flow per unit area	Percent of catchment cleared of trees and brush	U.S.A. Virginia	Riggs (1965)
Annual sediment yield	Precipitation Intensity Average annual runoff Erosion factor Annual precipitation Area Average slope	U.S.A.	Kohler (1954)
Reservoir sedimentation	Slope factor Age Gross erosion Capacity—inflow ratio Non-incised channel density Watershed shape	U.S.A. Illinois	Stall and Bartelli (1959)

number of gauged watersheds to give a form applicable to ungauged watersheds but for which basin characteristics and climatic variables are known. Gray (1962) reviewed the existing methods available and employed a two parameter gamma distribution to derive synthetic unit hydrographs from data from 42 watersheds including expressions of drainage area, length of main stream, length to centre of area, slope of main stream and mean land slope.

Although relations between input and process and between process and basin character have been derived for a large number of areas, the significance of certain parameters does not appear to obtain in the same way universally. This result may be expected partly from the nature of different areas and the inputs they experience. In Finland for example Mustonen (1967) has suggested that climatologic variables such as seasonal precipitation and mean annual temperature, are more significant than basin characteristics in determining stream flow parameters and he ascribed this to the fact that heavy rainfalls are seldom experienced in Finland. This type of variation is one reason for spatial variations (Chapter 6) but a further consideration

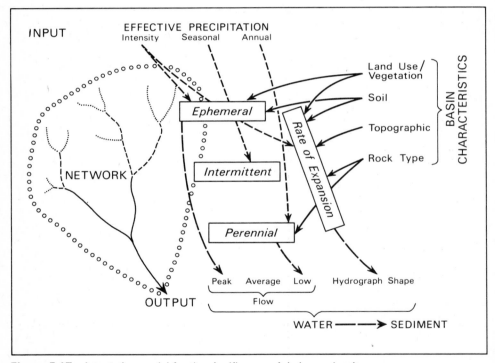

Figure 5.17 A tentative model for the significance of drainage density
The diagram endeavours to indicate how the drainage network may be expressed in several ways, each way having a particular significance for basin process, and each affected by input and by other basin characteristics. The several expressions of drainage density are also related to particular indices of output from the drainage basin and some of these are indicated. No attempt is made to indicate all possible linkages and a similar model could be designed for all other basin characteristics in view of the dynamic nature of the drainage basin (pp. 86, 286).

is that certain basin characteristics influence some indices of basin response more than others. Thus rock type has a profound influence upon low flows whereas vegetation and soil type can exercise an influence particularly upon peak streamflow and upon sediment yield. Thus all drainage basin characteristics may affect all aspects of process but the rank order of the significance of the characteristics varies according to the particular process parameter being considered. For geomorphological purposes this can be illustrated by the drainage network which is composed of several elements, indicated as perennial, intermittent and ephemeral in Figure 5.17, although the pipe network could also be included. Some methods of expressing drainage density (pp. 45–8) approximate to the total network while others are closer to the perennial net and a further characteristic of the net arises from its dynamic nature and can be expressed as Rate of Expansion (Figure 5.17). The major ways in which indices of input, measures of basin characteristics, and parameters of basin response are related to these elements of the drainage network are outlined in Figure 5.17 and the same type of model could be constructed to explain the significance of other basin characteristics. Indeed Walling (1971) in a study of small catchments in south-east Devon has distinguished dynamic characteristics and has also suggested the characteristics which influence the various expressions of streamflow (Table 5.12).

Table 5.12 Streamflow parameters for instrumented catchments in southeast Devon and their associated dominant influencing catchment characteristics (*after Walling, 1971*)

Streamflow parameter	Dominant influencing catchment characteristics
Water balance components	
a Annual runoff	Topographic aspect, vegetation
b Monthly runoff	Rock type, vegetation
Flow duration curves	
a Variability	Rock type, vegetation
b Minimum levels	Rock type, vegetation, soil
c Maximum levels	Rock type, topography, soil
Runoff components	
a Component values	Rock type, vegetation, soil
b Response ratio	Rock type, vegetation, soil
c Quickflow duration curves	Rock type, aspect
Depletion curves	Rock type, soil
Rainfall/runoff relations	
a Quickflow volume	*Dynamic characteristics, Rock type, soil, vegetation
b Hydrograph rise	*Dynamic characteristics, topography, soil
Hydrograph form	*Dynamic characteristics

*Dynamic characteristics connote the way in which varying proportions of the catchment influence stream discharge; the extent of the contributing area will depend upon preceding and prevailing moisture conditions

Such conclusions arise from the fact that a drainage basin is not easily and realistically represented as the sum total of a number of average basin characteristics such as area, relief, shape, soil and vegetation. Rather it is composed a number of response units each possessing particular characteristics and each contributing to water and sediment production according to these characteristics; depending upon the position of the response unit in the watershed, the preceding conditions, and the nature of the input received during a specific time period. Appreciation of the significance of the response units depends upon the way in which water and sediment is produced in the drainage basin. As noted elsewhere (pp. 28- 31) not all of the basin produces water and sediment and the T.V.A. (1965) showed how the runoff-producing area varied during a storm and this variation is partly reflected in the changing drainage density.

Throughflow is now recorded from many basins and this may be one reason for the dynamic nature of basin characteristics but in addition runoff has been seen as the product of expansion of saturated zones along valley floors and over the lower portions of adjacent slopes. This mechanism was supported by Dunne and Black (1970a, b) who concluded that the major portion of storm runoff was produced as overland flow in small saturated areas close to streams, whereas the remainder of the catchment area acted mainly as a storage reservoir during storms, supplied base flow between the storm events, and maintained the wet areas that produce storm runoff. A slightly different view proposed by Ragan (1967) visualised runoff as due to two main mechanisms. Firstly, near the stream, patches of wet ground show discharge from seeps during storm events as a result of direct interception of rainfall and an acceleration of seepage rates; secondly a sudden temporary rise in the water table near the stream network promotes an extra thickness of saturated flow entering the channel as seepage and this in turn increases the velocity of groundwater flow into channels.

Some theories of drainage basin dynamics have emerged from studies of the inter-relationships between form and process and Strahler (1958) outlined the seven variables controlling drainage density and reduced them to four by use of the pi theorem (see pp. 360–2). In a different way Melton (1958), from a study of 15 morpho-metric, surficial and climatic variables of small basins in the south-west U.S.A. grouped the variables as a basis for the identification of variable systems. Such systems expressed the pattern and nature of interrelationships and involve a view of the drainage basin in a steady state such that an input to the system will be accommodated to minimise the effect of the change. Such a view of adjustment is paralleled by the quasi-equilibrium developed in the case of hydraulic geometry and by the concept of least work advocated in the study of channel patterns. It thus appears that drainage basin form and process relations have much in common with, and cannot be regarded in isolation from, relations between form and process in the component channel pattern and channel cross-section subsystems. The pattern of world variation (Chapter 6) and of changes in time (7) need to be viewed against this background.

Selected reading

Stream Channel processes and fluvial sediments are reviewed in:

J. R. L. ALLEN (1965), A review of the origin and characteristics of recent alluvial sediments, *Sedimentology*, **5**, 89–191.
M. A. CARSON (1971), *The mechanics of erosion*, 15–63.

Fundamental reviews of hydraulic geometry and of channel patterns are contained in:

L. B. LEOPOLD and T. MADDOCK (1953), The hydraulic geometry of stream channels and some physiographic implications, *U.S. Geol. Surv. Prof. Paper*, **252**, 1–56.
L. B. LEOPOLD and M. G. WOLMAN (1957), River channel patterns, *U.S. Geol. Surv. Prof. Paper*, **282B**.

Some aspects of drainage basin characteristics are included in:

S. N. DAVIS and R. J. M. DE WIEST (1966), *Hydrogeology.* (564 pp.)

K. J. GREGORY and D. E. WALLING (1968), The variation of drainage density within a catchment, *Int. Ass. Sci. Hydrol. Bull.*, **13**, 61–8.

N. HUDSON (1971), *Soil Conservation.* (320 pp.)

M. J. KIRKBY and R. J. CHORLEY (1967), Throughflow, overland flow and erosion, *Int. Ass. Sci. Hydrol. Bull.*, **12**, 5–21.

H. L. PENMAN (1963), *Vegetation and Hydrology*, Technical Communication no. **53**. Commonwealth Bureau of Soils, Harpenden.

Many relevant aspects are contained in:

Proceedings International Hydrology Symposium, September 6–8 1967, Fort Collins, Colorado, U.S.A.

6   Variations in space

The universal curiosity of man about the world beyond his immediate horizon, a world known to differ in varying degrees from the home area, is the foundation of all geography. *R. Hartshorne, 1959*

Consideration of spatial variation in the operation of fluvial processes necessarily involves areal extrapolation of many of the concepts and ideas concerning the influence of drainage basin characteristics on catchment response, presented in Chapter 5. The juxtaposition of catchments of contrasting topography, rock type, land use and soil characteristics, possibly subject to different meteorologic inputs, will result in areal variations in catchment dynamics. Furthermore, fluvial processes will operate differentially even within an individual watershed, and on a wider scale, climate exerts a general and pronounced influence over the basic pattern of catchment response. Spatial variations can therefore conveniently be considered at several levels, namely, the micro-scale, the meso-scale and the macro-scale—representing local, regional, and continental and worldwide contrasts respectively. The magnitude and nature of any variation is most conveniently and best assessed by consideration of several indices of catchment dynamics which are themselves surrogates for, or indicative of, the pattern of operation of the drainage basin system. Parameters describing the output of runoff, sediment and solutes from a drainage basin have primarily been used for this purpose. In many cases suggestions can be made concerning the factors governing the areal variation of these parameters, but it should be accepted that it is difficult, if not almost impossible, to isolate individual controls within the drainage basin system. Moreover, a distinction must be made between natural variations and those which are primarily man-induced, and this discussion will deal initially with the former and then consider the implications of the latter.

6.1   The micro-scale

Early views of drainage basin dynamics were predominantly based upon the concept of a lumped system in which streamflow was generated uniformly over the watershed, and where the source area was synonymous with the catchment area. In this context, fluvial processes could justifiably be viewed as operating uniformly over the watershed, although channel processes would necessarily be restricted in occurrence. However, observation and experiments have tended to refute such theories and the concepts of Unit Source Areas and Partial and Variable Source Areas have been developed (p. 30). These concepts have particular relevance to the dynamics of runoff production, but they also have important implications for a consideration of the areal variability of drainage basin processes. If runoff is produced preferentially from certain zones of a catchment, and if these zones vary in extent through time, then fluvial processes must operate differentially across a watershed.

In studying small catchments at the Coshocton Experimental Station in Ohio, U.S.A., Amerman (1965) concluded that any watershed larger than 1·2 to 2·0 hectares must be viewed as physically complex and as containing a number of Unit Source Areas, each exhibiting a different response. The subdivisions proposed by him for a 30·8 hectare watershed are indicated in Figure 6.1A. In Figure 6.1B, the isolines of Annual Hydrologic Response (the ratio of direct runoff to annual precipitation) for a 2243 hectare drainage basin in the Coweeta Hydrologic Laboratory, U.S.A., further demonstrate the variations in storm runoff production that may occur within a single basin. The Response varies fivefold across the basin, with increased values reflecting the steeper terrain and thinner soils found at higher elevations.

Contrasts in catchment response between drainage basins contained within a small area have been demonstrated by Walling (1971) in a study of five adjacent watersheds in south-east Devon (Figure 6.1C). Because of the close proximity of the basins, it was assumed that contrasts between them were primarily the result of differences in catchment characteristics rather than meteorological inputs. The range of variation in total runoff, flow components, peak flows, minimum flows and suspended sediment and dissolved load concentrations and yields are portrayed in Figure 6.1C. Detailed analysis of the factors responsible for these contrasts indicated that topography, rock type and land use and vegetation characteristics exerted strong influences and that the individual response parameters were dominated by different controlling factors (Table 5.12).

6.2   The meso-scale: runoff dynamics

At the meso-scale, variations in drainage basin dynamics can be viewed as reflecting the interaction of both catchment characteristics and meteorological factors such as the input of precipitation and solar radiation. The appreciable spatial variations that can be found may be reviewed by considering, firstly, runoff parameters and, secondly, indices of suspended sediment and solute production (Section 6.3).

6.2a   Runoff volumes

The total volume of runoff (mm) or the mean discharge (m³/s) from a catchment is controlled by the relative magnitude of the water balance components within that

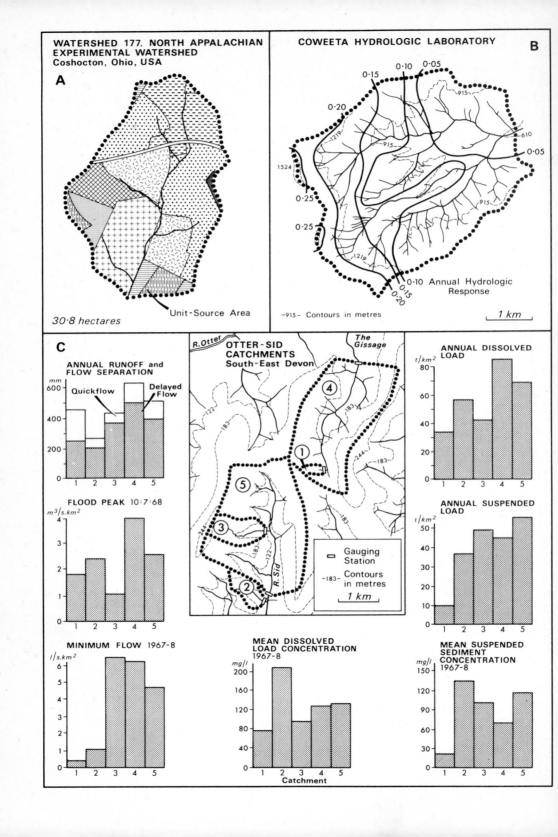

WATERSHED 177. NORTH APPALACHIAN EXPERIMENTAL WATERSHED
Coshocton, Ohio, USA

A

30·8 hectares

Unit-Source Area

COWEETA HYDROLOGIC LABORATORY

B

0·15 0·10 0·05
0·20
0·25
0·25
0·10 Annual Hydrologic Response
0·15
0·20

-915- Contours in metres

1 km

C

ANNUAL RUNOFF and FLOW SEPARATION

mm
600
400
200
0

Quickflow   Delayed Flow

1 2 3 4 5

FLOOD PEAK 10·7·68

m³/s.km²
4
3
2
1
0

1 2 3 4 5

MINIMUM FLOW 1967-8

l/s.km²
6
5
4
3
2
1
0

1 2 3 4 5

R.Otter

OTTER-SID CATCHMENTS
South-East Devon

The Gissage

④
①
⑤
③
②

R.Sid

☐ Gauging Station
-183- Contours in metres

1 km

ANNUAL DISSOLVED LOAD

t/km²
80
60
40
20

1 2 3 4 5

ANNUAL SUSPENDED LOAD

t/km²
50
40
30
20
10

1 2 3 4 5

MEAN DISSOLVED LOAD CONCENTRATION 1967-8

mg/l
200
160
120
80
40

1 2 3 4 5
Catchment

MEAN SUSPENDED SEDIMENT CONCENTRATION 1967-8

mg/l
150
120
90
60
30

1 2 3 4 5

area, or more specifically by the magnitude of precipitation input and evapotranspiration losses. In many areas, including the British Isles, the magnitude of the precipitation and evapotranspiration components closely reflect variations in altitude, with increased altitude resulting in increased precipitation and decreased evapotranspiration consequent upon the reduced temperatures. This occurrence is illustrated clearly and in detail by the work of Maderey (1971) in the Rio Tizar Basin in Mexico (Figure 6.2) and a more general example is provided by the maps of average annual precipitation, evaporation, runoff, and relief for Poland, reproduced in Figure 6.2. Within Poland, the average mean annual runoff of 5·5 l/s.km² comprises extremes greater than 30 l/s.km² in the Tatra Mountains and as low as 1–2 l/s.km² in the central lowland areas.

Although much of the areal variation in runoff volumes can be ascribed to meteorologic controls, which are themselves closely related to altitude, detailed study of the magnitude and extent of the contrasts often reveals the influence of many other physiographic factors which combine with the former to produce intricate patterns of variability in runoff depths. Multivariate statistical analysis has been used in several studies to explain these areal contrasts, by determining the influence and relative importance of a series of independent variables or parameters representing the controls thought to be important. The equations provided by multiple regression analysis (Table 6.1) are analogous to maps in that they present a quantitative description of the magnitude and pattern of variation in the dependent variable. If the extent of spatial variations of the independent variables are known, then inferences about the magnitude of the dependent variable can be made. The various sets of independent variables listed in Table 6.1 range from predominantly meteorological parameters (Equation 1) to one including several physiographic indices (Equation 4).

At a more local scale, vegetation and particularly forest cover exert considerable influence over the total volume of runoff from a catchment. There is abundant evidence, particularly from the United States (Sopper and Lull, 1967), to indicate that the presence of forest is marked by lower runoff. In a classic study at the Stocks Lysimeter in Yorkshire, Law (1956) demonstrated that runoff from a spruce plantation was only 66 per cent of that from adjacent rough pasture. Work in certain areas of Russia has, however, demonstrated the reverse occurrence, and it would appear that under the particular conditions associated with the steppe and forest-steppe zones of this country, the presence of a forest cover can give rise to increases in total runoff. Bochkov (1959) has argued that large forested catchments may show an increase in mean long-term runoff of 20 to 60 per cent on unforested basins, and Rakhmanov (1962) has accounted for such increases in terms of reduced evaporation and increased precipitation beneath the forest (Table 6.2). The general character of the areal pattern of annual runoff volumes is primarily governed by the meteorological inputs into the drainage basin system, whilst the detail is superimposed by the influences of the various physiographic characteristics of an area.

Figure 6.1 Variations in space : the micro-scale
This figure illustrates the unit source areas contained within a 30·8-hectare watershed at Coshocton, Ohio (A) ; the variation in Annual Hydrologic Response across a 2243-hectare drainage basin in the Southern Appalachians, U.S.A. (B) ; and the contrasts in response exhibited by five adjacent small catchments in east Devon (C). (A, after Amerman, 1965 ; B, after Hewlett, 1967).

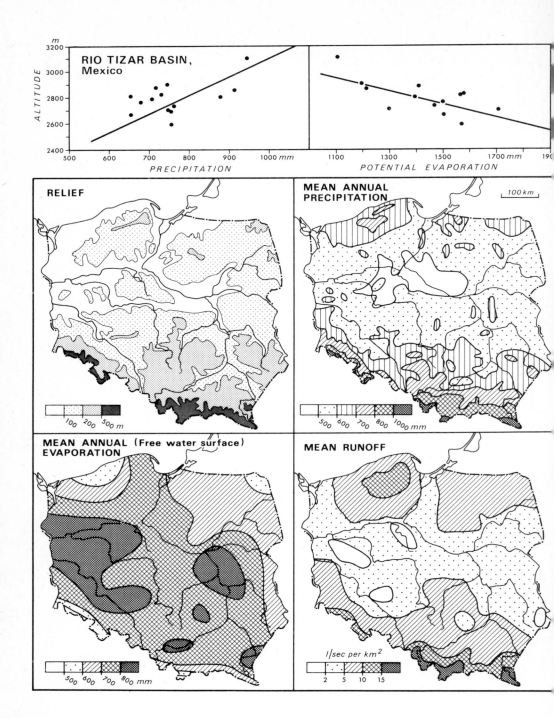

RIO TIZAR BASIN, Mexico

ALTITUDE

PRECIPITATION

POTENTIAL EVAPORATION

RELIEF

100 200 500 m

MEAN ANNUAL PRECIPITATION

100 km

500 600 700 800 1000 mm

MEAN ANNUAL (Free water surface) EVAPORATION

500 600 700 800 mm

MEAN RUNOFF

l/sec per km²

2 5 10 15

Table 6.1 Multivariate relationships between annual runoff volume and mean annual discharge and controlling factors

Worker	Region	Equation
Mustonen (1967)	Finland	$R_a = -11 + 0.83\ P_w + 0.73\ P_f + 0.57\ P_s - 0.21\ \text{PET} - 21T + 0.29\ \Delta SM - 0.99\ FD - 0.77\ VFS + 0.86\ CS$ $(R = 0.94)$
Busby (1964)	Conterminous United States	$R_b = 150 + 0.42\ P - 2.23\ T + 0.083\ S - 0.38\ W + 0.071\ D_p + 0.054\ D_t - 0.008\ d$
Thomas and Benson (1970)	Potomac Basin U.S.A.	$Q_m = 2.89 \times 10^{-4} . A^{1.06} . S^{0.10} . P^{1.87} . S_n^{0.18}$
Taylor (1967)	New Zealand catchments	$Q_m = -11.398 + 0.74\ A - 3.42\ L_b + 0.028\ H + 0.363\ P$ $(R = 0.923)$

where R_a = mean annual runoff (mm)
P_w = winter precipitation (mm)
P_f = autumn precipitation (mm)
P_s = summer precipitation (mm)
PET = potential evapotranspiration in summer (mm)
T = average annual temperature (°C)
ΔSM = change in soil moisture deficit during the year (mm)
FD = frost depth 31 March (cm)
VFS = volume of forest growing stock (m³/ha)
CS = percentage of area with coarse soils

R_b = mean annual runoff (inches)
P = mean annual precipitation (inches)
T = mean annual temperature (°F)
S = mean annual snowfall (inches)
W = average wind velocity (m.p.h.)
D_p = average number of days with measurable precipitation
D_t = average number of days with temperatures of 90°F or more
d = average heating degree days

Q_m = mean annual discharge (cfs)
A = drainage area (mi²)
S = main-channel slope (ft/mi)
P = mean annual precipitation (inches)
S_n = mean annual snowfall (inches)

Q_m = mean annual discharge (cfs)
A = catchment area (mi²)
L_b = maximum basin length
H = total relief
P = perimeter

Figure 6.2 Spatial variation in water balance components
This figure illustrates the variation according to altitude of precipitation and potential evapotranspiration within the Rio Tizar catchment, Mexico (after Maderey, 1971); and the general relationship of relief to the pattern of mean annual precipitation, mean annual evaporation and mean runoff in Poland (after Mikulski, 1963).

Table 6.2 Mean values of the water balance components of forested and non-forested drainage basins in European U.S.S.R., 1949–54

Regions	Maximum degree of forestation (%)	PRECIPITATION (mm) Open terrain	Forested	EVAPORATION (mm) Open terrain	Forested	RUNOFF (mm) Open terrain	Forested
Smolensk	75	550	630	380	340	170	290
Kirov	100	480	550	345	300	135	250
East of the Volga*	30	375	400	300	285	75	115

*Only the central and northern areas.
Source: Rakhmanov (1962)

6.2b   *Flood magnitude*

Many suggestions as to the factors controlling variations in flood intensity in both space and time have appeared in the literature (for example, International Association of Scientific Hydrology, 1969; Rodda, 1969). The very existence of, and need for, these suggestions is ample testimony to the considerable variations that may occur within a region. In a study of the magnitude of floods in the Appalachian Region of the United States, Schneider (1965) produced a map of the magnitude of the 50-year flood adjusted to a catchment area of 259 km², and the values shown ranged from 140 m³/s up to 625 m³/s. The pattern shown by the map (Figure 6.3) indicates a distinct contrast between the Appalachian Plateaux to the north-west and the mountainous Ridge and Valley and Blue Ridge provinces in the south-east. Flood intensities in the former are greater because of the orographic influence of the mountains on the westward movement of moisture over the region and the dendritic character of the Plateaux drainage which results in more rapid flood concentration than in the elongated valleys of the southeast. Superimposed on this basic contrast is a general southerly increase in flood magnitude, such that in the Appalachian Plateaux there is a threefold increase along a transect from Pennsylvania to Alabama, primarily as a result of an increase in annual precipitation from 1000–1100 mm to 1300–1400 mm.

The Appalachian Region provides a useful indication of the manner in which the spatial pattern of flood intensity over a region will reflect both meteorologic and physiographic factors, the former being concerned primarily with the intensity and amount of rainfall and the latter with the storm runoff-producing potential of the watersheds and the routing properties of the channels. A further example is provided by the provisional regional flood frequency map of England and Wales (Figure 6.4), produced by Cole (1966). The regional contrasts in flood magnitude shown on this map can be explained partly in terms of geology and topography, with the northern and western areas or Highland Britain possessing older and more impermeable rocks and greater relief, and partly in terms of the general pattern of storm rainfall which emphasises the former effects. Storm rainfall, as indexed by the 24-hour rainfall total with a recurrence interval of two years (Figure 6.4), ranges from in excess of 76 mm in the western areas to less than 38 mm in the south-east.

Multiple regression equations have been developed by several workers in attempts to relate flood data to controlling independent variables, and examples are listed in Table 6.3. These equations demonstrate the importance of both meteorologic and physiographic variables, and in some areas vegetation cover, particularly the presence or absence of forest cover, can be seen to exert an influence on flood intensity. In large basins the character of the main channel and the presence of lakes and swamps are important in controlling downstream transmission of the flood peak.

Finally, the limited areal extent of many individual flood events must be mentioned. The causative intense rainfall can often be highly localised, and because of this the occurrence of high discharges and the associated geomorphic effects may be restricted in extent. A generalised map of the daily rainfall values associated with the severe flooding of 23 and 24 May 1966, in the Marlborough region of New Zealand is presented in Figure 6.5. In this particular example (New Zealand Ministry of Works, 1966) the intense rainfall lasted 11 to 15 hours and the totals for the associated 24

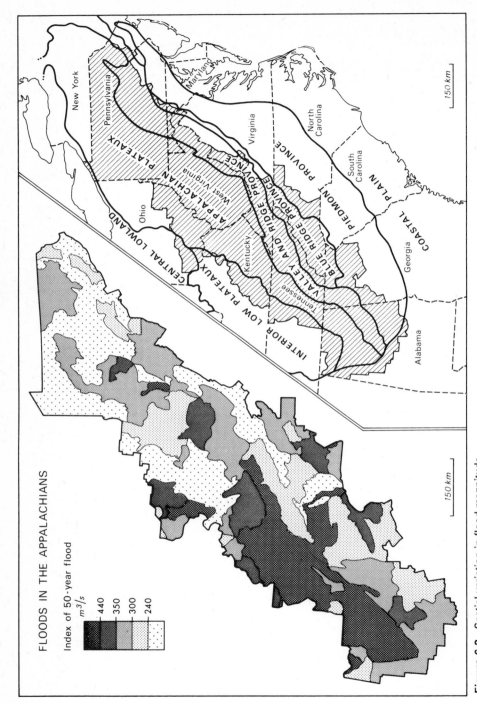

Figure 6.3 Spatial variation in flood magnitude

Flood intensity, expressed as the magnitude of the 50-year flood adjusted to a catchment area of 259 km², varies considerably over the Appalachian Region of the United States, and the pattern partly reflects the different characteristics of the various physiographic provinces (based on Schneider, 1965).

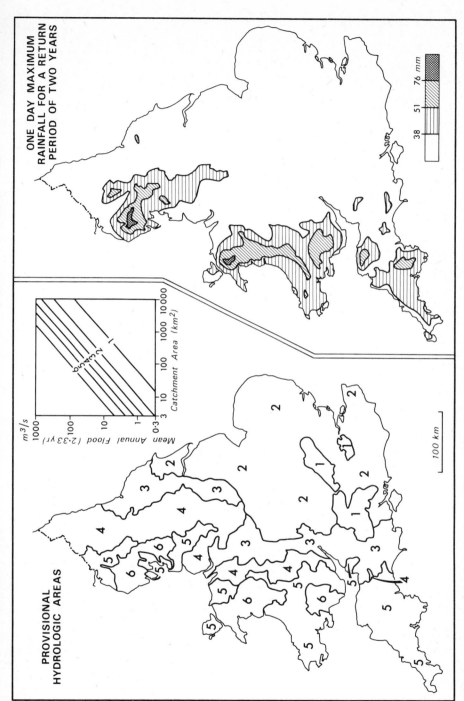

Figure 6.4 Flood and storm rainfall magnitude in England and Wales
Variations in flood magnitude demonstrated by Cole (1966), using regional flood frequency analysis, closely reflect the pattern of storm rainfall prescribed by Rodda (1967).

Table 6.3 Multivariate relationships between flood discharges and controlling factors

Worker	Region	Equation
Thomas and Benson (1970)	Central Valley, California	$Q_{50} = 2 \cdot 31\ A^{0 \cdot 84} \cdot E^{-0 \cdot 48} \cdot F^{1 \cdot 20}$
Wong (1963)	New England	$Q_{2 \cdot 33} = 0 \cdot 096\ S_L{}^{1 \cdot 29} \cdot S_a{}^{0 \cdot 97}$
Nash and Shaw (1966)	Great Britain	$\bar{Q} = 0 \cdot 009\ A^{0 \cdot 85} \cdot R^{2 \cdot 2}$
		$\bar{Q} = 0 \cdot 074\ A^{0 \cdot 75} \cdot S^{1 \cdot 0}$
Rodda (1969)	United Kingdom	$Q_{2 \cdot 33} = 1 \cdot 08\ A^{0 \cdot 77} \cdot R_{2 \cdot 33}{}^{2 \cdot 92} \cdot D^{0 \cdot 81}$

where Q_{50} = annual peak discharge with a recurrence interval of 50 years (cfs)
A = drainage area (mi^2)
E = average basin elevation in 1000 feet above sea-level datum
F = forested area, in per cent of total drainage area

$Q_{2 \cdot 33}$ = mean annual flood (cfs)
S_L = length of main stream (mi)
S_a = average land slope (ft/mi)

$\bar{Q}$ = mean annual maximum discharge (cfs)
A = catchment area (mi^2)
R = mean annual rainfall in inches
S = catchment slope in parts per 1000

$Q_{2 \cdot 33}$ = mean annual flood (cfs)
A = catchment area (mi^2)
$R_{2 \cdot 33}$ = mean annual daily maximum rainfall (inches)
D = drainage density (mi/mi^2)

hour period ranged from a maximum in excess of 450 mm down to only 25 mm, within a distance of 75 km. The major cells of intense rainfall were only about 30 km in diameter. Within the zone affected by the storm, flood discharges were as high as 4·7 m³/s.km² and most of the rivers experienced peaks in excess of 2·2 m³/s.km² and there were reports of considerable streambank erosion, floodplain scouring, gravel movement, and landslips on steep slopes. The Hapuku River changed course and covered 16 hectares of good farmland with gravel. Nevertheless, 80 km away the impact of the storm would have been minimal.

6.2c Minimum flows

Spatial contrasts in low flows are controlled primarily by the rainfall input and storage characteristics of individual drainage basins. The latter influences the extent to which the basin can store water for release during drought periods, and the former deter-mines the volume of water available to recharge this storage, and the magnitude and frequency of drought events. Storage properties are governed by rock type and to a lesser extent by soil type, and can vary markedly within a small region. An indication of the range of low flow discharges that may be encountered within a region is provided by Figure 6.6 which portrays, firstly, the minimum recorded discharge levels for a portion of the county of Devon and, secondly, the average annual low flows of streams in the Appalachian Region of the United States. The flow values range from

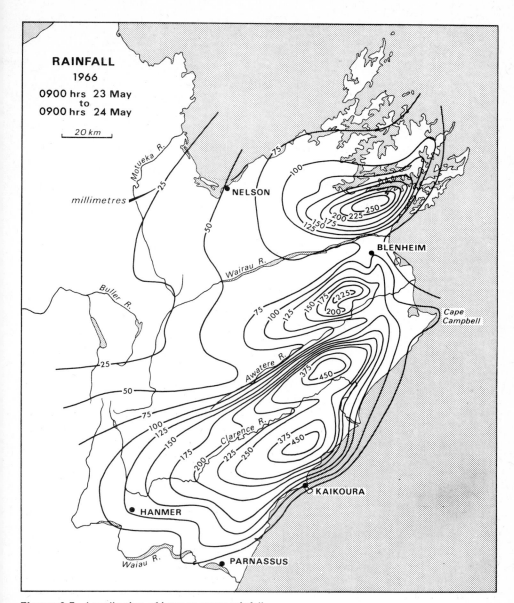

Figure 6.5 Localisation of intense storm rainfall
This figure demonstrates the localised pattern of the storm rainfall responsible for the severe flooding of 23 and 24 May, 1966, in the Marlborough district of New Zealand (based on New Zealand Ministry of Works, 1966).

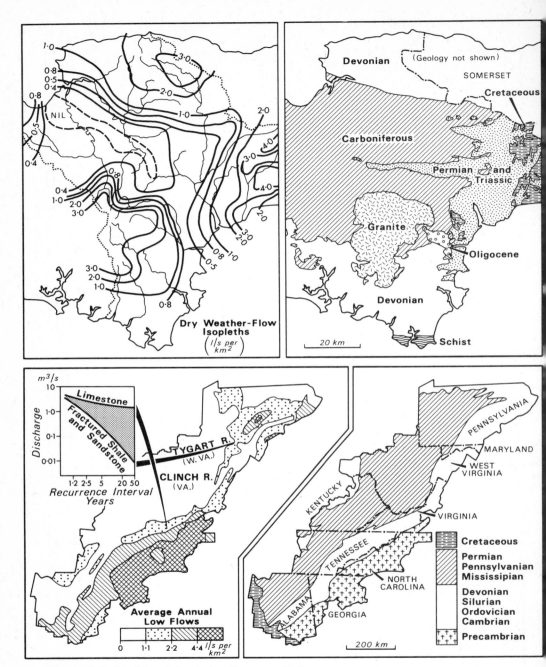

Figure 6.6 Spatial variation in low flow magnitude
The spatial pattern of minimum recorded flows within part of Devon (after Hall, 1967) is closely controlled by the underlying geology. Rock type exerts a similar influence over both average annual low flows and low flow frequency characteristics of streams within the Appalachian Region of the United States (based on Schneider and Friel, 1965).

zero to > 3·0 l/s.km² in Devon (Hall, 1967) and, within the Appalachian Region, from near zero in parts of Kentucky to 13·25 l/s.km² for the French Broad River Basin in North Carolina (Schneider and Friel, 1965).

In Devon, the pattern of low flows reflects the interaction of precipitation, rock type and soil cover. Because rock type is reflected by soil type and relief, which in turn influences the distribution of precipitation, a clear correspondence can be seen between the maps of geology and low flows. The minimum flow levels occur in the lowlands of the south-east and north-west where the annual precipitation is at its lowest (750–950 mm) and where the impermeable rocks sustain very low ground-water contributions to streamflow. Higher values appear in the extreme south-east due partly to the increased rainfall (1000 mm) and, more important, to the outcrop of permeable rocks which include several good aquifers. The maximum flow levels in the south-west and north-east correspond to the upland areas of Dartmoor and Exmoor, where increased rainfall up to 2200 mm and 1800 mm respectively provides a greater source of supply. Similarly, in the Appalachian Region areal contrasts in low flows are largely a reflection of the geologic units. The general pattern is one where the highest yields occur in the areas underlain by Precambrian rocks, moderate yields are provided by the Early Palaeozoic strata, and minimum yields mark the outcrop of Middle and Late Palaeozoic rocks. In this region, rock type also influences the form of the low flow frequency relationship, and in Figure 6.6 a comparison is made between the relationship for the Clinch River, Virginia, which is predominantly underlain by limestones, and that for the Tygart River in W. Virginia, which drains fractured shales and sandstones. A consideration of the pattern of drought flows with a high recurrence interval could be expected to demonstrate further contrasts within this region.

Studies of low flow variability have been carried out in many other areas, and the work of Schneider in the Upper Little Miami Basin, Ohio (1957) and the Swatara Creek Basin, Pennsylvania (1965), Waugh (1970) in Northland, New Zealand and Wright (1970) in the Lothians area of Scotland (Figure 5.14) provide valuable examples. In all these studies geology was found to be of prime importance in explaining the range of minimum flows. However, many other catchment characteristics can be expected to exert an influence on low flow volumes, adding to the complex spatial pattern, and it is again of value to present several multiple regression equations expressing a relationship between minimum flow and certain controlling variables (Table 6.4). Furthermore, it should be noted that in many regions the variations in low flow discharge will be closely related to those of flood flows because, as Giusti (1962) has shown, areas with the lowest minimum flows are frequently those with the highest flood discharges.

6.2d   Runoff components

Although many studies have considered spatial patterns of total runoff, and maximum and minimum flows, much less attention has been given to runoff components or the magnitude of runoff volumes from different origins. This facet of runoff character can be extremely valuable in the context of the operation of fluvial processes within the drainage basin because it provides an indication of the relative importance of surface and subsurface processes. Woodruff and Hewlett (1970) have constructed a

Table 6.4　Multivariate relationships between low flow discharges and controlling factors

Worker	Region	Equation
Thomas and Benson (1970)	Potomac River Basin	$M_{7,2} = 2\cdot74 \times 10^{-8} . A^{1\cdot08} . P^{3\cdot93} . F^{-0\cdot61} . S_i^{2\cdot08}$ $M_{7,20} = 3\cdot43 \times 10^{-4} . A^{0\cdot58} . S^{-1\cdot23} . S_n^{1\cdot53} . S_i^{4\cdot57}$
Thomas and Benson (1970)	Central Valley California	$M_{7,2} = 5\cdot04 \times 10^{-8} . A^{1\cdot31} . I_{24,2}^{6\cdot96} . S_w^{1\cdot00}$ $M_{7,20} = 1\cdot84 \times 10^{-9} . A^{1\cdot68} . I_{24,2}^{6\cdot89} . P^{4\cdot47} . F^{-3\cdot17}$

where $M_{7,2}$ = minimum annual 7-day average flow with a recurrence interval of 2 years (cfs)
　　$M_{7,20}$ = minimum annual 7-day average flow with a recurrence interval of 20 years (cfs)
　　　　A = drainage area (mi²)
　　　　P = mean annual precipitation (in)
　　　　F = forested area, in per cent of total drainage area
　　　　S_i = soils infiltration index (in)
　　　　S = main channel slope (ft/mi)
　　　　S_n = mean annual snowfall (in)
　　$I_{24,2}$ = intensity of 24-hour, 2 year rainfall (in)
　　　　S_w = average 1 April water content of snowpacks (in)

Hydrologic Response map of the eastern United States (Figure 6.7). The Hydrologic Response of a watershed is defined as

$$\frac{\text{Annual quickflow volume}}{\text{Annual precipitation volume}} \times 100\,\%$$

and is therefore indicative of the importance of storm runoff. Within the eastern United States the values exhibit an appreciable range from less than 4 per cent to in excess of 24 per cent. Although the pattern shown on the map parallels the trend of the major topographic and lithologic features within the region, Hewlett and Woodruff were unsuccessful in an attempt to relate the response ratios of ninety constituent basins to a selection of fifteen independent variables describing catchment morphometry and land use. Nevertheless, the values must reflect a complex interaction of morphometric and storage parameters, the latter being very difficult to quantify.

Sharov (1965) has carried out a similar analysis of runoff data for European Russia, although he was primarily concerned with mapping and analysing variations in the coefficient of underground flow defined as

$$\frac{\text{Volume of underground flow}}{\text{Volume of precipitation}} \times 100\,\%$$

The values plotted ranged from 0·2 per cent to 40 per cent, and several trends were distinguished. A general reduction in values from 5 to 10 per cent to 1 per cent in a north-west to south-east direction was ascribed to the reduced precipitation input, and the mountainous zones and areas of karst were associated with increased values. More locally, the depth of dissection and the presence and nature of permafrost exerted an influence.

Other runoff indices could be selected in this attempt to demonstrate spatial contrasts in catchment dynamics, but those which have been considered represent

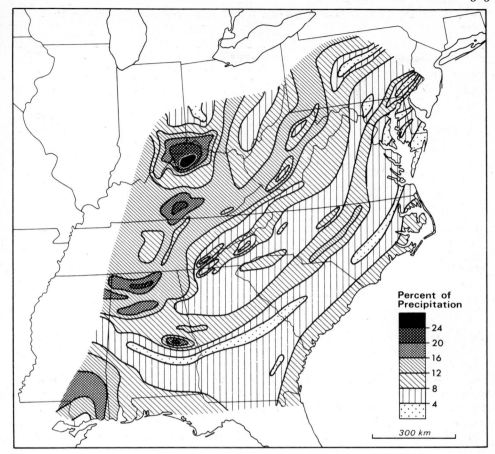

Figure 6.7 Average hydrologic response map of the eastern United States
The Annual Hydrologic Response is defined as the ratio of the annual quickflow volume to the
annual precipitation volume (based on Woodruff and Hewlett, 1970).

the major facets and provide a preliminary and clear indication of the extent of
regional variations. It is more useful to extend this review to include sediment and
solute dynamics.

6.3   The meso-scale: sediment and solute dynamics

Parameters of sediment and solute production reflect the operation of fluvial processes
in a drainage basin much more directly than indices of runoff, because they are
themselves measures of the intensity of denudation. The detailed pattern of sediment
and solute production within an area is the resultant of numerous meteorologic and
physiographic factors, often termed active and passive controls respectively, and
marked spatial contrasts can occur.

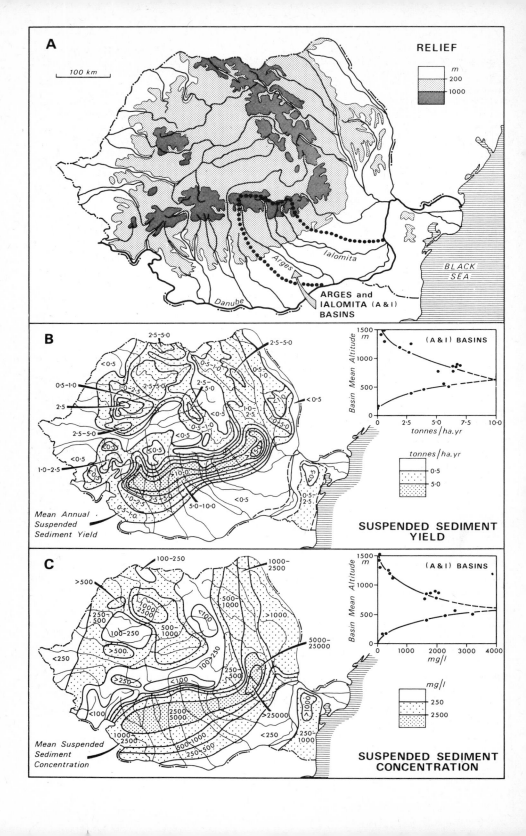

A

RELIEF

100 km

m
200
1000

Ialomita

Arges

Danube

BLACK
SEA

ARGES and
IALOMITA (A & I)
BASINS

B

2·5-5·0

2·5-5·0

<0·5

0·5-1·0

0·5-1·0

0·5-1·0

2·5-5·0

2·5·5·0

2·5

0·5-1·0

1·0-2·5

2·5

<0·5

<0·5

1·0-2·5

0·5-1·0

<0·5

2·5-5·0

<0·5

<0·5

<0·5

<0·5

<0·5

5·0

10·0

1·0-2·5

5·0-10·0

0·5-1·0

2·5-5·0

<0·5

2·5-5·0

Mean Annual
Suspended
Sediment Yield

(A & I) BASINS

Basin Mean Altitude

1500
m

1000

500

0   2·5   5·0   7·5   10·0
tonnes/ha.yr

tonnes/ha.yr

0·5
5·0

**SUSPENDED SEDIMENT
YIELD**

C

100-250

1000-
2500

>500

1000-
2500

250-
500

500-
1000

<100

100-250

500-
1000

>1000

>500

100-250

5000-
25000

<250

250-
500

>250

<100

2500-
5000

>25000

<100

1000-
2500

<100

250-
1000

1000-
2500

500-1000

<250

250-500

Mean Suspended
Sediment
Concentration

(A & I) BASINS

Basin Mean Altitude

1500
m

1000

500

0   1000   2000   3000   4000
mg/l

mg/l

250
2500

**SUSPENDED SEDIMENT
CONCENTRATION**

6.3a Suspended sediment transport

A general example of the extent of countrywide variations in suspended sediment production is provided by a study of the sediment loads of Roumanian rivers carried out by Diaconu (1971). Data from over two hundred sampling stations was analysed, and the mean annual loads were found to vary between 0·02 and 55·4 tonnes/hectare whilst mean concentrations ranged from 25 to 47,200 mg/l; a range of over 1000-fold in both cases. The generalised maps of mean concentration and annual loads demonstrate a clear interrelationship between sediment production and relief (Figure 6.8), although this relationship does not involve a simple linear function. Minimum loads and concentrations occur in both the lowlands and the areas of high altitude, and the maximum values occur in the intermediate areas between 200 and 600 metres, particularly on the southern slopes of the Carpathians. This trend was explained in terms of reduced sediment production occurring in the mountains due to resistant rocks and increased vegetation cover, and in the lowland areas due to the low relief. The detailed relationships between altitude and sediment yield and concentration for the Arges and Ialomita basins, on the southern flank of the Carpathians, are shown in Figure 6.8. The precise form of the relationship varies across the country in response to other controlling factors and Diaconu delimited eleven major regions each with its own sediment-altitude function.

The contrasts in sediment yield between several physiographic areas of the Midwest, U.S.A., and the range of values within the individual areas are shown in Figure 6.9A. In many of the areas, the range is up to 100-fold and clearly demonstrates the influence of meteorologic and physiographic controls in providing contrasts in sediment production. Scale is a very important factor in this context because maps or data for large areas are by necessity generalised and reflect the broad influence of meteorologic conditions and drainage basin characteristics. Within smaller areas detailed study will often reveal a more intricate pattern of variation in response to local conditions. The influence of land use on sediment yield in the Potomac Basin has been demonstrated by Wark and Keller (1963). In this catchment of nearly 40,000 km², areas of forest cover exhibit lower loads than cultivated areas, and from the general relationships developed (Figure 6.9B), an increase in forest cover from 20 per cent to 80 per cent was shown to be associated with a decrease in sediment yield from more than 100 to less than 20 tonnes/km². Similar local variations in sediment transport were described by Schumm (1954) in the San Rafael Swell, Utah and in Land Use District 18 of the Navajo Indian Reservation in Arizona and New Mexico, although in this case sediment production, as measured by accumulation in small stock reservoirs, was found to be clearly influenced by geology (Table 6.5), and relief.

An indication of the several other factors that can influence the spatial pattern of sediment production within a region, and of the extent of the areal variation, is provided by the multiple regression equations obtained by several studies which have

Figure 6.8 The pattern of suspended sediment transport within Roumania
The maps of mean annual suspended sediment yield (B) and mean suspended sediment concentration (C) demonstrate the general influence of relief (A) on suspended sediment production, and the Arges and Ialomita basins provide a particular example of the interrelationship (based on Diaconu, 1971).

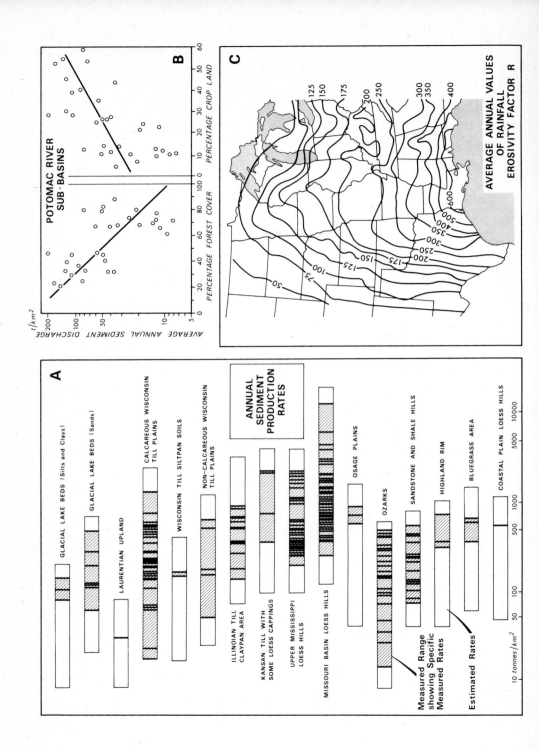

A

ANNUAL SEDIMENT PRODUCTION RATES

GLACIAL LAKE BEDS (Silts and Clays)

GLACIAL LAKE BEDS (Sands)

LAURENTIAN UPLAND

CALCAREOUS WISCONSIN TILL PLAINS

WISCONSIN TILL SILTPAN SOILS

NON-CALCAREOUS WISCONSIN TILL PLAINS

ILLINOIAN TILL CLAYPAN AREA

KANSAN TILL WITH SOME LOESS CAPPINGS

UPPER MISSISSIPPI LOESS HILLS

MISSOURI BASIN LOESS HILLS

OSAGE PLAINS

OZARKS

SANDSTONE AND SHALE HILLS

HIGHLAND RIM

BLUEGRASS AREA

COASTAL PLAIN LOESS HILLS

Measured Range showing Specific Measured Rates

Estimated Rates

10 tonnes/km²   50   100   500   1000   5000   10000

B

POTOMAC RIVER SUB-BASINS

AVERAGE ANNUAL SEDIMENT DISCHARGE   t/km²

200   100   50   10   5

PERCENTAGE FOREST COVER   100   80   60   40   20   0

PERCENTAGE CROP LAND   0   10   20   30   40   50   60

C

AVERAGE ANNUAL VALUES OF RAINFALL EROSIVITY FACTOR R

50   75   100   125   150   175   200   250   300   350   400   600

500   400   350   300   250   200   175   150   125

Table 6.5 The influence of rock type on values of annual sediment production estimated from studies of sedimentation in small reservoirs

Lithology	Sediment loss (m³/km²) Utah	New Mexico–Arizona
Resistant: Conglomerate, limestone and resistant sandstone	143	95–143
Medium: Friable sandstone	571	523
Soft: Shale and gypsum	1237	761

Source: Schumm (1954)

Table 6.6 Multivariate relationships between sediment production and controlling variables

Worker	Region	Equation
Anderson and Wallis (1963)	Pacific Coast, western Oregon and California	Log SS = −4·721 +1·244 log MQ +1·673 log FQ +0·116 log A +0·401 log S +0·0486 SC +0·482 SA + 0·942 R +0·0086 RC +0·0280 BC −0·0036 OC
Roehl (1962)	South-eastern Piedmont, U.S.A.	Log DR = 4·5005 −0·2304 log A −0·5102 colog RL −2·786 log BR
Striffler (1965)	Northern Michigan, U.S.A.	Log SD = 3·831 +1·190 log Q +0·134 log EB −0·003 SA −0·007 SB − 0·003 SC −0·765 RF.

where SS = suspended sediment discharge (tons/mi² . yr)
MQ = mean annual runoff (cfs/mi²)
FQ = discharge peakedness
A = watershed area (mi²)
S = slope of stream of 1 mile mesh length (ft/mi)
SE = silt and clay fraction of topsoil (%)
SA = surface aggregation ratio (%)
R = portion of catchment covered by roads (%)
RC = portion of catchment cutover in last ten years (%)
BC = portion of catchment in bare cultivation (%)
OC = portion of catchment in other cultivation (%)

DR = sediment delivery ratio in per cent of annual gross erosion
A = watershed area (mi²)
RL = ratio of basin relief to average stream length
BR = bifurcation ratio

SD = sediment delivery rate (lb/mi² . day)
Q = stream discharge (cfs/mi²)
EB = total length of eroding banks (ft ×10²)
SA = portion of watershed covered by Soil Association (a)
SB = portion of watershed covered by Soil Association (b)
SC = portion of watershed covered by Soil Association (c)
RF = rising or falling stage (rising = 1, falling = 2)

Figure 6.9 Spatial variation in suspended sediment production
(A) indicates the extent of variation in suspended sediment yields both within and between several physiographic areas of the midwestern United States; (B) demonstrates the influence of land use on the annual sediment discharge of the Potomac River sub-basins; and (C) illustrates the areal variation of rainfall erosion potential over part of the midwestern United States. (Based (A) on Brune, 1951; (B) on Wark and Keller, 1963; and (C) on A.S.C.E., 1970.)

attempted to derive statistical relationships between sediment yields and influencing factors. Several of these equations are listed in Table 6.6. Methods designed for evaluation or prediction of sediment yield from a semi-quantitative consideration of terrain characteristics (for example, Shown, 1970) also provide useful background material.

6.3b   Soil loss

Detailed studies have also been undertaken on spatial contrasts in the rate of soil loss by sheet erosion from small areas, particularly erosion plots. The work of the United States Department of Agriculture, Soil and Water Conservation Research Division is outstanding in this context in that it has collected data from throughout the United States and has used this to develop the Universal Soil Loss Equation (F.A.O., 1965). This equation incorporates over 10,000 plot years of data from 1200 field plots located at 47 research stations in 24 states, and relates rate of soil loss to several controlling factors through an easily-applied prediction formula, which takes the form:

$$\text{where} \quad A = R.K.L.S.C.P.$$
$$A = \text{average annual soil loss}$$
$$R = \text{rainfall factor}$$
$$K = \text{soil erodibility factor}$$
$$L = \text{length of slope factor}$$
$$S = \text{steepness of slope factor}$$
$$C = \text{cropping and management factor}$$
$$P = \text{supporting conservation practice factor.}$$

Again, the controls can be divided into two major types, firstly, meteorologic or active influences (R) and, secondly, local condition or passive influences (K, L, S, C and P).

The rainfall factor, R, is a measure of the rainfall energy or the capability of the local rainfall to erode soil from an unprotected field, and has been described in detail by Wischmeier and Smith (1958). It is derived on a storm basis as the product of the kinetic energy of the storm and its maximum 30-minute intensity, and annual values are obtained by summing the individual values for all storms with rainfall in excess of 12 mm (p. 188). Research has shown that this factor alone explains from 72–97 per cent of the variation in individual storm soil losses from cultivated fallow soils in several U.S. states. The annual R values range from less than 50 in the western semiarid plain areas to more than 600 in the south-eastern Gulf States and values for part of the midwestern United States are shown in Figure 6.9C. The five remaining factors in the formula incorporate the influence of the physical characteristics of the land surface on the response to the erosion potential of the rainfall, by expressing the ratio of expected soil loss for a particular condition to that from an arbitrary selected standard. The general approach offered by the Universal Soil Loss Equation in describing and accounting for areal variations in soil loss has also been applied successfully in Roumania (Mircea, 1970) and in Czechoslovakia (Pretl, 1970).

6.3c   Dissolved load transport

There are many examples of the extent to which dissolved loads and solute concentrations of streams can vary within a region, and the rivers of Kazakhstan in the U.S.S.R.

studied by Pavelko and Tarasov (1967) may usefully be cited. Analysis of data from 290 stations where samples had been taken for over 25 years indicated that the weighted mean concentrations ranged between 50 and 1400 mg/l whilst the maximum annual mean values ranged between 100 and > 8000 mg/l. More specifically, variations can also be found in the ionic composition of the solutes and Douglas (1968), studying 24 catchments in north-east Queensland, found that, whereas mean total dissolved solids contents varied from 29–69 mg/l, a range of approximately 2·3, individual ions exhibited much greater contrasts. For example, calcium concentrations varied from 0·34 to 3·09 mg/l and silica from 3–11 mg/l. In general, however, solute loads and concentrations exhibit less areal variation than suspended sediment values.

The detailed pattern of solute transport reflects many controls, but the precipitation input and the geological and pedological characteristics of a drainage basin would seem to be the major influences. The nature of the precipitation input is important both from the viewpoint of quantity and quality. In many catchments, an increase in precipitation could be thought of as diluting the solutes released by weathering processes to provide lower stream solute concentrations. Quality aspects, however, exert a more direct influence in that material dissolved in the precipitation can contribute directly to the solute load of a stream, and in areas of low stream solute loads it can provide a major source of dissolved material. Furthermore, the chemical quality of incident precipitation must be important in the context of the nature and intensity of chemical weathering processes.

The content of cyclic salts in precipitation at a single station can vary in response to air mass conditions (Gorham, 1955), and Kriventsov (1961) working in the Rostov Province of the U.S.S.R. between 1952 and 1955 reported variations from 16·5 to 194 mg/l, depending on wind direction. Similarly, marked spatial contrasts in chemical quality of precipitation can occur. Matveyev and Bashmakova (1967) produced maps of the average solute content of precipitation over European Russia and these showed a range of 10–60 mg/l with the associated precipitation of dissolved material varying from 5–50 tonnes/km². A more detailed study of the content of particular ions in precipitation has been made by Junge and Werby (1958) for the United States. Maps of the annual average concentration of various inorganic ions in rainwater were produced. Chloride concentrations were shown to be as high as 8·0 mg/l near the coast, to decrease rapidly inland, and to level off about 800 km inland to values as low as 0·2 mg/l. Calcium concentrations exhibited a reverse trend with maximum concentrations of about 3·0 mg/l occurring in the central and south-west inland areas and minimum values of 0·3 mg/l occurring on the eastern and western coasts. The pattern of chloride concentrations was ascribed primarily to marine influences because a large proportion of the content of this ion is derived from sea salts and because atmospheric vertical mixing reduces the concentration of sea spray in the lower troposphere over inland areas. The increased calcium concentrations in inland areas were the result of dry ground surface conditions and the occurrence of dust storms in the dry interior regions.

It has long been recognised that the chemical and physical properties of the rock and soil beneath a drainage basin exert a strong control over the solute content of streamwater and as early as 1924 Clarke (1924) produced convincing figures from Bohemian rivers and streams which demonstrated the contrasts in solute content

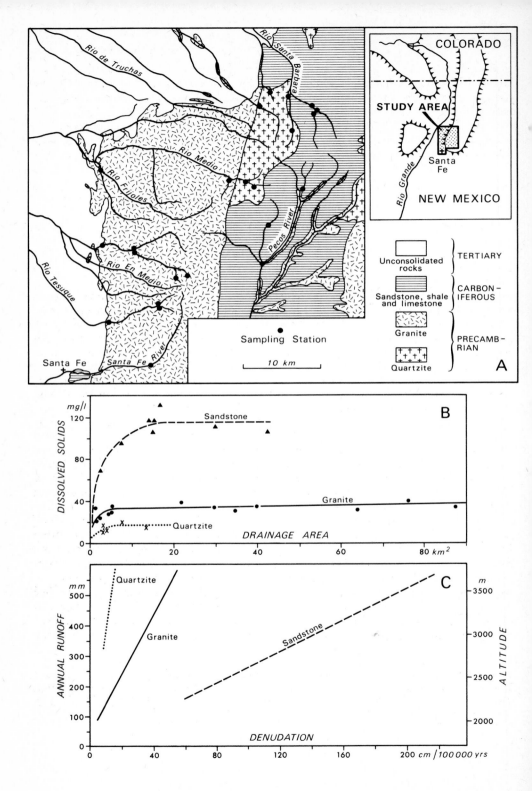

Rio de Truchas

Rio Santa Barbara

COLORADO

STUDY AREA

Rio Medio

Rio Friijoles

Santa Fe

NEW MEXICO

Rio Grande

Pecos River

Rio En Medio

Rio Tesuque

Santa Fe   Santa Fe River

Sampling Station

10 km

Unconsolidated
rocks } TERTIARY

Sandstone, shale
and limestone } CARBON-
IFEROUS

Granite } PRECAMB-
RIAN

Quartzite

A

mg/l

120

80

40

0

DISSOLVED SOLIDS

Sandstone

Granite

Quartzite

DRAINAGE AREA

0   20   40   60   80 km²

B

mm

500

400

300

200

100

0

ANNUAL RUNOFF

Quartzite

Granite

Sandstone

DENUDATION

0   40   80   120   160   200 cm/100 000 yrs

m

3500

3000

2500

2000

ALTITUDE

C

resultant upon rock type (Table 6.7). A more recent study is that carried out by Miller (1961) in the Sangre de Cristo Mountains of New Mexico. Within this area he isolated streams draining three major rock types, namely, granite, metaquartzite and sandstone (Figure 6.10A) and collected samples from streams draining these

Table 6.7 The influence of rock type on the chemical composition of stream water in Bohemia

Rock type	Ion concentration (mg/l)						
	Ca	Mg	Na	K	HCO$_3$	SO$_4$	Cl
Phyllite	5·7	2·4	5·4	2·1	35·1	3·1	4·9
Granite	7·7	2·3	6·9	3·7	40·3	9·2	4·2
Mica schist	9·3	3·8	8·0	3·1	48·3	9·5	5·4
Basalt	68·3	19·8	21·3	11·0	326·7	27·2	5·7
Cretaceous rocks	133·4	31·9	20·7	16·4	404·8	167·0	17·3

Source: Clarke (1924) after Gorham (1961)

individual outcrops. Because this was largely a wilderness area, the variations in solute concentrations between outcrops were not influenced by complications of man-made pollution. Figure 6.10B shows a consistent pattern of contrasts between the three rock types in terms of total dissolved solids content and further differences were found for individual ion concentrations. The average dissolved solids concentrations of streams draining quartzite, granite and sandstone respectively were in the proportion 1:2.5:10. By estimating annual runoff at each of the sampling points, from a general relationship between mean annual runoff and altitude, approximate denudation rates were calculated (Figure 6.10C). The peak denudation rates from the sandstone were over eleven times those from the quartzite and four times those from the granite. Studies by Hack (1960) in the Shenandoah Valley of Virginia and Pasternak (1963) in Poland have demonstrated similar contrasts in solute dynamics between different rock types, and the conclusion reached by Conway (1942), that igneous and metamorphic rocks usually produce stream solute concentrations less than 50 mg/l whilst sedimentary rocks exhibit values greater than this, would seem to be generally acceptable.

Vegetation can also exert a significant, if rather minor, influence on the solute transport from a drainage basin. Several studies have demonstrated that the solute content of rainfall can increase markedly after percolation through a forest canopy and Tamm (1953) has cited increases in potassium, calcium and sodium content of up to 18-fold, 4-fold and 3-fold respectively. These increases are related partly to leaf excretion and partly to entrapment of aerosols from dry air by the vegetation and subsequent removal by the rain. Furthermore, the production and movement of solutes within a drainage basin should not be viewed as a purely mechanical process, because it is intimately connected with the nutrient cycling of vegetation, with plants

Figure 6.10 Spatial variation in dissolved load production
The dissolved solids concentration of stream samples from the Sangre de Cristo Mountains of New Mexico (A) closely reflect the three major rock types of the area (B). Denudation rates estimated from the general relationship of runoff to altitude are similarly closely controlled by rock type (C). (Based on Miller, 1961.)

both consuming and releasing solutes to the system (Table 4.12). The chemical character of water percolating through the soil can change significantly during the period of leaf fall. In addition, animal life can contribute to the biological nutrient budget.

In conclusion of this brief treatment of spatial variations in the operation of fluvial processes at the meso-scale it is useful to refer again to the concept of the drainage basin presented in Figure 1.3. Because the drainage basin functions as a system transposing input of precipitation and solar radiation into output of water and sediment yield, it is inevitable that differences in the input and in the internal characteristics of the system will result in numerous contrasts in drainage basin dynamics at the regional level. Variation at the continental and world level, as primarily conditioned by climate or the general setting within which the system operates, must also be considered.

6.4   The continental and world scale

Consideration of spatial variations in fluvial processes on a continental or world scale follows logically from the input/process and form/process relationships considered above (Sections 6.2 and 6.3), and must be concerned with general trends in the intensity of the processes under the influence of climatic controls. The development of studies of climatic geomorphology has stressed the dependence of geomorphic processes and land form evolution on climate, and the world has been divided into a series of morphoclimatic and morphogenetic zones (for example, Figure 1.5). Peltier (1950) has attempted to distinguish the major processes operating in different zones, but although he proposed a general scheme for variations in stream erosion in accordance with mean annual precipitation and temperature (Figure 1.4A), no detailed and precise work has as yet appeared on the intensity of fluvial processes in different climatic zones. Indeed, conflicting views have sometimes been produced (Corbel, 1964; Fournier, 1960). Much of this present uncertainty results from the lack of reliable and long-term hydrologic data for many areas of the world, and the international cooperation and data collection schemes inspired by the International Hydrological Decade, will doubtless go a long way towards remedying this situation in the near future. Following the theme already developed within this chapter, spatial variations in parameters of runoff and sediment and solute production at the world level can be reviewed in the light of current data availability and knowledge.

6.4a   Runoff

The average annual runoff of the land surface of the earth is 267 mm (Barry, 1969), although there is imbalance between the continents, with the annual runoff from Australia totalling only 60 mm whilst that from South America is 490 mm. More detail on the pattern of world runoff is provided by the map produced by Lvovitch (1958) and shown in Figure 6.11. On this map, annual runoff values range from less than 50 mm to in excess of 1000 mm and the pattern closely reflects that of world climates which themselves control the all-important components of rainfall and evapotranspiration loss. A general relationship between mean annual runoff and mean annual precipitation for different temperature conditions has been proposed by

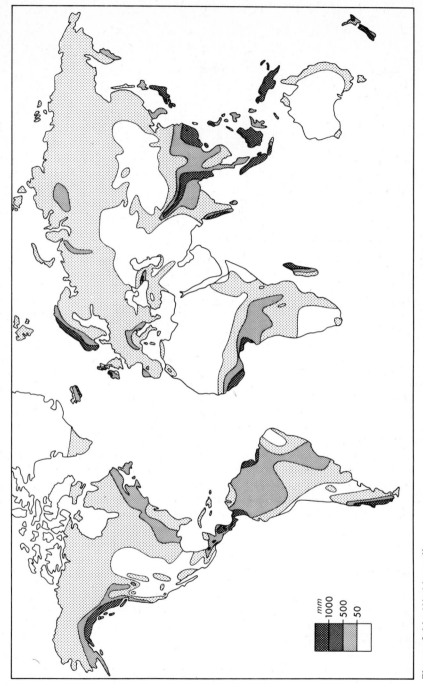

Figure 6.11 World runoff
In this simplified runoff map, based on the work of Lvovitch (1958), runoff is expressed as the equivalent depth in millimetres.

mm
1000
500
50

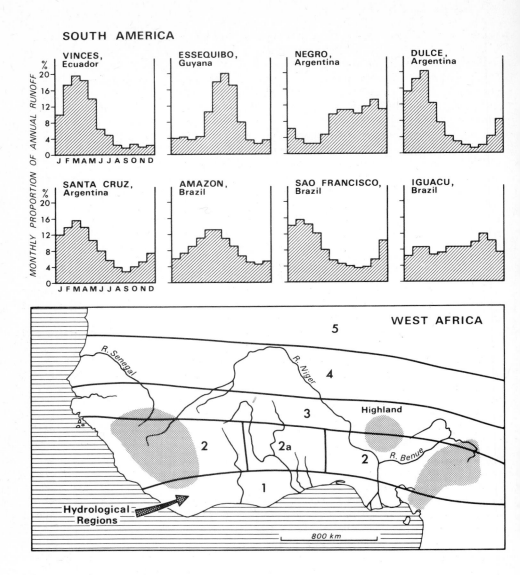

Figure 6.12 Regime diagrams and a hydrological regionalisation of West Africa
Dimensionless regime diagrams are presented for several contrasting South American rivers (based on data obtained from Oltman, 1968 and UNESCO, 1969). The hydrological regionalisation of West Africa illustrated is that suggested by Ledger (1964) and the major hydrological characteristics of the individual regions are listed in Table 6.8.

Langbein and others (1949) (Figure 1.4B) and this further demonstrates the influence of the evapotranspiration component which is directly related to temperature.

However, a single value of total runoff can conceal considerable contrasts in catchment runoff response and the analysis of river regimes, as proposed by Pardé (1955), provides deeper insight into spatial patterns of runoff by demonstrating the influence of climate on time distribution of streamflow. In some rivers the total runoff will be fairly uniformly distributed throughout the year whilst in others, particularly those influenced by snowmelt, there may be a marked concentration of flow into a short period. In Figure 6.12 dimensionless regime diagrams are presented for eight South American rivers. Contrasts are obvious, and the almost uniform regime of the Iguacu River, rising in a zone of Humid Subtropical climate is markedly different from those of the Dulce and Essequibo rivers influenced by Steppe and Desert, and Tropical Rain-forest and Tropical Savanna climates respectively. Ledger (1964) has constructed a division of West Africa into hydrological regions based upon regime characteristics and the proposed regionalisation (Figure 6.12) follows closely the major climatic zones of the area. Some of the contrasts in response characteristics between the different regions are listed in Table 6.8. All values show a general decrease from south to north, conditioned by decreasing rainfall depths.

To consider floods in more detail, it is clear that climate must exert an important influence over flood intensity in different areas of the world by its control over rainfall magnitude and intensity, although relief and other catchment and channel characteristics are important in transforming the potential of the storm rainfall into the flood event. It is difficult to cite values of world maximum floods, partly because of lack of discharge records on many rivers and partly because of the influence of catchment area on flood magnitude. Pardé (1961) reports several extreme flood discharges; one on the Bunton Branch River in Texas (10·6 km^2) of 37·4 m^3/s.km^2, another on the Mangakotukutuku River in New Zealand (18·7 km^2) of 36·4 m^3/s.km^2 and one on the Pecos River in Texas (75,000 km^2) of 2·93 m^3/s.km^2, and these would seem to be strong contenders in this class. The general influence of climate on flood potential is shown more clearly by studying rainfall data, and Figure 6.13 is a precipitation depth-duration diagram showing the world maximum recorded values for a given duration, and equivalent data from several stations in the United States. The U.S. data indicates the variations in rainfall flood potential that can occur over a large area. In the world data, the various short duration record values are ascribed to several different places, but the maximum depths reported for durations between 9 hours and 8 days are those from La Réunion, an island in the Indian Ocean, where values of 1087 mm and 4130 mm respectively have been recorded. For durations greater than 8 days, the maximum depths are those from Cherrapunji in India where as much as 9300 mm has been recorded in a single month. Contrasts in the U.S. data can be viewed both in terms of the values of precipitation for a given duration and of the form of the relationship for individual stations. The maximum values and the relationships with the steepest slopes occur in the subtropical areas (Stations A and B), whilst the minimum values are those from the subarctic (Station I). The stations in arid and semi-arid zones (C, F, H and J) provide an interesting contrast, for they exhibit a marked reduction in the rate of increase in depth versus time for longer durations. In these areas, flash storms of short duration are the most significant for flood production.

Table 6.8 Regional hydrological characteristics of West African rivers * (cf. Figure 6.12)

Region	Runoff (%)	Mean annual runoff 1/s . km²	Flood magnitudes 1/s . km²		Minimum flows 1/s . km²
			$Q_{2\cdot33}$	Q_{10}	
1	15–30	5–20	40–50	70	0·05–2·0
2	25–35	10–21	50–90	70–120	0·05–2·0
2ᵃ	6–10	2–4	15–50	40–100	0–0·50
3	c10	2–8	10–80	40–120	0–0·10
4	1–3	0·20–0·50	—	5–10	0
5	(5–10)†	—	—	(100–200)‡	0

*Catchments 10,000–20,000 km²
†Runoff limited to small mountainous catchments, for example, 5000 km²
‡Mountainous catchments of 3000–5000 km²
Based on Ledger (1964)

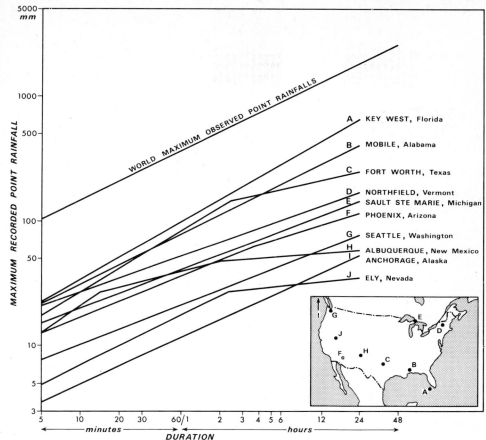

Figure 6.13 Spatial variation in precipitation depth-duration characteristics
This figure illustrates the depth-duration precipitation characteristics of several locations in the United States and the maximum recorded values for the world. (Data obtained from Todd, 1970.)

Climatic conditions are also important in influencing the detailed functioning of the runoff process and contrasts in the relative magnitude of different runoff components can occur over wide areas. Russian workers (for example, Glushkov, 1933) have extended the zonal concept applied to soils and vegetation and based on the work of Dokuchayev (1899) to the runoff process. Muraveyskiy (1956) classified runoff as one of the three basic geographical factors which, along with climate and topography, form geographical complexes. Detailed quantitative description and comparison of the runoff process has, however, only been made possible with recent advances in geochemical analysis of stream waters (Voronkov, 1963). Chemical analysis of the water enables runoff from different zones to be distinguished and the relative importance of individual components to be determined.

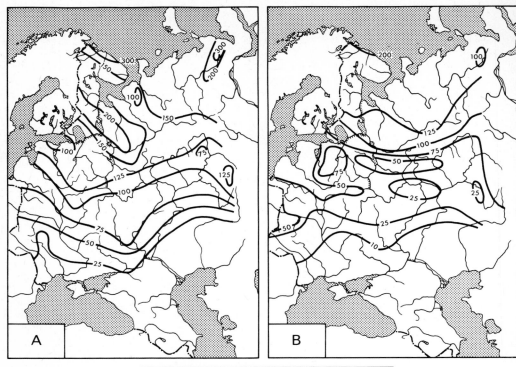

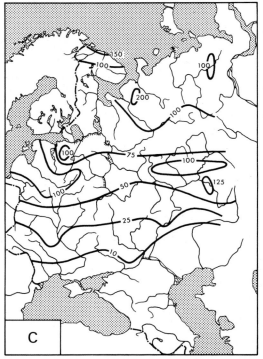

Skakalskiy (1966) has produced generalised maps of European Russia showing the magnitude of the major runoff components, namely, slope, topsoil-ground, and ground-water runoff which broadly represent surface runoff, throughflow and inter-flow, and groundwater flow respectively (Figure 6.14). For each component, the predominant characteristic of the distribution map is the general southerly decrease in runoff volumes in response to the decreasing precipitation and increasing eva-potranspiration in this direction. Further detail of the pattern of variation is provided by the values of the relative proportions of the different components comprising the total runoff of individual zones (Table 6.9). In all zones frost and snow are important

Table 6.9   Runoff sources in the natural zones of European Russia

| | Proportion of total annual runoff (%) | | |
Zone	Slope waters	Topsoil-ground water	Ground-water
Tundra	45–50	30	20–25
Taiga	40–60	30–40	20–30
Mixed forest	40–50	25–35	20–30
Forest steppe	65–80	10–15	10–20
Steppe	60–80	15–30	15–30

Source: Skakalskiy (1966)

controls, particularly in the Tundra zone where permafrost exists. The Steppe zone is the most favourable for the occurrence of slope runoff, partly because the increased evapotranspiration losses reduce the contributions from the two other sources and partly as a result of conditions of soil freezing, ice crusting and surface compaction which are conducive to high rates of surface runoff. Ground-water contributions from the Steppe are the lowest of the five zones. The Taiga areas exhibit the highest topsoil-ground runoff contribution because of the high permeability of the upper soil horizons which reduces surface runoff and the existence of a compact illuvial layer at slight depth which promotes lateral flow through the soil. This contribution decreases southwards towards the Forest Steppe as a result of the drier ground and the greater depth to the compact illuvial horizon, but it remains high in the Tundra areas to the north because of the abundant subsurface drainage associated with the active layer during the period of melt.

6.4b   Suspended sediment transport

The maximum values of suspended sediment concentration reported in the literature are probably those cited by Beverage and Culbertson (1964) for the Paria River at Lees Ferry, Arizona. An instantaneous concentration of 646,000 ppm and a mean daily value of 411,000 ppm have been reported from this measuring station. Two New Mexican rivers, the Rio Puerco and the Rio Salado, were also cited as having provided instantaneous concentrations of 418,000 and 405,000 ppm respectively.

Figure 6.14   Runoff component maps of European Russia
These maps based on the work of Skakalskiy (1966) depict the magnitude of the slope (A) topsoil-ground (B), and ground-water (C) runoff components within European Russia.

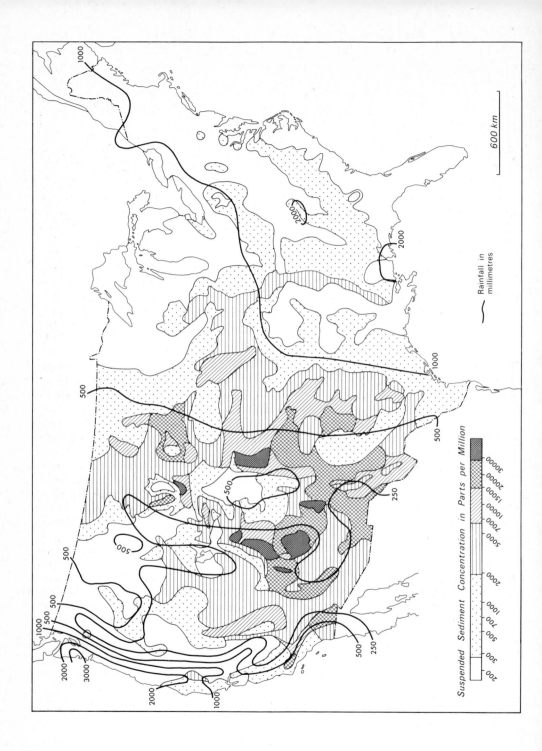

Suspended Sediment Concentration in Parts per Million

200 300 500 700 1000 2000 5000 7000 10000 15000 20000 30000

Rainfall in
millimetres

600 km

The precise distinction between a stream laden with these concentrations of sediment and a mudflow is difficult to define, but a threshold value for mudflows of 800,000 ppm would seem generally applicable. The tributaries of the Rio Grande and the Colorado are well known for their high concentrations with water 'too thick to drink and too thin to plough'. But other rivers of the world must also be mentioned. Todd and Eliassen (1938) have reported concentrations of 400,000 ppm in the Yellow River, China, and Pardé (1960) refers to a report of concentrations of 460,000 ppm in the River Esk near Hawkes Bay, New Zealand, during the extreme flood of April 1938. The New Zealand example can be distinguished from those cited previously because, whereas they occurred in semi-arid areas where conditions of dry soil and sparse vegetation cover provide a ready source of sediment during an intense storm, in New Zealand the high concentration was attributed to extreme flood conditions and steep unstable slopes.

On a world scale, therefore, suspended sediment concentrations range between something approaching zero to values in excess of 900,000 mg/l (600,000 ppm). The map of mean annual suspended sediment concentrations for the United States (Figure 6.15) exhibits a range of over 100-fold from less than 200 ppm to over 30,000 ppm. Because concentration values are controlled by many factors associated with supply conditions and transport capacity, it is difficult to isolate general trends, especially with respect to climate. Langbein and Schumm (1958) have, however, attempted to relate mean annual concentration to annual precipitation, using data from small drainage basins in the United States and they found a clear inverse relationship (Figure 6.16A). They argued convincingly that a decrease in annual precipitation is associated with a decrease in vegetation cover density from forest through grasslands to desert scrub, providing an increased supply of sediment. The pattern shown by Figure 6.15 substantiates these conclusions in that minimum and low concentrations generally occur in areas with an annual precipitation greater than 500 mm, whilst maximum values occur in drier areas.

The relationship between annual runoff and mean annual suspended sediment concentration plotted for a limited set of world data in Figure 6.16B might be expected to exhibit a similar trend to the precipitation/concentration relation of Figure 6.16A. However, although there is a definite inverse trend in the upper part of the plot, this is reversed in the lower part and it could be argued that both maximum and minimum concentrations occur at low runoff, whilst intermediate values occur at higher runoff levels. It is probable that the restricted range of climatic conditions in the United States only provides a partial picture of variations at the world scale. Looking at the data in Figure 6.16B in more detail, it is apparent that the low values of concentration for minimum runoff are provided by temperate catchments, whilst the high concentrations are associated with subtropical and tropical areas. Furthermore, the former values are from large catchments of low relief whereas the latter are from watersheds of more marked relief. More data and detailed multivariate analysis are required if definite world-wide patterns and trends are to be isolated.

Figure 6.15 Mean annual suspended sediment concentration of rivers in the United States This map portrays the average annual discharge-weighted mean suspended sediment concentrations of rivers in the United States. The discharge-weighted mean concentration is derived as the total load for a given year divided by the volume of water discharge for the same period (based on Rainwater, 1962).

The increasing amount of data available on sediment transport by rivers also serves to indicate the considerable range of suspended sediment yields occurring throughout the world. Some of the maximum reported values are listed in Table 6.10, and for comparative purposes, the yields of several of the world's major drainage basins are presented in Table 6.11. Two well-known maps of world sediment yields have also been produced (Fournier, 1960; Strakhov, 1967) and Strakhov's map is reproduced in Figure 1.2. There have been several attempts at explaining the world-wide pattern of suspended sediment yields, and the conclusions of Strakhov (1967), Fournier (1960), Langbein and Schumm (1958), Douglas (1967) and Fleming (1969) can be reviewed, although it will be apparent that no consistent and generally-accepted conclusion exists.

Strakhov (1967) proposed a simple distinction between two major and parallel contrasting zones of the earth's surface. The first, the Temperate Moist belt of the Northern Hemisphere where annual precipitation ranged from 150 to 600 mm and whose southern boundary was formed by the annual +10 °C isotherm, exhibited low annual values of suspended sediment yield averaging about 10 tonnes/km². The second zone was confined between the +10 °C annual isotherms on either side of the equator, and there yields were generally between 50 and 100 tonnes/km², rising in places to above 1000 tonnes/km². Arid areas were not included in this world classifica-

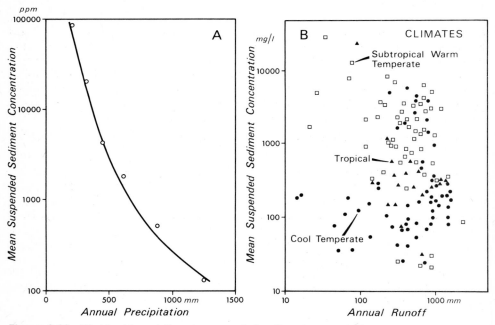

Figure 6.16 World-wide variations in suspended sediment concentration
(A) demonstrates the relationship between annual precipitation and mean suspended sediment concentration obtained by Langbein and Schumm (1958) for several groups of small catchments in the United States. In (B), world data published by Fournier (1969) has been used to produce a plot of mean suspended sediment concentration versus annual runoff for catchments classified into three major climatic groups.

Table 6.10 World maximum-recorded suspended sediment yields

River	Location	Average annual yield (tonnes/km²)	
Ching	Changchiashan, China	8040	a
Lo	Chuantou, China	7922	a
Waipaoa	Kanakanaia, New Zealand	6983	b
Tjatjabon	Java	6250	c
Lo-ho	Loyang, China	6068*	d
Marecchia	Pietracuta, Italy	4570*	b
Semani	Uraqe Kucit, Albania	4150	b
Soldier	Pisgah, Iowa, U.S.A.	4072	e
Shkumbini	Papër, Albania	3590	b
Kosi	Chatra, India	3130	a
Yellow	Shenhsien, China	2957	a
Indus	Kalabagh, West Pakistan	2498	a
Santa Anita Creek	Arcadia, Calif., U.S.A.	2374	e
Eel	Scotia, Calif., U.S.A.	2292	a

*1 year of record
Sources: a Holeman (1968)
b Fournier (1969)
c Douglas (1967)
d Fournier (1960)
e Brown (1950)

tion, presumably because the lack of runoff was assumed to be associated with low or negligible sediment yields, although those areas in the United States have been shown to possess high yields of over 200 tonnes/km².yr.

Fournier's (1960) classic study differed in approach in that he carried out detailed statistical analysis of the sediment yield data in an attempt to explain the variations and then used the general relationships obtained to construct the world map. Using data from ninety-six basins he tried to find an index of climate that would best account for the variation in the values of sediment yield, and the $\frac{p^2}{P}$ index was finally selected

Table 6.11 Suspended sediment yields of selected major drainage basins

River	Location	Average annual yield (tonnes/km²)
Ganges	delta, Bangladesh	1568
Yangtze	Chikiang, China	549
Indus	Kotri, West Pakistan	510
Mekong	Mukdaham, Thailand	486
Colorado	Grand Canyon, Arizona, U.S.A.	424
Missouri	Hermann, Missouri, U.S.A.	178
Mississippi	Louisiana, U.S.A.	109
Amazon	mouth, Brazil	67
Nile	delta, Egypt	39
Danube	mouth, U.S.S.R.	27
Congo	mouth, Congo	18
Rhine	mouth, Holland	3·5
St Lawrence	mouth, Canada	3·1

Source: Holeman (1968)

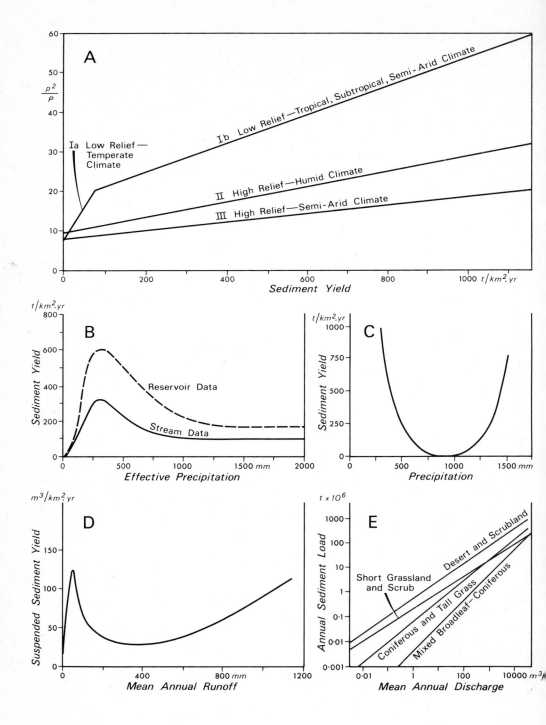

(p = rainfall in wettest month in mm, P = mean annual rainfall in mm) because it reflected seasonality as well as absolute depth of the precipitation. When plotted against this variable, the sediment data exhibited four distinct groupings and trends which were related to relief and other climatic influences (Figure 6.17A), although in all cases an increase in the $\frac{p^2}{P}$ index, reflecting increases in total rainfall and seasonality of the regime, was associated with increased sediment yield. This occurrence can be explained in terms of the increased erosion potential of high rainfall totals, particularly when the rainfall is concentrated into a short time period, and the decrease in vegetation density and associated increased sediment supply in areas with a seasonal drought. Subsequently, Fournier (1960) devised a relief factor which could be used with the precipitation index to produce a single empirical equation for the prediction of sediment yields:

$$\text{Log DS} = 2.65 \text{ Log} \frac{p^2}{P} + 0.46 \text{ Log} \frac{H^2}{S} - 1.56$$

where   DS = supended sediment yield (t/km².yr)
              H = mean relief of the basin or the difference between the mean altitude and the minimum altitude (m)
              S = catchment area (km²)

The map of world sediment yields produced by extrapolation of these relationships exhibits maximum production rates in the seasonally humid tropics, with yields decreasing towards the equatorial regions where seasonal effects are lacking and towards the arid zones where runoff is low. Values rise again in the Mediterranean areas where a seasonal rainfall regime occurs and finally decrease over the temperate and cold regions. Fournier's map is essentially similar in general trends to that produced by Strakhov but there are discrepancies of up to a whole order of magnitude in the absolute magnitude of the values depicted for certain areas. It would seem that much of Fournier's original data reflects increased yields consequent upon human activity rather than natural rates of sediment production.

A more specialised study of the influence of mean annual precipitation on suspended sediment yields has been undertaken by Langbein and Schumm (1958). By using data only from the United States and from a restricted range of catchment area, they hoped to obtain precise results. Values of sediment yield were obtained from ninety-four stream sampling stations with catchment areas averaging 3885 km² and from 163 reservoir surveys in watersheds averaging 78 km², and the standardised precipitation values were estimated from runoff data by using a general graphical relationship

Figure 6.17 World-wide variations in suspended sediment yield
Various attempts at explaining world patterns of suspended sediment yield are presented. (A) illustrates the relationships between sediment yield and an index of precipitation magnitude and seasonality derived by Fournier (1960); (B) depicts the relationship between effective precipitation and sediment yield obtained for the United States by Langbein and Schumm (1958); (C) presents the relationship between sediment yield and annual precipitation obtained for world data by Fournier (1960); (D) illustrates the relationship between sediment yield and annual runoff produced by Douglas (1967) and (E) demonstrates the relationships derived by Fleming (1969) between sediment yield and mean annual discharge for several different vegetation zones.

between mean annual precipitation and runoff, adjusted to a standard temperature of 10 °C. The grouped data was used to construct the generalised graphical plots shown in Figure 6.17B. Both sets of data exhibit the same trend, and the higher values associated with the reservoir data are the result of the smaller catchment areas. The detailed shape of the curves was explained convincingly in terms of the erosion potential of precipitation and its interaction with the protective capacity of a vegetation cover. Maximum sediment yields occur at an annual precipitation of about 300 mm; below this there is too little rain to cause a significant erosive effect, and above this the vegetation cover is too dense to permit increased erosion despite the increased erosion potential of the rainfall.

The argument presented by Langbein and Schumm is applicable primarily to the United States and that country only represents a limited number of the climatic zones of the world. Tropical areas are not represented. Fournier (1960) attempted to define a relationship between annual precipitation and sediment yield for world data and the result is shown in Figure 6.17C. Langbein and Schumm (1958) refer to this result and suggest that the falling limb represents the right hand side of their curve (Figure 6.17B) because the yields must decrease again towards zero with near-zero precipitation. The rising limb is to be expected from the previous discussion of Fournier's work because the tropical monsoon areas with high annual rainfall are also the areas with high $\frac{p^2}{P}$ indices and maximum sediment yields. It seems reasonable to suggest that Langbein and Schumm's curve would exhibit a similar rising limb for maximum levels of precipitation if it included data from tropical areas. This suggestion is further supported by the result obtained by Douglas (1967) in his analysis of world sediment yield and runoff data (Figure 6.17D). This plot shows two sediment yield maxima, one at low values of runoff which essentially mirrors the peak on Langbein and Schumm's graph (Figure 6.17B) and one at high runoff levels which is similar to the rising limb on Fournier's plot (Figure 6.17C).

Fleming (1969) has demonstrated in a different manner the importance of vegetation as a control over the pattern of world sediment yields. He considered the general relationship of suspended sediment yield to mean annual discharge (m³/s) and found that the scatter could be reduced by isolating four individual relationships corresponding to four major vegetation zones (Figure 6.17E). Maximum sediment yields were associated with the desert and scrubland zone and minimum yields with the mixed broadleaf-coniferous zone. This result generally concurs with the conclusions of Langbein and Schumm but it provides for a more accurate prediction equation for suspended sediment yield.

The above studies clearly demonstrate the influence of annual precipitation and runoff and also relief and vegetation on world suspended sediment yields, and indicate that maximum yields are associated with semi-arid and tropical areas with seasonal precipitation regimes. However, care should be exercised if graphs and relationships are extrapolated as general relations applicable to all areas of the world, because of the multivariate nature of the controls and because they are often based on restricted samples which can be biased by results from areas where human interference is important. Furthermore, although variations in yields and concentrations are the major facet of suspended sediment production that has been studied and analysed, there is considerable scope for other analysis. Regime studies could be extended to

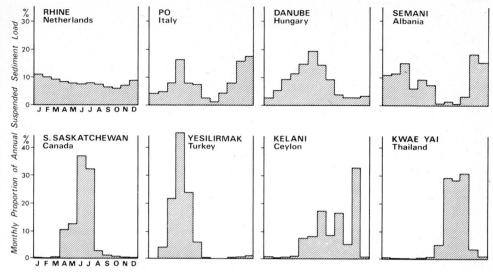

Figure 6.18   Suspended sediment regime diagrams
This figure illustrates the contrasts in seasonal pattern of sediment production exhibited by several world rivers (based on data in Fournier, 1969).

sediment transport because climate also affects the time pattern of delivery, and in Figure 6.18 graphs of the average monthly proportions of sediment transport are shown. These graphs are directly analogous to the runoff regime diagrams in Figure 6.12 and indicate appreciable contrasts between the several catchments in response to seasonal climatic influences.

6.4c   Dissolved loads

According to Livingstone (1963), the mean solute content of the rivers of the world is 120 mg/l and the mean values suggested for the individual continents are listed in Table 6.12. Absolute values do not exhibit as great a range as suspended sediment concentrations but Livingstone reports values of less than 10 mg/l from the northeastern Highlands of Victoria, Australia, and as high as 5457 mg/l and 7900 mg/l for the Pecos River, Texas and the Kalaus River, U.S.S.R. respectively. Pavelko and Tarasov (1967) have also described concentrations of 6–7000 mg/l in the rivers of Kazakhstan, U.S.S.R. An example of contrasts in dissolved load concentrations at the continental scale is provided by the generalised map of stream solute concentrations in the conterminous United States (Figure 6.19). Rock type is important in controlling spatial contrasts in solute concentrations (Section 6.3c), but climate exerts a strong, if not a dominant, influence at this scale. There is a general increase of levels of concentration towards semi-arid and arid areas in Figure 6.19, and this trend is further demonstrated for the United States by the data presented by Durum, Heidel and Tison (1960) for 8 major drainage areas and by Langbein and Dawdy (1963) for 170 sampling stations (Figure 6.20A). In both cases there is a decrease in

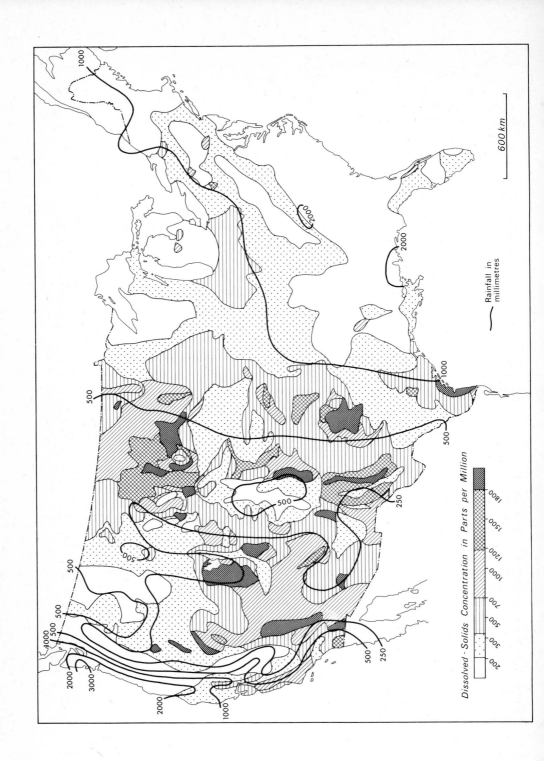

Rainfall in
millimetres

Dissolved - Solids Concentration in Parts per Million

200 300 500 700 1000 1200 1500 1800

Table 6.12   Mean dissolved load concentration of river waters of the world

Continent	Mean solute content (mg/l)
North America	142
South America	69
Europe	182
Asia	142
Africa	121
Australia	59

Source: Livingstone (1963)

concentration with increased runoff similar to that exhibited by suspended sediment concentrations in Figure 6.16A. Maximum solute concentrations found in arid areas are the result of the small volumes of precipitation and runoff which become highly charged with solutes from salt accumulations in the soil and which are further concentrated by evaporation effects. Furthermore, the concentration of cyclic salts in the incident precipitation are also often at their highest over arid areas and Matveyev and Bashmakova (1967) present maps of the chemical content of atmospheric precipitation over European Russia which show, for example, how along meridian 40 °E concentrations increase from about 10 mg/l at 65 °N to approximately 60 mg/l at 45 °N. This increase is associated with contributions from soil dust in dry areas, an occurrence which is substantiated by a change in the dominant chemical composition of the solute content from sulphate and sodium ions in the north to sulphate, bicarbonate and calcium ions in the south. Langbein and Dawdy (1963) (Figure 6.20A) are specific about the nature of the decrease in solute concentrations of streams from arid to humid conditions and argue that the decrease from an average of 800 mg/l is gradual up to a threshold of an annual precipitation of about 250 mm and thereafter more rapid in the form of a straight dilution effect. Different climatic conditions can also be expected to cause differences in the chemical constituents of dissolved load, because of their influence on the nature of the weathering process, soil character and runoff dynamics (cf. p. 329).

The dissolved loads of world rivers do not vary as greatly as suspended sediment yields and Strakhov (1967) reports a maximum range of 70-fold from 3.9 to 290 tonnes/km² for the Yana and Sulak rivers respectively, which contrasts with a maximum range of over 500-fold for sediment yields. The magnitude of the dissolved load of a drainage basin will be influenced by the underlying rock, but several workers have also isolated a definite relationship between dissolved load magnitude and runoff depth on a world scale. The relationships between annual dissolved load and runoff suggested by Livingstone (1963) for the world, Langbein and Dawdy (1963) for the United States and Van Denburgh and Feth (1965) for the Western United States shown in Figure 6.20B exhibit the reverse trend to that shown for concentration (Figure 6.20A). This is because the decrease in concentration with increasing runoff is more than offset by the increase in runoff volumes. The three plots shown in

Figure 6.19   Mean annual dissolved solids concentration of rivers in the United States
The values of dissolved solids concentration mapped are the prevalent or modal concentrations (based on Rainwater, 1962).

Figure 6.20B are approximately similar in the range 5–1000 mm of runoff, but whilst Langbein and Dawdy indicate a gradual flattening of the curve at high runoff due to an increased dilution effect, and both they and Van Denburgh and Feth portray similar load values at high runoff, Livingstone indicates substantially greater values. This distinction may be because the U.S. data only represents a limited range of climatic conditions and because the world data used by Livingstone includes tropical areas. All three relationships, however, clearly demonstrate that in general solute loads are least in arid areas and greatest in areas of maximum precipitation.

To conclude this brief review of variations in sediment and solute production at the world scale, it is useful to compare the relative efficacy of these two components of denudation in different areas of the world by comparing the graphs of dissolved load and suspended load versus runoff produced by Langbein and Dawdy (1963) and Livingstone (1963), and Langbein and Schumm (1958) and Douglas (1967) for the United States and the world respectively (Figure 6.20C). Although indicating certain detailed contrasts, these two sets of relationships serve to indicate the general world pattern of the two components of stream load. In arid areas, solute loads contribute only a small proportion of the total load, but solute transport increases in magnitude with increase in annual runoff and approaches equality with sediment transport where runoff exceeds 600 mm. At greater runoff levels, it is likely that both exhibit a similar increase. It is also possible to distinguish a trend within the U.S. data for an increase in dissolved load to be associated with a decrease in suspended sediment load and this has also been demonstrated for the United States by Judson and Ritter (1964) (Figure 6.20D). This trend is not so apparent for the world situation (Figure 6.20C) in which both suspended load and dissolved show increases at runoff levels less than 75 mm and above about 750 mm. Indeed, the data presented by Strakhov (1967) on the relative importance of the two load components at the world level clearly indicates this latter tendency for both suspended and dissolved loads to increase together (Figure 6.20D). Strakhov (1967) demonstrates the influence of relief in addition to climate in this context by distinguishing between Temperate

Figure 6.20 (A and B) World-wide variations in dissolved solids concentration and load
The relationship between annual runoff and mean dissolved solids concentration proposed by Langbein and Dawdy (1963) is presented in (A), and a similar trend is exhibited by a plot of the data cited by Durum, Heidel and Tison (1960). The relationships between annual runoff and dissolved load proposed by Livingstone (1963) for the world and by Langbein and Dawdy (1963) for the United States and Van Denburgh and Feth (1965) for the western United States are illustrated in (B).

(C and D) The relative importance of sediment and solute loads
Comparison of the relative importance of sediment and solute loads is afforded by (C) which presents the relationships between dissolved load and annual runoff proposed by Langbein and Dawdy (1963) and Livingstone (1963) (dashed), and the relationships between suspended sediment load and annual run-off proposed by Douglas (1967) (dashed) and derived from the data cited by Langbein and Schumm (1958). A more direct comparison is illustrated by (D) in which the data cited by Judson and Ritter (1964) for the United States and by Strakhov (1967) for several world climate and relief zones have been plotted.

(E) The world pattern of denudation
By combining the relationships between annual runoff and sediment and dissolved load proposed for the United States by Langbein and Dawdy (1963) and Langbein and Schumm (values of effective precipitation converted to runoff) and for the world by Livingstone (1963) and Douglas (1967) general relationships between total denudation and annual runoff have been constructed.

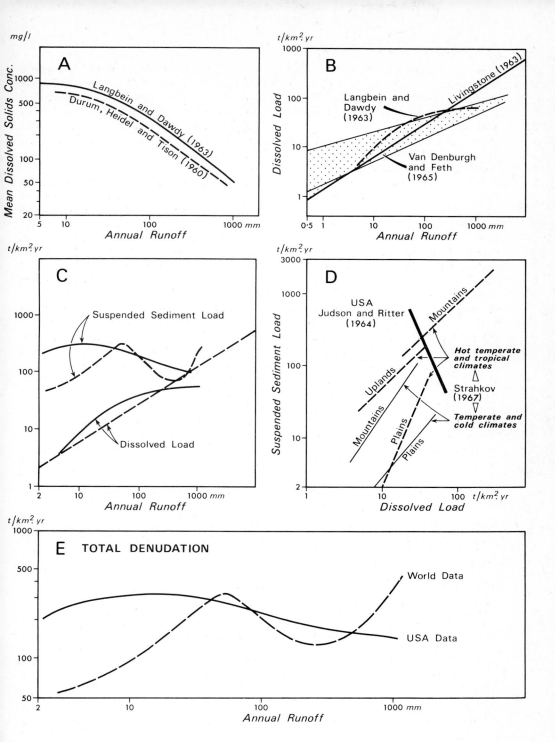

and Cold climates, and Hot Temperate and Tropical climates and mountain and plain areas. According to his data (Figure 6.20D), loads in mountain areas are appreciably greater than in plain areas and dissolved loads are greater than suspended loads in plain areas under both sets of climatic conditions, whilst suspended sediment loads are greater than solute loads in mountain zones. The overall increased magnitude of suspended sediment yield over dissolved load is shown by the figures for the denudation of the continents in Table 6.13.

With the greater availability of data to be expected in the near future, more definite conclusions concerning the world pattern of denudation, as indicated by the combined sediment and solute loads of rivers, will be possible. For the present, a useful 'best-estimate' can be obtained by combining, firstly, the results of Langbein and Dawdy (1963) and Langbein and Schumm (1958) concerning the relation of solute and suspended sediment loads to annual runoff in the United States, and, secondly, the equivalent results of Livingstone (1963) and Douglas (1967) for the world. The resultant plots of total denudation versus annual runoff are shown in Figure 6.20E. If the restricted climatic range of the U.S. data is accepted, the pattern shown is one of two denudation maxima, the first in semi-arid areas with runoff between about 15 mm and 60 mm and a second in tropical areas with runoff in

Table 6.13   Denudation of the continents*

Continent	Annual suspended sediment discharge†		Annual dissolved load discharge‡	
	tonnes/km²	10⁹ tonnes	tonnes/km²	10⁹ tonnes
North America	96	1·99	33	0·68
South America	63	1·22	28	0·54
Africa	27	0·55	24	0·49
Australia	45	0·23	2·3	0·012
Europe	35	0·33	42	0·40
Asia	600	16·16	32	0·86

*Values cited are with reference to total yield to the oceans
†Based on Holeman (1968)
‡Based on Livingstone (1963)

excess of approximately 1000 mm. This pattern mirrors that suggested by Fournier (1960) and Douglas (1967) for suspended sediment production and provides an interesting contrast with the ideas of Corbel (1964) and Peltier (1950) (Figure 1.4A) who stress the dominance of temperate areas and zones of intermediate precipitation magnitude respectively in the world distribution of denudation intensity. There is clearly need for further detailed research in this context.

6.5   The influence of man

Few, if any, areas of the world can be said to be unaffected by man, but the extent of this influence varies appreciably. In many areas man exerts an important control over the spatial pattern of fluvial processes, and this is particularly so at the meso-scale where, for example, contrasts in agricultural practice or between rural and urban areas can be more important than differences in geology and topography or meteorological conditions. The magnitude and nature of these effects can be considered either in terms of their detailed interaction with process dynamics or their influence on the various parameters selected as indicative of drainage basin response. The latter approach is more relevant in the context of this analysis of areal variations in the operation of fluvial processes, and a convenient distinction can be made between those effects associated with rural areas and those related to the urban environment and also including areas of mining, building activity and road construction. Furthermore, it is only worthwhile to consider those effects for which quantitative data is available. Many of the examples must necessarily be drawn from the United States, because an awareness of the impact of man on the environment in this country has instigated the setting up of several studies to monitor and assess the precise magnitude of these effects.

Plate 19 Incipient gully development
Heavy rainfall on this bare soil has caused rill erosion and the early stages of gully development.

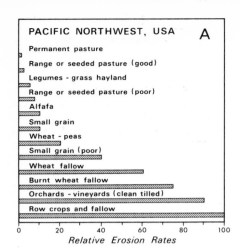

PACIFIC NORTHWEST, USA — A

Permanent pasture
Range or seeded pasture (good)
Legumes - grass hayland
Range or seeded pasture (poor)
Alfafa
Small grain
Wheat - peas
Small grain (poor)
Wheat fallow
Burnt wheat fallow
Orchards - vineyards (clean tilled)
Row crops and fallow

0   20   40   60   80   100
Relative Erosion Rates

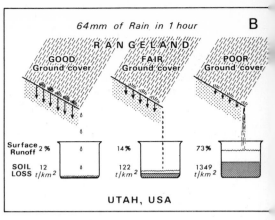

B

64mm of Rain in 1 hour

RANGELAND

GOOD Ground cover	FAIR Ground cover	POOR Ground cover
Surface Runoff 2%	14%	73%
SOIL LOSS 12 t/km^2	122 t/km^2	1349 t/km^2

UTAH, USA

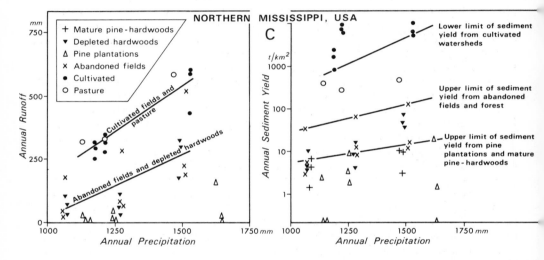

NORTHERN MISSISSIPPI, USA — C

mm

+ Mature pine - hardwoods
▼ Depleted hardwoods
△ Pine plantations
× Abandoned fields
● Cultivated
○ Pasture

Cultivated fields and pasture

Abandoned fields and depleted hardwoods

Annual Runoff

1000   1250   1500   1750 *mm*
Annual Precipitation

t/km^2

Lower limit of sediment yield from cultivated watersheds

Upper limit of sediment yield from abandoned fields and forest

Upper limit of sediment yield from pine plantations and mature pine - hardwoods

Annual Sediment Yield

1000   1250   1500   1750 *mm*
Annual Precipitation

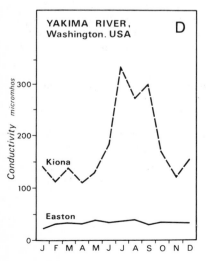

YAKIMA RIVER, Washington. USA — D

Conductivity micromhos

Kiona

Easton

J F M A M J J A S O N D

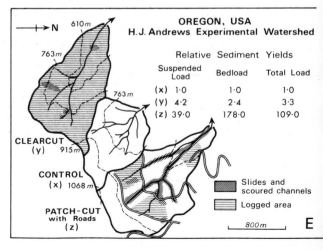

E

→N   610m

763m

763m

OREGON, USA
H.J. Andrews Experimental Watershed

Relative Sediment Yields

	Suspended Load	Bedload	Total Load
(x)	1·0	1·0	1·0
(y)	4·2	2·4	3·3
(z)	39·0	178·0	109·0

CLEARCUT (y)   915m

CONTROL (x)   1068m

PATCH-CUT with Roads (z)

■ Slides and scoured channels
▦ Logged area

800m

6.5a Rural areas

In rural areas, agricultural activity can greatly modify soil and vegetation character-istics and therefore runoff and sediment production, and the general implications of these changes have been demonstrated by Strakhov (1967) in his evaluation of the world map of sediment yield. According to this map, the plains region of Europe is designated as a zone with annual sediment yields of 30–50 tonnes/km², but he argues that in their natural state, uninfluenced by agriculture, they would produce less than 10 tonnes/km². Similarly, the Mississippi region of the southern United States is mapped as a zone of 150–230 tonnes/km², whereas essentially similar though little altered areas of South America only exhibit values of 50–100 tonnes/km². Changes in the pattern of runoff must also be associated with these changes in sediment yield, especially where the stream network has been modified by the construction of ditches and subsurface drains. Johnson (1966) has estimated that in England and Wales, where average drainage densities are 1 to 3 km/km², there is an average of 3 km of ditch per square kilometre. More specific and local examples of contrasts in catch-ment response conditioned by agricultural activity are numerous and are best considered at three levels; firstly, contrasts that can occur within a particular general land-use type; secondly, those that occur between land-use classes; and thirdly, those consequent upon changes in the pattern of land-use.

A clear example of the variation that may arise within a particular land-use class, in this case cultivated land, is provided by Figure 6.21A, which shows the relative erosion rates associated with different crop types in the Pacific north-west of the United States. In this region there is a range of up to 15-fold between erosion rates on hayland and fallow and those associated with row crops. More detailed data is provided by Table 6.14 where a comparison between conditions of clean tillage and dense crop cover is made for five locations in the United States. At these sites, the ratios of soil loss and runoff from clean tilled land to those from dense crop cover are as high as nearly 1:200 and 1:33 respectively. However, tilling of the soil surface does not in every case increase runoff and sediment production, because there are several reports from Russia in which the reverse occurs, and where autumn ploughing can cause a decrease in surface runoff. Kuznik (1954) presents results from the Irgiz and Maly Uzen basins of the Transvolga region of the U.S.S.R. which indicate that run-off, caused in this case by spring snowmelt, is 3·3–13 times greater from unploughed stubble than from ploughed land. Shevchenko (1962) found a similar situation in the central Chernozem zone of the U.S.S.R. (Table 6.15) and also demonstrated that the precise type of tillage influenced the magnitude of the reduction in runoff.

Figure 6.21 The influence of man : rural areas
This figure demonstrates the influence of crop type on erosion rates in Pacific north-west United States and of range condition on runoff and sediment yield in Utah (A and B). (C) illustrates the influence of land use on annual runoff and sediment yield from watersheds in the hill land of the upper coastal plain in northern Mississippi and (D) indicates the effects of irrigation return water on the dissolved load of the Yakima River, Oregon, by comparing the conductivity of water at Easton and Kiona, upstream and downstream respectively of the irrigation return water inflow. The effects of logging activity on sediment yield is demonstrated in (E) which compares the sediment yield of a clearcut, a patchcut with roads, and an untouched control watershed (based A on Brown, 1950; B on Noble, 1965; C and D on Ursic and Dendy. 1965; and E on Fredriksen 1970).

Table 6.14 Comparison of runoff and sediment yield from a clean-tilled crop and a dense cover crop at five locations in the United States

Location	Average annual precipitation (mm)	Clean-tilled crop		Dense-cover crop	
		Annual soil loss (tonnes)	Annual runoff (%)	Annual soil loss (tonnes)	Annual runoff (%)
Bethany, Missouri	884	69·88	28·31	0·29	9·30
Tyler, Texas	1037	28·40	20·92	0·13	1·15
Guthrie, Oklahoma	838	24·68	14·22	0·033	1·23
Clarinda, Iowa	681	19·12	8·64	0·061	0·97
Statesville, N. Carolina	1149	22·94	10·21	0·012	0·33

Source: American Society of Civil Engineers (1962)

Table 6.15 The influence of type of tillage on spring runoff from experimental plots in the Talovskiy region of Voronezh province, U.S.S.R.

Type of tillage	Water equivalent of snow at the beginning of melt + precipitation during the runoff period (mm)	Depth of runoff (mm)	Runoff coefficient (%)
Unploughed with stubble	105·5	74·5	71
Ordinary ploughing to a depth of 20–22 cm with basin listing	118·5	16·2	14
Ordinary ploughing to a depth of 20–22 cm	89·7	5·6	6
Ploughing with chisels to a depth of 37 cm	98·5	22·6	23
Ploughing without a moldboard to a depth of 35–40 cm	101·7	6·5	6

Source: Shevchenko (1962)

Marked contrasts in catchment response indices can also occur within the general category of pasture and rangeland as a result of the intensity of grazing. Lusby (1970) has demonstrated the influence of grazing on runoff and sediment yield within the Badger Wash basin, Colorado, U.S.A. by comparing the response of four pairs of grazed and non-grazed catchments. The non-grazed basins exhibited 30 per cent less runoff and 45 per cent less sediment yield than the grazed watersheds because the increased vegetation cover density reduced erosion and increased infiltration. The influence of grazing on reduction of vegetation is felt most strongly in areas where plant regeneration is slow, and in these areas excessive grazing can destroy the vegetation cover. Noble (1965) has indicated the effects of reduction in ground cover from good (60–75 per cent cover) through fair (37 per cent cover) to poor (10 per cent cover) on runoff and sediment production in intermontane Utah (Figure 6.21B) and Copeland (1965) cites the classic case of the Parrish Creek basin in Utah. Overgrazing drastically depleted the natural vegetation of this 558 hectare watershed and this was associated with a series of debris floods in 1930 during the summer of which erosion rates were as high as 8000 m³/km². The response of Parrish Creek contrasted strongly with the nearly undamaged Morris Creek where peak erosion rates of only 0·13 m³/km².yr occurred.

Most of the variations in catchment response cited above are conditioned indirectly by man's agricultural practices. Mention must also be made of conservation measures by which man attempts to directly modify the operation of fluvial processes. Rates of runoff and sediment yield for a particular crop or husbandry can be significantly reduced by the application of conservation measures and the Universal Soil Loss Equation (p. 318) makes provision for these effects on rates of soil loss. Crop rotations and careful planning and management of cultivated areas can reduce soil losses by as much as 70 per cent and the use of contour farming, strip cropping and terracing can provide further reductions. Baird (1964) compared two small watersheds in the Texas blacklands, one with good conservation practices and one without conservation, and found that sediment yields of the former were reduced to only 12 per cent of those from the untreated basin.

Plate 20 Gully development
Overgrazing has caused the slopes of this field to be dissected by small gullies.

Contrasts in catchment response between different land-use types may be even more significant than those already described and their extent has been admirably demonstrated by Ursic and Dendy (1965) for the hill lands of the upper coastal plain in northern Mississippi, U.S.A. By comparing individual watersheds under various land-use practices (Figure 6.21C), they showed that sediment yields from cultivated catchments were over 10 times greater than from areas of abandoned fields and forest covers and over 100 times greater than from pine plantations and mature pine-hardwoods. Runoff rates exhibited similar contrasts (Figure 6.21C), with annual runoff rates from cultivated and pasture areas being as much as double those from abandoned fields and woodland. The greatest contrasts usually occur between cultivated areas and forest and the magnitude of the increased sediment yields associated

Table 6.16 Comparison of suspended sediment yields from forested and cultivated tropical catchments

	Sediment yields (m³/km² . yr)	
Locality	Forested	Cultivated
Mbeya Range, Tanzania	6·9	29·5
Cameron Hills, Malaysia	21·1	103·1
Tjiloetoeng, Java	900·0	1900·0
Barron, Queensland	5·7	13·6
Millstream, Queensland	6·2	12·3
Northern Ra., Trinidad	1·8	16·0
Apiodoume, Ivory Coast	97·0	1700·0

Source: Douglas (1969)

with cultivated land in tropical areas is indicated in Table 6.16. The tendency for increased runoff from cultivated areas is not universal, because Lvovitch (1958) has cited Russian data which shows that cultivation of former forest areas has cut by half the streamflow of the rivers of the Transvolga and other areas of the dry steppe zone in European Russia. He has also estimated that flood discharges in the Don (378,000 km²), of which 70 per cent is under cultivation, have decreased by 25–30 per cent as a result of agricultural activity and total yields have decreased by 16–18 mm or 20–22 per cent.

Fertilizer application associated with agricultural activity can affect the solute loads of streams so that they are unrepresentative of natural conditions and contrast strongly with areas where fertilizer application is absent. These effects are most striking in irrigated areas where return water is generally very high in dissolved solids due to leaching of fertilizers and accumulations of salt within the soil. Sylvester and Seabloom (1963) report the case of the Yakima River basin in Oregon, U.S.A. where irrigation schemes have been extensively developed. The average solute concentration of water applied during the growing season was 85 mg/l whilst return water exhibited concentrations of 292 mg/l in the surface drains and 435 mg/l in the subsurface drains. The precise effects of this irrigation return water on the Yakima River was evaluated by studying changes in solute loads between Easton and Kiona, stations upstream and downstream respectively of the receipt of irrigation return flow (Figure 6.21D). Specific conductance, for example, was found to increase from less than 50 micromhos to over 300 micromhos, although the exact extent of the increase depended on season. Irrigation practice can also modify the discharge of a river when large volumes of water are diverted. It has been estimated that irrigation in India and China takes over 6–8 per cent of the annual streamflow, whilst in some rivers such as the Nile and Syr Darya this figure rises to nearer 50 per cent (Lvovitch, 1958).

The process of land-use change and modification can give rise to very marked contrasts in catchment response because the process equilibrium is often drastically upset. Areas of forest clearance and logging activity with their associated track and road construction, heavy machinery operation, and often nearly complete removal of the vegetation cover, have been found in many regions to exhibit increased flooding and particularly increased transport of debris and suspended sediment. Scientifically planned cutting and clearing operations can reduce these effects considerably and the influence of two types of logging practice on the response of small catchments in a Douglas Fir stand in western Oregon has been studied by Fredriksen (1970). He compared the response of three small watersheds in the H. J. Andrews Experimental Forest; one patch-cut and with roads constructed for timber removal, one clearcut and with the logs removed by aerial ropeways, and the third an undisturbed control basin (Figure 6.21E). Contrasts in sediment yields between the three catchments were very significant (Figure 6.21E), and the presence of logging roads was found to be the dominant control over increased yields because whereas vegetation removal was more complete in the clearcut watershed it exhibited a total sediment yield only 3 per cent of that from the patch-cut catchment where roads had been constructed. The forest roads were associated with severe mudflow and landslide activity which scoured the stream channels and which provided a massive source of sediment and debris. There were also contrasts in the relative proportion of suspended load and bedload comprising the total load because whereas the ratio was 1:1 and nearly 2:1 in the control and clearcut catchments respectively it changed to 1:4·5 in the patch-cut watershed where the total sediment yield was 3130 tonnes/km². The increase in the relative importance of bedload in the patch-cut basin was attributed to the landslide activity.

Although the construction of forest tracks and roads can be seen to be of prime importance in altering the response of a catchment during logging activity, the removal of the vegetation can completely alter other aspects of drainage basin dynamics, but on a less spectacular scale. Many studies have demonstrated that forest clearance is associated with increased runoff of up to 47 per cent (Sopper and Lull, 1967) and Brown and Krygier (1967) have shown how clear cutting can alter the temperature regime of small streams. They monitored the influence of clear-cutting on water temperatures in the forested Coast Range near Toledo, Oregon. The daily change in temperature and mean monthly maximum temperatures increased by up to 7 °C. Where herbicides are used to completely devegetate a catchment even greater changes can occur, particularly in the nutrient and solute budgets. The extent of these changes has been demonstrated by Pierce et al. (1970) and Hornbeck et al. (1970) for the Hubbard Brook Experimental Basin, a 15·6 hectare watershed in central New Hampshire, U.S.A. The catchment was completely cleared of all woody vegetation and treated with herbicides for three years. Streamflow volumes increased from 240 mm to 346 mm, and very marked changes in solute loads occurred. Nitrate concentrations increased by an average of 50 times and the major cation concentrations rose between 3- and 20-fold.

Fires are often associated with vegetation clearance and land-use change, although they can occur naturally. They can cause major changes in catchment response, especially increased flooding and sediment transport. Brown (1972) studied the effects of a bushfire which covered about 725 km² of the lower Tumut Valley in south-east New South Wales, Australia, and found that hydrograph shapes altered appreciably as a result of the rapid runoff from the burnt areas. Substantial increases in sediment loads also occurred and the highest observed sediment concentrations and sediment loads in Wallaces Creek were 7052 ppm and 272 tonnes/day before the fire and 143,000 ppm and 118,000 tonnes/day after the event. However, it is to the contrasts between rural and urban areas that we must turn to find some of the most striking spatial variations that can occur.

6.5b   Urban areas

The hydrological environment of urban areas stands in complete contrast to that in rural areas. Because large areas are covered by buildings, concrete and tarmac and vegetation is sparse, evapotranspiration is generally reduced. Gleason (1952) has estimated that evapotranspiration losses in urban areas of Pasadena, California, amount to only 125 mm from streets and pavements, and 610 mm from estates and residential areas, whilst the near-natural areas of lawns and trees exhibit values of 915 mm. Infiltration rates are decreased because of the existence of impervious surface coverings and groundwater levels are consequently lower. Runoff dynamics are drastically altered because of changes in the water balance components, different infiltration and storage conditions and modifications to the stream network. Small stream tributaries are often infilled, whilst others may be diverted into pipes or concrete lined channels, and the network can be extended by the construction of surface water drains, street gutters and even roof gutters. Reduction in infiltration capacities will cause increased storm runoff volumes and reduced base flow and the generally reduced roughness of the catchment surface and channel boundaries and the increased density and efficiency of the drainage network will increase storm runoff intensity, causing increased flood peaks. Protection of the catchment surface and channel boundaries with concrete and similar materials will considerably reduce the supply of sediment to the streams and in the absence of man-made pollution, dissolved loads will be lower because of the reduced runoff contribution from the soil and rock.

The precise magnitude of these changes in response can be demonstrated by reference to several specific examples where urban watersheds have been compared with rural areas, or where the changes induced by the development of an urban area have been monitored. The impact of urban development on the response of East Meadow Brook, a small stream on Long Island, New York, has been studied by Sawyer (1963) and evaluated by Cohen et al. (1968). Urbanization of the drainage basin with the construction of residential housing increased rapidly after 1952, and storm drains were built to carry runoff to the stream. Some of the resultant changes in the pattern of runoff are indicated in Figure 6.22A which shows that annual direct runoff volumes increased by up to 250 per cent after 1951 and that individual storm hydrographs

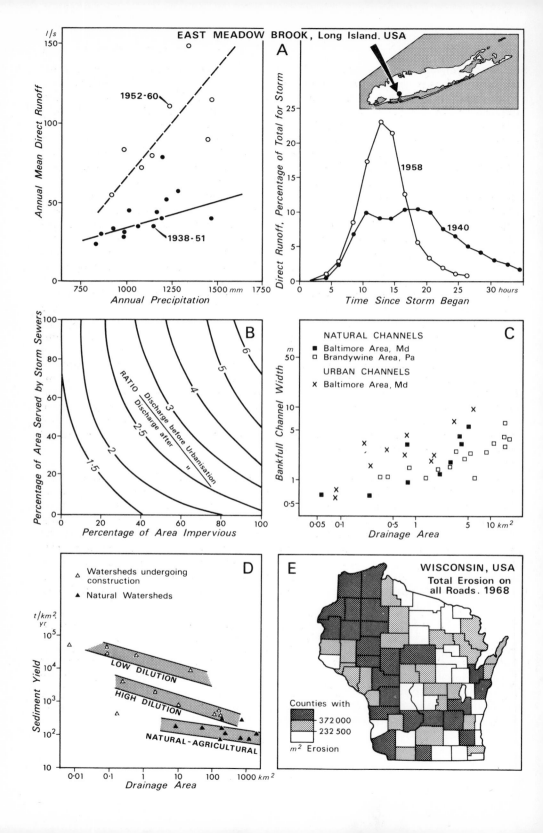

exhibited more pronounced peaks, reduced lag times and more rapid recession limbs in the period after urbanization. Increases in direct runoff volumes and hydrograph peakedness are associated with increased flood intensity and the flood frequency and magnitude characteristics of an urban area will contrast with those from a rural area. Wilson (1967) compared the flood flows of streams draining natural and urban areas around Jackson, Mississippi, and found that the mean annual flood for a totally urbanised catchment was about 4·5 times that from a similar rural basin, whilst the 55-year flood was increased by about 3-fold. A similar study was conducted by Carter (1961) for suburban areas in the vicinity of Washington D.C., and he found a maximum increase in flood peaks of various recurrence intervals from catchments larger than 10 km² of 1·8 times. The exact increase in flood magnitude will depend upon the extent of the impervious area which produces increased storm runoff, and upon the density and degree of development of the storm sewer network which concentrates the flow. Leopold (1968) analysed the results obtained by seven individual studies in order to derive a general relationship between the magnitude of the increase in the mean annual flood and the percentage of impervious area and the proportion of the catchment served by storm sewers (Figure 6.22B). Increase in flood magnitude is associated with increase in the frequency of flooding, and the bankfull capacity of an unmodified channel may be exceeded more often than the average frequency of 1·5–2·0 years. Leopold (1968) has calculated that a catchment with 50 per cent of its area sewered and 50 per cent of its area impervious would exhibit a 4-fold increase in the number of floods equal to, or exceeding, bankfull capacity. Streams will tend to adjust their channel dimensions to such changed flow frequency characteristics, and channels will be enlarged. Wolman (1967) has compared the relationship between bankfull channel width and drainage area for natural and urban channels, and demonstrated that many urban channels exhibit greater widths than their rural counterparts (Figure 6.22C). This tendency for channel scour is also attributable to the decrease in availability of sediment in an urban watershed.

Although it could be argued that dissolved loads of urban streams should be reduced, pollution will in most cases provide for increased solute transport. Koch (1970) compared the dissolved loads of streams draining urbanized and natural watersheds on Long Island, New York, and found, for example, that the average nitrate content of urban streams was about fourteen times greater than that of rural watersheds and that average dissolved solids contents were about three to four times greater. In this particular example the increases were primarily attributable to the large number of cesspools and septic tanks discharging waste into the shallow groundwater reservoir which in turn discharged into the streams. Water temperatures may

Figure 6.22   **The influence of man: urban areas**
The influence of urbanisation on the runoff pattern of the East Meadow Brook is demonstrated in (A) by comparing the relationships between annual precipitation and storm runoff volumes, and the form of the storm hydrograph for two similar storms, before and after urbanisation (based on Cohen *et al.,* 1968). A general relationship between degree of urbanisation of a catchment and the magnitude of the mean annual flood for a basin of 2·59 km², proposed by Leopold (1968), is presented in (B). The influence of urbanisation on channel form is demonstrated in C (based on Wolman, 1967) and the effects of building construction on sediment yield are indicated in D (based on Wolman and Schick, 1967). (E) depicts the magnitude of roadside erosion in Wisconsin (after Briggs *et al.,* 1969).

also differ between natural and urban streams, and Leopold (1968) cites an investigation carried out on Long Island which demonstrated that streams most affected by man's activities exhibited temperatures 5·0 to 8·0 °C above a natural control in summer, and 3·0 to 5·5 °C below in winter. Solute and sediment loads of urban streams can also be influenced by atmospheric pollution and dry fallout, although these effects will also operate over surrounding areas. Weibel *et al.* (1964), for example, measured annual dustfall contributions to the land surface of Cincinnati, Ohio, of as much as 1·4 tonnes/hectare.

Areas of building construction constitute a special, though transitory, zone of urban areas where the removal of vegetation, earth moving and excavation, and the operation of heavy machinery can completely disrupt the surface of a drainage basin. These conditions are associated with increased runoff and, more particularly, increased erosion and suspended sediment transport. The onset of building activity monitored by the authors in the catchment illustrated in Figure 1.10 (Walling and Gregory, 1970) has caused suspended sediment concentrations to increase by as much as 10-fold and storm peaks have increased in magnitude by up to three times. However, much greater contrasts between rural areas and building sites are found in south-eastern U.S.A., especially in the metropolitan district of Washington, where vast areas are cleared for subsequent residential development. Guy (1965) has suggested that the extent of erosion and sediment transport associated with this building activity is sufficient to cause as much as 10 tonnes of sediment to be transported to the Potomac estuary for each person added to the metropolitan area of Washington. Furthermore, Thompson (1970) has estimated that the 2·1 per cent of the metropolitan area of Detroit which was under construction produced as much sediment as the remaining 97·9 per cent of the area. Peak suspended sediment transport is not always associated with storm events, because the operation of machinery within a stream channel can produce the same effect. To cite some general values, Wolman (1967) has estimated that erosion rates on very small areas affected by construction activity may exceed by 20,000 to 40,000 times the amount eroded from farmland and woodland in an equivalent period of time, and a general relationship between suspended sediment yield and drainage area (Figure 6.22D) produced by Wolman and Schick (1967) for natural streams and streams subject to building activity in Maryland and adjacent areas indicates that yields from areas undergoing construction can be as high as 38,000 tonnes/km².yr and up to 200 times those from rural areas. The solute loads of streams draining construction areas can also be expected to be increased because of the disturbance of the soil mantle and the exposure of bedrock, and those of the catchment illustrated in Figure 1.10 have shown a general increase from about 150 mg/l to over 220 mg/l and values in excess of 600 mg/l have been recorded on occasions.

Road and highway construction is closely analogous to general building activity and can give rise to important contrasts in catchment response in otherwise rural areas. The construction of cuts and fills and the disposal of surface runoff into roadside ditches can constitute a severe erosion problem in many areas and the magnitude of erosion rates that can occur on areas of road construction have been monitored by Diseker and Richardson (1962) on six plots of varying slope in Georgia, U.S.A. During 1960, the erosion rates averaged 250 tonnes/hectare on gently sloping plots, 349 tonnes/hectare on medium sloping plots and 489 tonnes/hectare on steeply

sloping plots, and these rates were more than fifteen times those on the steep cultivated farmland of surrounding areas. The influence of the construction of a highway
across a drainage basin on its response has been studied by Vice *et al.* (1969) in the
Scott Run Basin in Virginia. Eleven per cent of this 11·8 km² basin was affected by
road building activity and this small proportion of the total catchment contributed
85 per cent of the total sediment yield. The average yield from the construction area
was 158 tonnes/hectare and the yield of an average storm event in this area was about
10 times greater than from adjacent cultivated areas, 200 times greater than from
grass areas and over 2000 times greater than from forest areas. Roadside verges,
cuts and fills can continue to be an erosion problem long after construction, unless
they are well vegetated and stabilised, and a survey of roadside land in Wisconsin
(Briggs *et al.* 1969) showed that there were more than 29,500,000 m² or 2950 hectares
of active erosion bordering roads in this State. This area was calculated as being
equivalent to a strip of land 5 m wide stretching from Madison, Wisconsin to New
York and back across the continent to Los Angeles. The areal distribution of this
roadside erosion was not uniform and the north-western counties were found to be
particularly dominant (Figure 6.22E).

The development of urbanised areas and industrialisation can have a considerable
impact on surrounding areas in the development of mining activity. Underground
mining can influence subsurface dynamics, but the most important disruptive effects
are those associated with open-cast mining. There are nearly 4000 km² of open-cast
strip mining for coal in the Appalachian region of the United States, and the impact
of surface disturbance and vegetation removal is closely analogous to that of building
construction. Sediment yields are increased, and Davis (1967) has cited examples
where erosion losses from spoil heaps have been as high as 165 tonnes/hectare.yr
whilst losses from nearby undisturbed forest soil were less than 0·4 tonnes/hectare.yr.
Massive landslides resulting from an excess of internal water in a spoil bank can also
occur. Runoff dynamics are modified because excavation activity can seriously
disrupt the normal drainage pattern and the occurrence of bare mined areas and
spoil heaps is usually associated with increased storm runoff and higher flood peaks.
Reduced storm runoff can, however, occur where spoil material provides increased
storage capacity and where spoil banks act as flood retarding terraces. Base flows will
in some cases be reduced but in others, where increased storage capacity occurs, they
may be increased. Davis (1967) refers to a study in Indiana where 67·6 km² of mined
area within a 699 km² study area contributed an average base flow of 200 l/s during
October when streams in the adjacent areas were dry. Pumped drainage to a stream
can also modify the pattern of runoff by directly increasing the flow and by altering
the groundwater levels. Golf (1967) has described the process of draining open-cast
brown coal mining areas in Lower Lusatia, German Democratic Republic, in order
to maintain a cone of depression of the water table in the excavated area. In this area,
an average of 6·3 m³ of water is pumped and drained for every tonne of coal mined
and pumping commences 4 to 5 years before the coal is dug. The result is an increase
in runoff volumes, a more balanced flow regime, increased low flows and decreased
flood flows, but when mining and pumping ceases runoff volumes decrease as a result
of the recharging of the cone of depression. Golf studied the particular case of
two tributaries of the Spree River, the Dobra and the Vetschauer Mühlenfliess,
which represented mined and natural conditions respectively (Figure 6.23A) and

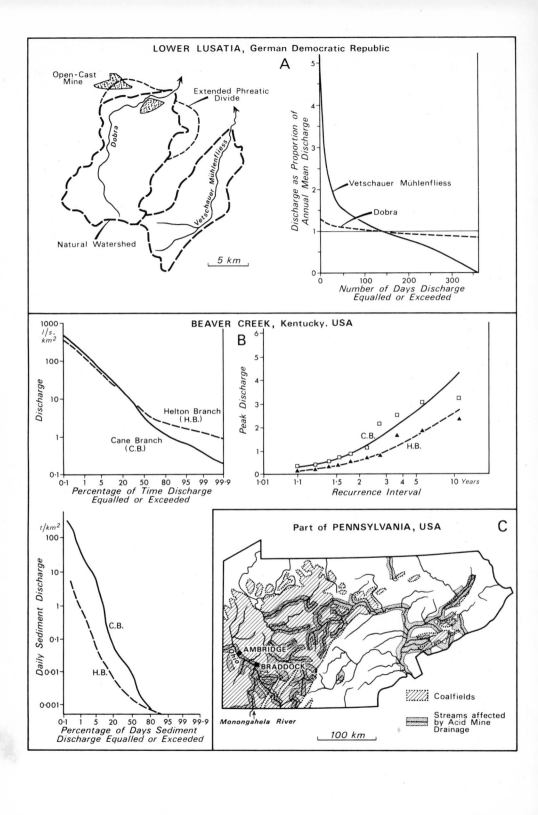

LOWER LUSATIA, German Democratic Republic

A

Open-Cast Mine

Extended Phreatic Divide

Dobra

Vetschauer Mühlenfliess

Natural Watershed

5 km

Discharge as Proportion of Annual Mean Discharge

Vetschauer Mühlenfliess

Dobra

Number of Days Discharge Equalled or Exceeded

BEAVER CREEK, Kentucky. USA

B

Discharge

l/s. km²

Helton Branch (H.B.)

Cane Branch (C.B.)

Percentage of Time Discharge Equalled or Exceeded

Peak Discharge

C.B.

H.B.

Recurrence Interval

Years

Daily Sediment Discharge

t/km²

C.B.

H.B.

Percentage of Days Sediment Discharge Equalled or Exceeded

Part of PENNSYLVANIA, USA

C

Ohio

AMBRIDGE

BRADDOCK

Monongahela River

Coalfields

Streams affected by Acid Mine Drainage

100 km

demonstrated the changes in the phreatic divide consequent upon groundwater pumping and the contrasts between the duration curves of the two streams (Figure 6.23A) which show decreased high flows and increased low flows for the mined catchment. Similar findings were obtained by Dobroumov (1967) who studied the hydrological consequences of mining in the Kursk iron mining region of the U.S.S.R. and particularly of the pumping required to maintain a cone of depression within the workings. The Mikhaylov mine exhibited a maximum drop in the water table of 60 m, a cone of depression 36 km in diameter and a discharge of 180 l/s into the drainage system and the flow of the Oskolets river was at one point 39 per cent higher than the presumed natural discharge.

Exposure of subsoil and overburden in strip mined areas also has considerable influence on chemical weathering and water quality. Most important are the effects of sulphide bearing minerals such as pyrite, and marcasite which form ferrous sulphate and sulphuric acid. Davis (1967) has estimated that one tonne of pyrite may ultimately produce more than 1·5 tonnes of sulphuric acid. The presence of this acid severely modifies the soil weathering processes and instances have been reported where the amount of dissolved material contained within acid spoil may be up to one thousand times greater than in natural soil. Struthers (1961) has aptly christened the 72,900 hectares of strip mining activity in Ohio as 'Ohio's largest chemical works'. Runoff from these disturbed areas and pumped drainage finds its way into streams and radically alters the pattern of solute transport within an area. Figure 6.23C indicates the extent to which the major rivers of the Appalachian area of Pennsylvania, U.S.A. are affected by acid mine-drainage water, and it has been calculated that the Ohio River at Ambridge and the Monongahela River at Braddock, Pennsylvania, transport annually the equivalent of 150,000 and 200,000 tonnes of sulphuric acid respectively (Musser, 1965).

The extent to which strip-mining can cause marked spatial contrasts in catchment response has been admirably demonstrated by a detailed study carried out within the Beaver Creek Basin of south-central Kentucky (Collier et al., 1964 and 1970). The response of the natural Helton Branch watershed was compared with the Cane Branch watershed which was strip-mined over 10·4 per cent of its area. Although it was not possible to isolate precisely the contrasts caused by mining activity from those which would have occurred naturally, considerable differences were found between the catchments. The mined Cane Branch exhibited higher flood peaks and a greater range of flows (Figure 6.23B), the reverse occurrence to that found in Lower Lusatia by Golf (1967) (Figure 6.23A). Dissolved loads per unit area were over 12 times greater from the mined watershed and suspended sediment yields were over 75 times greater. Analysis of the suspended sediment duration data showed that whereas storm concentrations and daily sediment discharges commonly exceeded 30,000 ppm

Figure 6.23 The influence of mining activity on catchment dynamics
The influence of open-cast mining on the phreatic divide of the Dobra catchment and the contrast in the form of the duration curve between the mined Dobra and the unmined Vetschauer Mühlenfliess basins are shown in (A, after Golf, 1967). In (B) the flow duration curves, flood frequency curves and suspended sediment concentration duration curves of the Cane Branch (mined) and Helton (natural) watersheds are compared (after Collier et al., 1964, 1970). The extent of river pollution by acid mine drainage in part of Pennsylvania, U.S.A., is shown in (C, based on Musser 1965).

and 6 tonnes/km² in the Cane Branch catchment, these were the maximum levels recorded in the Helton Branch catchment (Figure 6.23B). In this particular case, man's activity has caused immense changes in the dynamics of a small drainage basin.

Many other instances of man-induced contrasts in the pattern of fluvial processes could be cited, but the above examples provide an indication of the magnitude of these effects. As technology develops, some of man's disruptive activities may be mitigated by careful planning as his control over the environment will increase. Reservoirs already modify the flow of many rivers, and with the increasing demand for water it may be that natural patterns of runoff will disappear. Furthermore, the suppression of evaporation from water bodies by the use of monomolecular films and the artificial inducement of precipitation have already been shown to be practical possibilities.

Selected reading

Information on the pattern of runoff and sediment production of rivers in various areas of the world can be found in:

UNESCO 1969, *Discharge of selected rivers of the world*, Studies and Reports in Hydrology, **5**.
F. FOURNIER 1969, Transports solides effectués par les cours d'eau, *Bulletin of the International Association of Scientific Hydrology*, **14**, 3, 7–47.

Similar detailed data for the United States is contained in two periodic publications of the U.S. Geological Survey: (*U.S. Geological Survey Water Supply Papers*): *Surface Water Supply of the United States* (for example, USGSWSP, **1901**); *Quality of Surface Waters of the United States* (for example, USGSWSP, **2011**).

Examples of studies concerned with spatial variations in drainage basin dynamics at the world scale, the continental scale, and the micro-scale respectively are provided in:

R. P. BECKINSALE 1969, River Regimes. In R. J. CHORLEY (ed.), *Water, Earth and Man*, Chapter 10, 1, 455–71.
D. C. LEDGER 1964, Some hydrological characteristics of West African rivers, *Transactions of the Institute of British Geographers*, **35**, 73–90.
D. E. WALLING 1971, Streamflow from instrumented catchments in southeast Devon in K. J. GREGORY and W. L. D. RAVENHILL (eds.), *Exeter Essays in Geography*, 55–81.

7 Changes in time

... imperative in science, more so than in any other domain, is a careful study of the past in order to understand the present and to subjugate the future. *J. D. Bernal*

7.1   **Theory of drainage basin changes**
7.2   **Short-term changes:** *7.2a Channel and pattern changes; 7.2b Drainage basin changes*
7.3   **Intermediate-scale changes:** *7.3a Channel geometry changes; 7.3b Channel pattern changes; 7.3c Drainage basin changes*
7.4   **Towards the long-term**

The basis for the understanding of the present has been outlined in Chapters 5 and 6. The numerous instances of equilibrium which have been demonstrated between form and process in the drainage basin as a whole and in its component subsystems show that any change over time can lead to an adjustment of basin characteristics or of drainage basin process. Knowledge of present drainage basin form and process and of their inter-relations can provide the basis for an interpretation of the character of changes which may take place over time but in addition the analysis and dating of sedimentary deposits provides further assistance. Increasingly, in sedimentary geology a knowledge of present environments has been employed to enlighten past events and developments, but in fluvial geomorphology comparatively little effort has yet been devoted to the extrapolation of the present into the past and a recurring criticism of the Davisian approach has been that it gave little attention to a knowledge and understanding of present processes. More recently, the increased interest in palaeohydrology (for example, Schumm, 1965, 1967, 1969; Skarzynska, 1965), the study of water and sediment dynamics in the past, introduces a promising field for future investigation. Study of the past in the light of understanding of the present is necessary to develop adequate models of landform evolution and progress has been made in other fields such as coastal and glacial geomorphology. The methods of study must lean heavily upon the additional information provided by superficial deposits, and must sometimes employ spatial variations of the present (Chapter 6) as a basis for indicating how changes may take place over time. It could be argued that a chapter on the changes over time is, even now, premature but the following is presented as a basis for understanding and as an indication of the subjects awaiting investigation.

Changes take place over time as a result of modifications of inputs to the basin, or of transformation of the effects of these inputs on the basin, and of changes of the basin represented by changes of basin characteristics. Changes of input arise directly as a result of climatic change but indirectly an apparent change in input can be manifested through modification of the vegetation cover, for example. Changes of the basin can be achieved by earth movements, and changes of basin characteristics can be a response to these factors, to the effects of sea level fluctuations, or to the hand of man. During the last 50 to 70 million years, during which the land surface has been largely shaped, all these changes have been manifest but isolation of their effects has proved difficult because not all areas have responded in the same way to a particular factor, many areas have been influenced by several factors unevenly distributed over the basin in space and in time, and separation of the influence of a single factor is not easily accomplished. Such attempted separation is complicated by the numerous instances of feedback which are liable to occur. Thus a change of climate with an increased total rainfall could have led to an increase in runoff, but a concomitant increase in density of the vegetation cover could have encouraged a reduced runoff rate and a reduction in the density of the drainage network.

Extrapolation of the results of present studies to longer time periods is potentially subject to error (for example, Winkler, 1970) unless contemporary processes and rates of erosion are interpreted in relation to the present controlling conditions to ensure that grossly misleading estimates are not produced. However, it has been shown (for example, Young, 1969) that rates of erosion as indicated by contemporary measurements are broadly in sympathy with the long-term estimates of erosion based upon landform evidence. More particularly, changes should be apparent through adjustments in channel geometry, in channel patterns and in drainage basin characteristics as a result of development over time. As the channel geometry has been shown to be adjusted, in quasi-equilibrium, with the stream channel processes operating in the context of a particular drainage basin environment, it is inevitable that different stream channel processes in the past should have inspired a different channel geometry Similarly, in the case of channel patterns, development over time could see the replacement of a meandering by a braided channel pattern, and in the drainage basin as a whole, although less easily detected over short periods, changes of characteristics should have occurred, particularly in the case of the drainage network. These three types of adjustment are not necessarily exclusive because the adjustment to change could take one of several forms according to local conditions and basin characteristics. Thus in a particular basin a modified runoff regime following a change of climate or of vegetation cover could, in an area of permeable rocks, find expression in a drainage network of modified extent, whereas on an impermeable rock the network could remain virtually unchanged but the channel geometry or the channel pattern could be adjusted. The adjustment that occurred would therefore probably take place according to the principle of least work advocated in the analysis of channel patterns (p. 257). Furthermore, the drainage basin has been studied in different ways and at different time scales so that conflicts in interpretation have arisen according to the experience of different researchers. Hack (1960), for example, contended that a dynamic equilibrium approach to landscape did not require the recognition of fragments of 'stage' in the landscape, whereas Holmes (1964) cited cases where the present landscape can be interpreted in terms of stages of development in the past. It is, therefore, necessary to consider the drainage basin at several time scales as was recommended by Schumm and Lichty (1965) and included in Table 1.4. The *short term* changes are arbitrarily considered as longer than 50 years (less than 50 years discussed in Chapter 6) but generally not greater than 2000 years and thus this represents the period in which human activity has predominated and in which climatic change has been instrumentally recorded or traced from historical records and evidence. *Intermediate* scale changes may be regarded as those which occurred during the bulk of Quaternary time and therefore encompassed the period in which changes of climate and of sea level were paramount. The *long term* scale is that of the pre-Quaternary and this will be introduced very briefly because at that scale reconstruction of basin development requires different techniques, because it is of less significance for the future development of the basin and for the geographer's interest in it, and because virtually all present basins must have been indistinguishable at that scale. Illustration of these three time scales and their effects must logically be prefaced by outlines of the theories proposed to indicate the magnitude and direction of time-based changes.

7.1 Theory of drainage basin changes

One of the first recent attempts to indicate the way in which changes should take place was made by Lane (1955) in a paper which aimed to reconcile the previously largely independent approaches initiated by G. K. Gilbert and by W. M. Davis. He adopted an equation of equilibrium (Table 7.1A) and distinguished six categories of change which could occur in response to the changes of the four variables. Slope will increase (Class 1) following an increase of Q_s or of D or a decrease of Q_w and these could be achieved by a climate or vegetation change or by interference by man. Lane cited the bed of the Yuba river in northern California which was raised 6 m when hydraulic gold mining after 1850 discharged large quantities of gravel into the stream. A decrease of Q_s or D (Class 2) could be accomplished as a result of dam construction or of reservoir development which promote aggradation upstream and lead to a reduced sediment load downstream. A sudden increase of slope at a particular point (Class 3) may arise from dam construction, landslide, or lava flows. A lowering of stream base level will be associated with Quaternary sea level fluctuations (Class 4), and changes of base level without a change of elevation can be achieved by natural or artificial diversions (Classes 5 and 6).

This approach, therefore, focuses attention upon the fact that if adjustment between the form and process of river channels is expressed in simple terms then a change in one of the variables will necessarily lead to an adjustment of other variables in the system. More recently, Schumm (1969) has characterised the changes which take place in rivers over time as *river metamorphosis* and, from the empirical relations between the channel geometry and sediment characteristics of 36 stable alluvial rivers for which hydrologic data was available, he has derived four general relations (Table 7.1B). The empirical relations used were those which indicated that channel width (w), channel depth (d) and meander wavelength (M_L) were directly related to water discharge whereas slope was inversely related and so, on this basis, an approximation (i in Table 7.1) for mean annual discharge or mean annual flood

Table 7.1 Theoretical approaches to channel changes with time

---

A   E.W.LANE (1955)

$$Q_s D \simeq Q_w S$$
where Q_s = bed material load
D = particle diameter
Q_w = water discharge
S = stream slope

Categories of change
Class 1   For increase of single variable Q_s or D, or single decrease of Q_w, slope will increase
Class 2   For decrease of single variable Q_s or D, or single increase of Q_w, slope will decrease
Class 3   Sudden increase of slope at a particular point, Q_s will increase
Class 4   Lowering of stream base level—slope initially greater
Class 5   Change of base level without change of elevation—base level effectively moved down-
and 6   stream (5) or upstream (6)

B   S.A.SCHUMM (1969)

On the basis of empirical relations suggests, for stable alluvial channels in semi-arid and sub-humid
  areas

(i) $$Q_w \simeq \frac{w \cdot d \cdot M_L}{S}$$ where w = channel width
d = channel depth
M_L = meander wavelength
S = stream slope

and if Q_w is constant then

(ii) $$Q_s \simeq \frac{w \cdot M_L \cdot S}{dP}$$ where P = sinuosity

$$Q_w^+ \simeq \frac{w^+ d^+ M_L^+}{S^-}$$ e.g., diversion of water into system

$$Q_w^- \simeq \frac{w^- d^- M_L^-}{S^+}$$ e.g., diversion of water out of system

$$Q_s^+ \simeq \frac{w^+ M_L^+ S^+}{d^- P^-}$$ e.g., deforestation or increase in cultivated area

$$Q_s^- \simeq \frac{W^- M_L^- S^-}{d^+ P^+}$$ e.g., afforestation or modified cultivation pattern

If Q_w and Q_s do not change alone

(iii) $$Q_w^+ Q_t^+ \simeq \frac{w^+ M_L^+ F^+}{P^-} S^- + d^- + ;$$ (v) $$Q_w^+ Q_t^- \simeq \frac{d^+ P^+}{S^- F^-} w^{\pm} M_L^{\pm}$$

(iv) $$Q_w^- Q_t^- \simeq \frac{w^- M_L^- F^-}{P^+} S^- + d^- + ;$$ (vi) $$Q_w^- Q_t^+ \simeq \frac{d^- P^-}{S^+ F^+} w^{\pm} M_L^{\pm}$$

where Q_t = percentage of bed load
F = width depth ratio

(Q_w) was derived. Similarly, employing empirical relations and the assumptions that
the per cent silt-clay in the channel bed and banks reflects the sediment moving
through the channel, and that this percentage is inversely proportional to the bed
material load, an approximation for bed material load was derived (ii in Table 7.1).
This then permits the deduction of a number of situations, for example, where water
is diverted into (Q_w^+) or out of (Q_w^-) the river system, or where increased erosion is
induced by deforestation (Q_s^+) or erosion is decreased (Q_s^-) following conservation

measures or afforestation, for example. In many cases, however, Q_w and Q_s change simultaneously, being interrelated, and are dependent upon the same variables; it is not then possible to assume that the per cent silt-clay is inversely related to Q_s because discharge is varying. Therefore, Schumm employed per cent bedload (Q_t) to indicate those situations where, for example, water is diverted from a bedload to a suspended load channel (iii), where dam construction leads to a reduction in water and sediment (iv), or where increased water use or improved land use can give rise to reduced water discharge but relatively increased per cent bedload (vi).

At the level of the drainage basin fewer detailed analyses have been attempted but the general directions of change have been available since the work of Langbein and Schumm (1958). The relationships which they proposed between mean annual precipitation and runoff and between mean annual precipitation and sediment yield provided a basis for sketching the direction of changes of drainage basin processes accompanying changes of climate (Figure 1.4B, C). Dury (1967) has noted that these relationships are not based upon a large number of values and so may subsequently require refinement. An alternative approach to the basin was provided by Strahler (1956, 1964) who envisaged an equilibrium equation for the basin as

$$D_d = \frac{1}{H} f \left(Q_r K, \frac{Q_r \rho H}{\mu}, \frac{Q_r^2}{Hg} \right)$$

where $Q_r K$ = the Horton number expressing the relative intensity of processes where Q_r is runoff intensity, K is erosion proportionality factor representing mass rate of removal per unit area divided by force per unit area

$\dfrac{Q_r \rho H}{\mu}$ = a form of the Reynolds Number (see p. 236) where ρ is density of fluid, H is relief, μ is dynamic viscosity of fluid

$\dfrac{Q_r^2}{Hg}$ = a form of the Froude Number (see p. 236) where g is acceleration of gravity.

Therefore, upsets of steady state can be envisaged using this equation and, for example, removal of the forest cover and its replacement by intensive cultivation would lead to an increase of the Horton number through increased runoff or surface erosion or both, and this culminates in an increased drainage density often expressed in gullying.

7.2 Short term changes

Over the last 2000 years the influence of man has been very significant in the drainage basin. This has been noted increasingly since the early studies by George Perkins Marsh (1864) and by Sherlock (1922) and has prompted reactions like that of G. W. Lamplugh who in 1914 commented that 'I am constantly struck with the effect of human culture upon the streams. Hardly in any particular has Man in a settled country set his mark more conspicuously on the physical features of the land.' The significance of man's influence exerted through agricultural improvement was clearly recognised by Beardmore in his Manual of Hydrology in 1862 and in 1830 T. L. Lauder had appreciated that flooding must have been increased in Scotland as a consequence of man's improvement of the land. Further interest in the effects of man was encouraged by the spread of soil erosion and the necessity to develop methods of soil conservation (Bennett, 1939). In South Carolina in 1911 a soil survey of Fairfield County showed that approximately 36,000 ha of land formerly cultivated had been gullied so extensively that it had to be classed as rough gullied land and an additional 19,000 ha of rich bottom land had been converted into swampy meadow land because streams had exceeded the capacities of their previous channels. Study and measures for remedial action were therefore developed in the U.S.A. (Bennett, 1939), in New Zealand (Cumberland, 1944) and in the U.S.S.R. (Tricart, 1950). Although the impact of man's influence has been reported, at least in some form, from most of the areas of the world, the impact was perhaps greatest in those areas where land development has occurred over the last few centuries and also in areas with a climate which was subject to substantial short-term variations. The areas most prone to man's influence are perhaps suggested by the position of the peaks in the rainfall-sediment relation proposed by Schumm (see Figure 1.4C).

Some of the major ways in which man has modified the character and the function of the drainage basin are summarised in Table 7.2. These ways are principally those

Table 7.2   Examples of drainage basin changes effected by man

Many of these can instigate changes of stream flow and of sediment and solute production and subsequently result in modifications of channel geometry (g), channel pattern (p), drainage network (n)

	Direct changes	Form affected
Drainage network changes:	irrigation networks	n
	drainage schemes	n
	agricultural drains	n
	ditches	n
	road drains	n
	storm water sewers	n
Channel changes:	river regulation	g p
	bank stabilisation, protection	g p
Water and sediment balance:	abstraction of water	g
	return of water	g
	waste disposal	g
	Indirect causes	
Land use:	cropland	n p g
	building construction	p g
	urbanisation	n p g
	afforestation	n p g
	reservoir construction	p g
Soil character:	drainage	n
	ploughing	n p
	fertilisers	

in which man has modified processes directly by abstraction of water, by export of water from one basin to another, and by influencing the solutes in a river, for example. Indirect effects also arise from the influence which man has exercised upon the characteristics of the drainage basin (Table 7.2). The implication of such changes is that it is necessary to know the magnitude of man's influence upon present processes before the rates and character of present processes can be used to assist in the interpretation of past situations. This is particularly significant in the case of the flood hazard and the sediment yield from drainage basins. In central Wales Howe, Slaymaker and Harding (1967) analysed the streamflow record of several basins tributary to the Severn and showed that in the period 1911 to 1940 a flood height of 5·1 m was expected once every twenty-five years. In the period 1940–64 this flood height was attained once every four years. The 25-year flood was 5·1 m in the first period but was 5·94 m in the second. The possible explanations for this increased flood hazard were sought from increased rainfall and from changes in watershed characteristics by man. It was shown that daily rainfalls of at least 63·5 mm at two stations had increased markedly in occurrence during 1940–64 as compared with 1911–40. In addition drainage density had been increased particularly due to afforestation, and had increased by the order of 0·7 km per km², and so was also partly responsible for the increased flood heights.

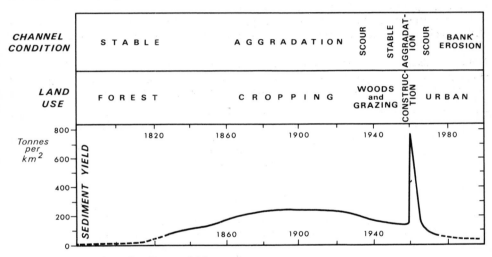

Figure 7.1 Variation of sediment yield over time
This model based upon Piedmont, U.S.A. was deduced by Wolman (1967).

Such increased flooding has been suspected in many other areas consequent upon man's activities in modifying land use, in regulating the drainage network, and in modifying river channels but general changes should also have occurred in the pattern and amount of sediment production. Thus in south-east Asia and in Australia Douglas (1967) has shown that sediment yields from areas substantially affected by man are greater than those from relatively undisturbed areas. In the middle Atlantic Piedmont region of the U.S.A. Wolman (1967) has proposed a sequence of changes which may have occurred through the phases of land use from pre-agriculture to a completely urbanised landscape. This model (Figure 7.1) suggests that prior to the advent of farming low sediment yields obtained but were increased substantially during the farming era, decreased slightly prior to urban development when farmland reverted to grazing or to woodland, increased dramatically during building construction to values of the order of 800 tonnes/km², and then finally in the urbanised phase sediment yield declined to values comparable to those of the initial pre-agricultural phase.

7.2a   Channel and pattern changes

As a consequence of man's effects or of climatic variation there should be detectable differences in the hydraulic geometry of stream channels. Thus Wolman (1967) incorporated channel geometry into his model (Figure 7.1). He indicated that changes would occur during the farming era due to aggradation; during the pre-urban phase through scour which was brought about by the reduced sediment load consequent upon land reversion; during aggradation in the urban phase due to the large quantities of sediment made available by urbanisation; and during scour and bank erosion in the subsequent urban phase. The latter situation arises because much of the catchment is unable to produce suspended and bed load material and so this encourages erosion of the channel bed and banks. Thus changes in hydraulic geometry are accomplished by bed and bank erosion in response to a reduction in sediment yield relative to water discharge, and by aggradation following an increase of sediment yield relative to water yield. Few specific studies have yet been made to indicate the exact ways in which channel geometry may be adjusted but Wolman (1967) tentatively compared the bankfull width of urban channels with the bankfull width of non-urban channels and showed that the former were substantially wider than the latter, presumably reflecting the increased runoff of urban areas (Figure 6.22). In the Centre Creek catchment covering 22 km² of South Island, New Zealand, substantial modification of the stream channels has been reported (O'Loughlin, 1969) following the accelerated erosion phase which was initiated by European-style farming in the mid-nineteenth century. Comparison of photographs of different dates indicated no perceptible changes over 15 years in the upper part of the catchment, a 5 per cent increase in streambed area in the middle catchment, and a 30 per cent increase of bed area in the lower catchment accompanying aggradation of the lower reaches.

The factors responsible for adjusted channel form over short time periods have also led to modifications of channel pattern in historic times. Such adjustments can have been occasioned by an increase or decrease of either water or sediment discharge provided to the channel reach. Subsequent to such changes adjustments of pattern may have occurred. Schumm and Lichty (1963) have documented the changes which occurred on the flood plain of the Cimarron river in south-western Kansas. Measurements made in 1874 showed that the channel ranged in width from 3 to 92 m and averaged 15·25 m in the 6 counties studied, it was narrow, meandering and probably stable, and the flood plain was grassed and provided good grazing. Beginning in 1914, and continuing until 1942, the channel widened until almost all of the flood plain was destroyed and the average channel widening over the period 1914–39 was 350·6 m. After 1942 the channel became narrower again and averaged 176·8 m. Whereas at some cross sections it continued to widen, at others it narrowed to 20 per cent or less of its former maximum width. The most recent period, from 1955 to 1960, may have been characterised by relative stability with some tendency to channel widening again. This channel behaviour (Plate 21) had originally been attributed to accelerated erosion prompted by overcultivation and the sequence of agricultural

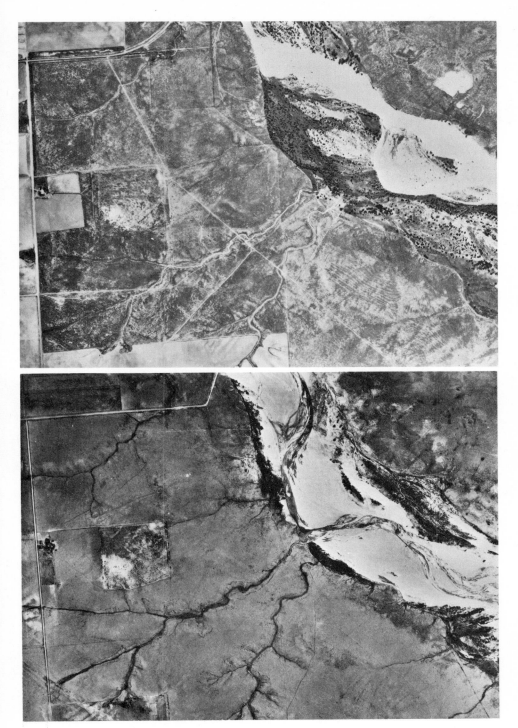

Plate 21 Cimarron River and flood plain, Kansas (*U.S. geol. survey*)
Above, taken in 1936; Below, taken in 1960, show the changes described along a portion of the
river.

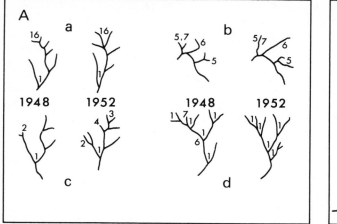
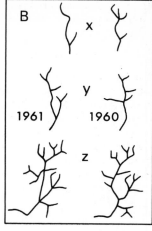

Figure 7.2   Detailed changes of drainage patterns
A is based upon drainage systems measured in 1948 and 1952 in badlands on an industrial dump
at Perth Amboy, New Jersey by Schumm (1956).
   B is based upon systems developed on an upraised lake shore of Hebden Lake Montana following an earthquake in August 1959 (Morisawa, 1964). Drainage patterns mapped after one (1960)
and two (1961) years of exposure showed an increase in the total number of streams, of stream
length accompanying elimination of tributaries but a steady state reflected by total stream length
was rapidly achieved and maintained.

history. However, Schumm and Lichty (1963) demonstrate that the period of channel
widening accompanied a period of below average precipitation and was marked by
floods of high peak discharge, and that the period of flood plain construction took
place during a period of above average precipitation with floods of lower peak
discharge. It thus transpires that not only do the periods of lower annual precipitation
sometimes include events which produce the most significant floods but also that
vegetation growth in the channel is dependent upon the discharge levels and that this
in turn affects river behaviour. In the wet years lower water levels allow a vigorous
growth of perennial vegetation which stabilises the existing deposits and encourages
further deposition. Comparison with similar sequences of floodplain development
suggest that floods in semi-arid and arid environments may be very destructive to the
channel and to the flood plain whereas in humid regions such rapid changes may not
occur.
   Close to the rivers in south-east Wales Crampton (1969) has identified fluvial
terraces composed of coarse material. Crampton has suggested that during the Iron
Age forest destruction probably led to extensive erosion which provided large
quantities of coarse debris for the rivers. The possibly greater runoff during the higher
rainfall and lower temperature conditions of the Atlantic and sub-Boreal periods may
have given the conditions for transport of the material and its deposition in the
valleys. Subsequently it appears that aggradation was succeeded by rivers cutting
into their flood plains and leaving the former deposits as low terrace fragments.
Such sequences of development have been little-reported but must be of wide occur-

rence in the humid lands of the world in view of the changes of climate and the man-inspired modifications which they have experienced.

7.2b Drainage basin changes

Distinction of channel and pattern changes (7.2a) from those taking place in the basin as a whole is artificial because channel and pattern changes are often caused by, and are associated with, changes at the level of the basin. Thus the changing regime of a stream expressed in erosion and aggradation, cut and fill, or in changes of channel pattern must be associated with changes in drainage basin characteristics. The most notable basin characteristic susceptible to change is the drainage network. In the short term numerous opportunities are provided to study drainage network changes and in badlands at Perth Amboy a map produced in 1948 was compared with a map of the same system four years later (Schumm, 1956) to indicate the changes which had occurred (Figure 7.2A). A similar opportunity was utilised by Morisawa (1964) after an earthquake at Hebden Lake, Montana in August 1959 had tilted the lake floor so that the northern shore was drowned and the raised south shore exposed some 2·4 to 6·1 m of former lake bottom and beach. On this emerged bottom strip the drainage pattern was mapped after one and two years respectively (Figure 7.2B). The changes from 1960 to 1961 involved an increase in the total number of streams and of stream length accompanied by elimination of tributaries, and it appeared that overall there was little change in total stream length and this could be ascribed to a rapid attainment of a steady state network.

Such opportunities provide the means to investigate the ways in which drainage networks develop but greater attention has necessarily been devoted to the modes of recent change of the drainage network. Such studies can be the basis for an understanding of drainage changes over longer time periods, they can indicate the controls which influence drainage networks, and they can suggest the trends which are characterising contemporary drainage networks in many parts of the world. Over time a drainage network can either expand or contract but the detection of either must be made bearing in mind the expansion and contraction of the drainage network which takes place at the present time (Figure 2.13). Therefore, change over time is really a change in the limits over which the network fluctuates and for this reason expansion is more easily recognised at this time scale than is contraction.

Distribution and pattern of gullying Expansion is most vividly apparent in gullies which have been widely described from areas including the south-west U.S.A., New Zealand, the steppes of the U.S.S.R., the mediterranean lands, South Africa, the Southern Uplands of Scotland, the New Forest in England and many other localities. The pattern of world distribution of gullies appears to be dominated by those arid and semi-arid areas where climatic changes are easily expressed in network changes, and also those areas where the influence of man has been substantial or rapid or both. Gullies have been denoted by regional names including arroyo in the south western U.S.A. and wawoz in Poland. Brice (1966) adopted the definition of a gully as a 'recently extended drainage channel that transmits ephemeral flow, has steep sides, a steeply-sloping or vertical head scarp, a width greater than 0·3 m and a depth greater than 0·6 m'. The general characteristics of gullies, implicit in the many alternative definitions, include the facts that they often have ephemeral streamflow, they are often incised into unconsolidated materials, and they may have a V-shaped cross section, where the subsoil is fine in texture and resistant to rapid cutting, but U-shaped in material like loess where the soil and subsoil are both equally susceptible to erosion. In size they are larger than rills, they are usually bordered by steep sides and heads which often have the appearance of erosional scarps, and they are usually so deep that restoration is impossible with normal tools and they cannot be crossed by a wheeled vehicle or eliminated by ploughing (Plates 20, 22).

Gully characteristics have now been described from many areas and Bennett in 1939 suggested that there were more than two hundred million active gullies in the U.S.A. From many areas it appears that gullies frequently occur in two main locations, in valley floors and on valley sides (Plate 7), that they may be discontinuous, and that they are often intimately associated with aggradation of valley floors. Thus in the Medicine Creek basin of Nebraska (Figure 7.3 A1) valley bottom gullies may develop from small depressions on the valley floor and are sometimes discontinuous. These have been distinguished from valley side and valley head gullies which are recognised by the shape of the gully head in plan view because this may be pointed, broadly lobed or complexly branching (Brice, 1966). Thus slope may be an important localising factor and in eastern Nigeria (Figure 7.3C) gullies were observed only on slopes greater than 5 degrees where the subsurface lithology was susceptible to erosion (Ofomata, 1967). On the volcanic plateau of North Island, New Zealand, Blong (1970) has described gullies, which are often discontinuous, on the floors of large flat-floored dry valleys which are cut in deposits of Taupo pumice sediments dated from an eruption of 130 A.D. (Figure 7.3B). Such discontinuous gullies may begin abruptly at head scarps but down gully the definition of the feature is less sharp and eventually it disappears as a detectable feature of recent erosion. Gullies are associated with aggraded valley floors because aggradation may be the lateral complement of gully erosion and also because gullies may be cut into aggraded valley floors. In western Iowa, Daniels (1966) distinguished entrenched streams, which are streams flowing in a steep-walled trench cut in alluvium, from valley slope gullies, which are small, steep-walled, steeply-incised, elongate depressions on the valley sides. In some studies it has been possible to quantify gully distribution or the rate of gully extension in terms of the controlling factors. Based upon gully activity at locations in Minnesota, Iowa, Texas, Oklahoma and Colorado, Thompson (1964) found an empirical relation which explained 77 per cent of the variance expressed as:

$$R = 0.15\ A^{0.49}\ S^{0.14}\ P^{0.74}\ E^{1.00}$$

where R = average annual gully head advance in feet

A = drainage area in acres

S = slope of approach channel in per cent

P = annual summation of rainfall from rains greater than 0.5 inches in 24 hours

E = clay content of eroding soil profile in per cent by weight.

Mechanics of gully development Some rates of gully development and extension are indicated in Table 7.3 and to appreciate the mechanism whereby gullies develop it is necessary to recall the continuum of erosion processes which occur on hillslopes

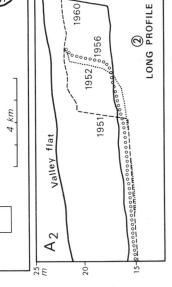

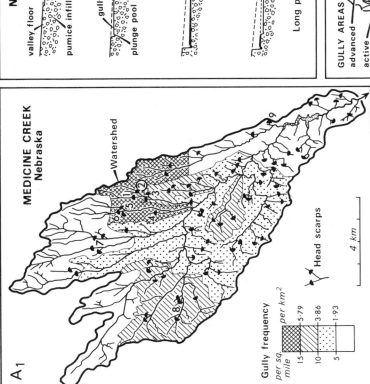

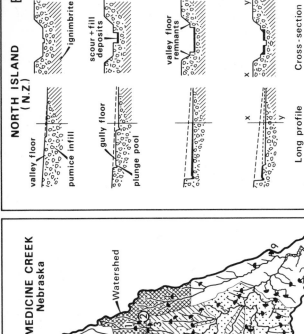

Figure 7.3 Network development and extension by gullying

A_1 shows the distribution and frequency of occurrence of gullies in a basin studied by Brice (1966) and A_2 indicates the rate of development of the long profile of one of the gullies 2 located on A_1.

A sequence of gully development proposed by Blong (1970) is illustrated in B and the distribution of different degrees of gully activity in eastern Nigeria is shown in C after Ofomata (1967).

NORTH ISLAND (N.Z.)   B

valley floor
pumice infill
ignimbrite

gully floor
plunge pool
scour + fill deposits

valley floor remnants

x ——— y

Long profile   Cross-section

GULLY AREAS
advanced
active
incipient

accelerated sheet erosion

ENUGU
CALABAR
PORT HARCOURT

EAST NIGERIA   C

100 km

MEDICINE CREEK
Nebraska

A_1

Watershed

Head scarps

4 km

Gully frequency
per sq. mile   per km²
15   5·79
10   3·86
5   1·93

A_2

Valley flat

1951   1952   1956   1960

25 m
20
15

LONG PROFILE ②

Table 7.3 Some rates of gully development

Area	Progress	Period	Total length	Approximate date of initiation	Source
San Simon Creek, S.W. Arizona	3·3 m	1944–6	104 km	1905	Peterson (1950)
Centennial Wash, near Salome, S.W. Arizona	15·2 22·9	1946–8 1947–8	16	1920 ?	Peterson (1950)
Deadman's Wash, near Shiprock, New Mexico	48·8	1944–8	3·2	1910	Peterson (1950)
Hogback Wash, New Mexico	61	1936–46	1·6	1920	Peterson (1950)
Near Shiprock, S.W. Arizona	45·7	1936–46	16	?	Peterson (1950)
Medicine Creek basin, Nebraska	4·6–6·1 228·6 106·7	4–5 July 1956 1937–52 1951			Brice (1966)
Steer Creek watershed, Western Iowa (loess)	26 lateral gullies	1915–63	16		Beer and Johnson (1965)
New Forest, Hampshire	64	1959–62 (2½ years)			Tuckfield (1964)
Mount Lofty Ranges, Australia	25% increase in length	1854–1957			Dragovich (1966)

and in drainage basins. This continuum can be visualised as ranging from sheet erosion, in which a sheet of water flows over the soil surface and can transport particles some of which may be dislodged by raindrop impact; to microchannel or rill erosion, which occurs when the soil surface is not uniform, the soils have high runoff characteristics and the upper horizons are easily eroded in areas subject to intense storms; and finally, to gully erosion when water is concentrated in definite channels and often succeeds the two previous stages. In eastern Nigeria (Figure 7.3C) Ofomata (1967) distinguished areas where erosion is very slight; in the eastern Highlands, the river valleys, swamps and delta region; areas where sheet erosion predominates, such as on the slopes bordering the Niger and western parts of the Cross river valleys; and areas where gullying occurs in various stages. Usually gullies are found on slopes greater than 5 degrees, are especially active during the rainy season, March to October, and are particularly well-developed on the margins of uplands composed of highly friable sandstones. Erosion of gullies is accomplished by scouring on the bottom or sides by running water, by mass movement of material into the gully from the sides, and by erosion over the well-defined headscarp. In valley floor gullies the the scarp may advance up-valley, facilitated by sloughing of material around the margins of the plunge pool, and in the process the scarp gradually increases in height. In this way the fusion of initially discontinuous gullies may take place and a general pattern of development (Figure 7.3B) has been sketched by Blong (1970). It is notable that development may not be uniform or continuous (Figure 7.3 A2) and Brice (1966) instanced one gully which extended 228 m between 1937 and 1952, 107 m of this occurring as a result of very high runoff in a single year, 1951.

The initial development of a gully in a valley floor location may focus upon one or more evenly shaped pits. In the New Forest Tuckfield (1964) observed the development of gullies from such pits and the deepening and coalescence of the pits during heavy rain. He showed that their coalescence gives rise to discontinuous gullies. Subsequently, rapid gully growth is achieved by fusion of discontinuous gullies rather than by headward erosion and, although deepening occurs by the action of running water, widening appears to be instigated chiefly by freeze-thaw action and by undercutting of the banks. Thus gullies may be initiated from small pits but in addition sinking of the ground and the presence of piping may have a significant influence upon the localisation of gullies. Terraces composed of silts near Kamloops, British Columbia, are dissected by numerous branching gullies and Buckham and Cockfield (1950) proposed that these gullies developed after infiltration, during storm and Spring melting, led to the flushing out of tunnels at levels below the surface. These tunnels collapsed to give funnel-shaped depressions which were subsequently developed into discontinuous gullies. A similar process has been described for the gullies of China (Plate 7) and more recently the localisation of gullies has been associated with piping. Pipes have now been reported in the western U.S.A., in Poland, New Zealand, Australia, China, Africa and the U.K. In Hong Kong, after deforestation, gullies have been recorded which extended by collapse and experienced little surface water flow even during the heaviest storms (Berry and Ruxton, 1960). In the Sudan subsurface channels have been described which have sometimes been used by rodents as burrows but are not exclusively of this origin. These pipes give rise to collapse features on the Nile terraces north of Khartoum (Berry, 1970), particularly on flat or very gently-sloping surfaces, often on land that has been irrigated.

Causes of gully development   The presence of small pits on the surface, or of pipes or sub-surface water movement below the surface may provide the locating factors for gully development but they cannot provide the explanation for why the gullies developed. Reasons for development of gullies include land use pressure; climatic fluctuation, which can be expressed in different runoff rates or in water table fluctuations; and more local factors, such as increased stream gradient due to local or differential uplift, or irrigation diversions which can initiate a gully cycle.

Land use pressure includes those cases where modification of the land use cover and of the surface is effected through a general change in land use. For example, a change from grassland with a dense matte of roots which inhibit surface runoff and promote transpiration to ploughed land which is susceptible to surface erosion, or from forest cover to ploughed land which can similarly instigate a drainage network adjustment. The impact of land use changes is most striking when they occur on high angle slopes which facilitate rapid runoff once the original cover is removed. In the Obervin-schgau of the European Alps gullies developed following the removal of trees on the lower parts of the sides of glacial troughs. Gullies also develop in areas with particular rainfall characteristics, especially the low annual totals and high intensities of semi-arid areas, and in localities with particular parent materials. Loess in China (Plate 22) and in the western U.S.A. has provided the situation for numerous gully systems (for example, Figures 7.3 A1). In New Zealand, gullies which developed after 1960 were associated with land development after 1950 (Blong, 1970) which involved compaction of the ground by cattle, and changes in pasture top soils, including an increase in organic matter and a decrease in infiltration characteristics. In South Africa, in the upper part of the basins of the Orange and Vaal rivers, soil stripping and gully cutting was stimulated by the overgrazing and burning of the degraded grassveld and its partial replacement by open shrub vegetation that provided little soil protection and gave increased runoff after 1880 (Butzer, 1971). There are numerous instances where specific gullies have been localised by vegetation or surface changes. Thus in New Zealand gullies developed after channelling alongside roads and tracks (Blong, 1970), in Nebraska many of the valley side and valley head gullies were associated with roads, buildings and fences and are almost exclusively confined (Figure 7.3 A1) to cultivated land (Brice, 1966). Of the six gullies studied by Tuckfield (1964) in the New Forest three developed along tracks, one along a fire break, one prior to laying a gas main, and one on a strip of land from which the vegetation had been cleared. Arroyos in the south-west United States have been attributed to overgrazing which modified the vegetation cover and the character of the surface, but this is not universally accepted as the complete explanation. The original view resorted to the high stocking densities of the late nineteenth century and in New Mexico 4 million sheep in 1880 were regarded as the reason for gully development. It has been demonstrated, however, that there were high stocking densities in the Rio Grande region in the 1820s accompanied by little or no gullying (Denevan, 1967).

Therefore, although overgrazing had been seen by some as the dominant, if not sole, cause of gullying in the south-west (for example, Antevs, 1952) other workers interpreted gullies as the consequence of *changes of climate* (for example, Bryan, 1928). The significance of past climates was indicated by the discovery of prehistoric gullies recorded in sediments, and by the appreciation of three broad stages in landscape evolution. These three stages were summarised by Yi Fu Tan

Plate 22 Gully development
Above, valley side gullies cut across varied rock types in Coverdale, Yorkshire.
  Below, aerial photograph of gullies developed on loess in China.

(1966) as, firstly, erosion of broad and deep valleys in bedrock of shale and sandstone or in basin fill; secondly, partial alluviation of these deep valleys; and, thirdly, epicyclic processes of erosion and deposition culminating in the widespread entrenchment of the later part of the nineteenth century and the present tendency towards widening and alluviation. Thus the existence of gullying prior to the impact of man necessitates consideration of the significance of changes of climate. Discussion has centred on how climatic fluctuation would be expressed in the vegetation cover and how it would influence runoff and erosion. Increasing aridity may or may not have been associated with an increase in intense rainfalls and accordingly two alternative models have emerged. Firstly, that erosion in prehistoric times was associated with drought because high-intensity storms under drought conditions had a substantial effect when the vegetation cover was reduced. Secondly, it has been argued that periods of intense summer rainfall saw phases of postglacial erosion and that periods of relatively light summer rainfall were characterised by alluviation. In New Mexico, Yi Fu Tan (1966) argues for a closer association of gullying with the intense storms of a wet season than with periods of pronounced drought characterised by small and frequent rains. This may be supported by analysis of rainfall at four stations in New Mexico (Leopold, 1951) which demonstrated a significantly greater number of large rains in the later nineteenth century than in the subsequent decades of the twentieth century and most of these were summer rains.

The occurrence of gullying prior to human influence upon drainage basins has also been noted by Vita-Finzi (1969) in the Mediterranean basin and more specifically by Harris and Vita-Finzi (1968) who showed that in Greece at Kokkinopilos, and in Epirus as a whole, incision has occurred since the deposition of the Red Beds (last glacial) except when interrupted by an aggradational phase that occurred between Roman times and the present. Gullying since Roman times was interpreted as a locally intensified manifestation of a general climatically-induced erosional phase, whereas biotic factors and human exploitation played a subordinate role in facilitating the adjustment of the drainage net to the climatic characteristics. In the study in New Zealand (Blong, 1970) at least one phase of gully activity was identified from sections in fill.

Views of gully development therefore indicate that this expansion of the drainage net is a response to increased runoff prompted either by a change of climate or by a modification of the vegetation cover. In each case the increased runoff is associated with an increased sediment yield achieved by gully development. A sequence of development of gullies may occur in some areas because initially increased runoff and erosion prompt channel widening and the mean depth of flow decreases for a given discharge. In unconsolidated sediments meandering may occur and King (1968) has suggested that widening and meandering lead to a decreased slope, and to reduced velocities, so that ultimately the channel attains a width at which the flow velocity is no longer competent to transport the sediment load so that the coarser fraction is deposited. Aggradation is slow at first, and subject to periodic scour, but in time the deposit can become increasingly stable, vegetated and able to absorb more water. The channel losses by seepage can then reduce water velocities further and encourage greater sediment accumulation.

Such changes over short periods of time have been noted also by Schumm and Hadley (1957) who proposed that discontinuous gullies often develop on areas of local steepening of valley fills, that aggradation downstream can lead to the dismemberment of the drainage net by isolating tributary channels, and that subsequently the drainage net is reintegrated after arroyo cutting dissects the fills. This underlines the fact that extension of the drainage net by gullying must be associated with aggradation for several reasons. Firstly, because gully erosion in one area is associated with aggradation down valley; secondly, because in many areas phases of incision or trenching have alternated with phases of aggradation or filling; and therefore, thirdly, because valley floor gullies are often cut into valley fills. Thus in South Africa active gullying in the Orange and Vaal basins was accompanied by alluviation of the major river channels, and in the Georgia Piedmont the Alcovy river is flanked by swamp and marsh which Trimble (1970) showed to be man-induced following accelerated erosion of the coarse soils in the upper parts of the drainage basins. The consequence of this accelerated erosion, which was initiated by land clearance in Georgia in the early nineteenth century and was locally intensified by plantation agriculture in the mid 1880s, was that sediment accumulation may have averaged almost one metre and was perhaps as much as 3 to 4 m in parts of the bottomland.

More generally, the record of sedimentation and erosion in drainage basins since late Wisconsin time has been studied as 'alluvial chronology', especially in the southwest of the U.S.A. The sequence of development usually contains evidence of a number of stages embracing both filling and cutting (Figure 7.3) which are depicted generally in Figure 7.4 and illustrated by Fivemile Creek in Wyoming (Hadley, 1960). Over wider areas in Wyoming, Leopold and Miller (1954) had demonstrated the presence of three terraces along many streams and major rivers; the highest is post-late Wisconsin, the middle terrace is generally a cut terrace and is developed in the material making up the youngest alluvium of the high terrace, and the lowest is a fill terrace with a surface only slightly higher than that of the present flood plain. This study was used to indicate, according to the sediment accumulation, the rates of sediment production at different times. Sediment production at the time of formation of the second terrace was comparable with that from contemporary streams in the south-west but was more than twice as great as that at the time of the development of the third terrace. This 3-fold sequence in Wyoming has been reviewed by Haynes (1968) and he has identified 3 units in the sedimentary record. The first was represented by the deposits of large streams in the late glacial which filled many of the pre-existing valleys (approximately 11,500 B.P.) and surmounted by weathering under conditions of low soil moisture (pre 6000 B.P.). The second usually occupies channels and includes arroyo cutting (after 7500 B.P.), channel filling (6000–4000 B.P.), soil formation with increased soil moisture (4500–3500 B.P.) and great thicknesses of sediments which include large proportions of locally-derived slope wash sediments. The third unit comprises alluvium and slope wash which overfilled channels (4000–2000 B.P.), erosion and arroyo cutting in some areas, and deposits formèd (post 1500 B.P.) prior to the modern arroyo cutting. The record from the study of alluvial chronology therefore illustrates the way in which changes in climate have been reflected in adjustments in the process, and therefore in the form, of the drainage basin and it also demonstrates the difficulty in distinguishing between the effects of climatic change and of man. Although most widely studied in arid, semi-arid, and mediterranean areas (for example, Vita Finzi, 1968), terracing (see Crampton, 1969) and incision have been recurrent features of drainage basin development during the post-glacial and are illustrated by the Holocene valleys of the Silesian upland (Figure 7.7).

7.3 Intermediate scale changes

Whereas the effect of human activity is notable at the short term, and this has usually produced changes of degree rather than kind except where deliberate human intervention has modified drainage basins directly, at the intermediate time scale more fundamental changes are apparent. Not only can changes of channel form and channel pattern and basin and network characteristics occur on a larger scale but in addition completely new basin forms may have been introduced. Changes of this magnitude have been inspired by changes of climate, of sea level, and of land level. These three categories have prompted two types of change in the drainage basin over Quaternary time. Firstly, direct, fundamental, and often widespread consequences were felt as a result of the shift of the world pattern of morphogenetic systems (Figure 1.5). The

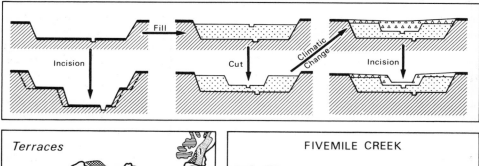

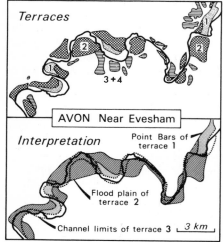

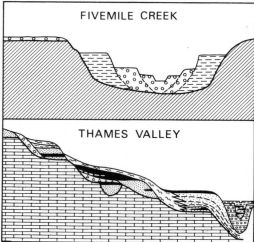

Figure 7.4 Terrace patterns
Several general mechanisms of terrace production by erosion and associated slope retreat and by cutting and filling are suggested above, the terrace pattern near Evesham was interpreted by Dury (1964) to signify the pattern shown and two vertical sequences of terraces are illustrated for Fivemile Creek, Wyoming (Hadley, 1960) and for the Thames valley (King and Oakley, 1936).

reaction to this has been shown by the study of glacial and periglacial geomorphology in middle latitudes and of landform inheritance, in Australia, for example. These changes, although varying in degree, were usually widespread and were categorised by W. M. Davis as climatic accidents. Secondly, there are consequences which have been more gradually effected or more locally distributed and these embrace the effects of sea level change, of land movement and of volcanic activity.

The number of Quaternary stages and their character is not exactly correspondent from various world areas (see Flint, 1971), but the analysis of deposits, of lake fluctuations, of pollen in deposits, and of deep sea cores has allowed the general trends to be established for particular areas. The general trends indicate that temperatures were generally reduced during the glacial periods of the Quaternary and that a difference of some 7 to 9 °C distinguished glacial and interglacial stages, although the temperature gradient would necessarily be steeper near the ice sheet margins. Thus in the location of New York the difference between glacial and present may have been of the order of 15 °C, compared with 11 °C in the Ohio valley and 7–9 °C in the Gulf of Mexico (Manley, 1955). In Europe the full glacial climate may have seen a mean annual temperature 9 °C lower than that of today in southern England (Manley, 1951) but between 5 and 7 °C in southern Europe. Accompanying these temperature changes were varying precipitation amounts and, although in northern latitudes precipitation may have been light in some areas near the margins of Pleistocene ice sheets, and largely confined to the summer months, elsewhere increases have been deduced and in the south-west U.S.A. may have been 25 cm greater during glacial times than at present. Temperature changes necessarily influenced evaporation rates and altered runoff amounts. The way in which the runoff amounts could change according to modification of mean annual temperature and precipitation have been suggested by Schumm (1965). In table 7.4 the change from present to glacial conditions in a Central Europe climate (A), a climate such as the British Isles (B), a semi-arid climate (C) and a tropical climate (D) are indicated together with the changes from the present to an interglacial climate in a temperate climate like B (E) and in a tropical

Table 7.4 Possible effects of climatic change upon drainage basin process (*based upon S. A. Schumm, 1965 and Dury, 1964*)

| | Mean annual temperature (°C) | | Mean annual precipitation (cms) | | Ratio of changed to present | | |
					Mean annual runoff	Mean annual sediment yield	Mean annual sediment concentration
	Present	Changed	Present	Changed			
A	10°	5°	50	75	5	0·5	0·1
B	10°	5°	75	100	3	0·8	>0·2
C	15°	10°	25	50	>20	>2·0	—
D	15°	10°	100	125	2	0·9	>0·5
E	10°	12·5°	75	62·5	0·5	1·5	3
F	15°	17·5°	75	62·5	0·3	1·7	7
G							

		Constant precipitation amount
−11° Evenly distributed	3·0 Evenly distributed
−11° Evenly distributed	2·5 Summer concentrated
−11° Evenly distributed	3·6 Winter concentrated
−11° Winter concentrated	2·1 Summer concentrated
−11° Summer concentrated	3·2 Summer concentrated

seasonal climate (F). These suggestions indicate the direction of changes in mean annual runoff and in mean annual sediment yield but they must be qualified by the facts that in some areas additional modification was occasioned by adventitious factors such as glacial meltwater or permafrost. In addition mean annual runoff may not be a particularly good index because basin elements may be adjusted to bankfull or to mean annual flood discharges which may not correlate directly with mean annual runoff. For example, a present climate such as B (Table 7.4) could have experienced a mean annual temperature 5 °C lower and a mean annual precipitation 25 cm lower during a glacial period and this (according to Figure 1.4B) could have produced a runoff 0·8 of that of the present time. However, this reduced runoff could have been achieved with a changed runoff regime so that the mean annual flood was, in fact, higher than that of the present. Dury (1964) has therefore estimated the changes which could arise from modifications of the mean annual temperature accompanied by a modification of the precipitation regime (Table 7.4G). This demonstrates that a change in favour of a winter concentration of precipitation would provide the greatest increases. Changes to summer concentrations were more usually experienced and such a change accompanying reduction of weighted mean annual temperature by 11 °C would give at least twice as much annual runoff. Speculation of this kind is a very necessary prelude to understanding processes of the past and it must be further developed in the future, in terms of peak process events and based upon a multivariate expression of drainage basin processes, for application to the problems of channel, pattern and basin development.

7.3a Channel geometry changes

River channels should reflect these past climates by being different in size and shape in the past but, as in the case of short term changes where the expectable channel changes are less dramatic, it is difficult in many cases to distinguish former channels from present ones. This difficulty does not arise, however, where deep channels have been infilled by sediments and are sometimes at levels lower than those of the present river. This type of situation was occasioned as a response to the low sea levels of the later part of the Quaternary and in many countries in middle latitudes deep buried channels have been identified beneath the flood plains of present rivers. Definition of the buried channel is difficult because the infilled portion may, in fact, be a buried valley floor and not entirely buried channel but Dury (1964) has identified large channels beneath the contemporary valley floors of many valleys of the English plain, the driftless area of Wisconsin, France and elsewhere. The form of the valley floor shows considerable variation in form and in the depth of the fill which obscures it (Figure 7.5C) but in general terms Dury has suggested that the ratio of buried channel width to present river channel width is of the order of 10:1. An example of a buried channel is provided by sections and boreholes at the sulphur mines of the Vistula valley south of Tarnobrzeg (Figure 7.7C). Dating of the deposits has demonstrated that a braided pattern, during which the gravel series was accumulated during the late Wurm, changed to the meandering stream pattern with finer deposits (Mycielska-Dowgiallo, 1969).

The Riverine Plain of New South Wales, Australia includes deposits and channels representing several phases of development and Schumm (1968) recognised three

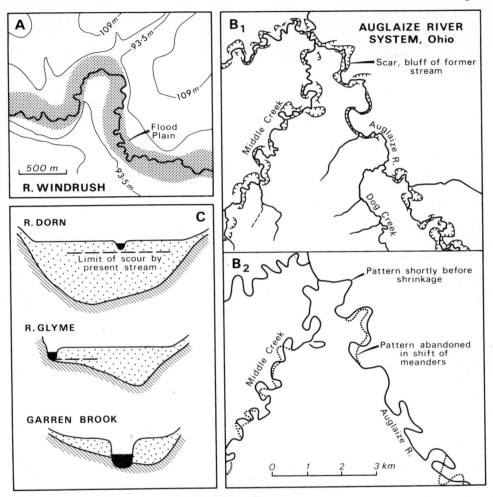

Figure 7.5 Underfit streams and meandering valleys
Illustrating the contrast between the underfit stream and its meandering valley (stippled) in A, showing three possible positions of present river channels in relation to buried channels in B, and demonstrating the present (B_1) and former (B_2) patterns of the Auglaize River. All based upon Dury (1964, 1965).

types of channels (Figure 7.6). The contemporary Murrumbidgee river channel is a suspended load channel which does not transport large quantities of sediment and reflects the modern climatic and basin characteristics. Depressions on the Riverine Plain and on the flood plain of the Murrumbidgee river are underlain by ancestral river channels which are similar to the contemporary river channels in shape, and therefore in width-depth ratio, sediment characteristics and pattern but they are much larger. Other channels termed Prior Stream Channels are relatively straight, wide and shallow and are filled with cross-bedded sands (Figure 7.6). Analysis of the geometry of these three types of channels in relation to present stream channel processes and in relation to sediment characteristics (Figure 7.6) indicated that the Ancestral channels, like the present river channel, were of the suspended load type (p. 243), whereas the prior stream channels were bedload channels. Bankfull discharges of the ancestral channels may have been five times greater than those of the present channels and the prior channel bankfull discharges were also of the same order of magnitude as the ancestral ones. In the light of these interpretations Schumm (1968) proposed a sequence of development which envisaged:

1 erosion of deep channels in Katandra sediment
2 deposition of Quiamong riverine sediments
3 erosion of prior stream channels—possibly associated with floods during a relatively dry climate
4 deposition of Mayrung riverine sediments
5 cutting of ancestral river channels possibly during more humid climate?
6 development of contemporary Murrumbidgee channel.

Similar examples have been identified elsewhere and in the Netherlands, for example, during the Wurm the Spring sno w melt gave high runoff over frozen sub-soil which prompted a characteristic braide d channel pattern. The numerous braided channels were subsequently covered, during the late Glacial, by a thin layer of clay as the sediment became finer, reflecting the replacement of tundra by forest vegetation and the disappearance of the permanently frozen ground. The reduced sediment load led the river to cut down into the sediments of its lower reaches and to replace the braided pattern by a meandering one (Poleman and Pape, 1967). Such studies indicate the potential and scope of palaeohydrological investigation when applied to the problems of past river development and they also indicate that changes in river channel cross section and in pattern cannot be isolated—such changes taking place over time have been referred to as river metamorphosis (Schumm, 1969).

7.3b Channel pattern changes

The buried valleys described from a wide variety of areas were often associated with a channel pattern different from that of the present river and thus in the Vistula valley (Figure 7.7) and in the Netherlands the buried channels were associated with a braided pattern but this was eventually replaced by the contemporary meandering one. Over a number of years the deep buried channels identified by Dury (1964) have been shown to be a widespread phenomenon and to be associated with channel patterns much larger than those of the present rivers. Thus contemporary meandering rivers may flow above a deep buried channel (Figure 7.5C) and be confined within a meandering valley (Figure 7.5A). Dury found that streams in such situations can have a meander wavelength much lower than that of the valley and so he termed such

Figure 7.6 Murrumbidgee River and Palaeochannels, Australia
The generalised section is based upon Schumm (1968) after B. E. Butler, Depositional Systems of the Riverine Plain in relation to soils *Commonwealth Scientific and Industrial Research Organisation*, Soil Pub. 10, (1958), 35 pp. The graphical relationships comparing the present, prior stream and ancestral channels are derived from Schumm (1968).

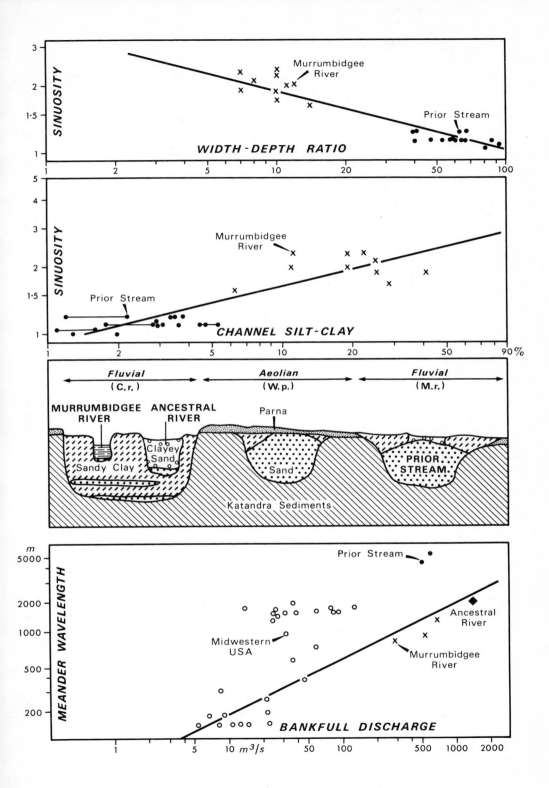

streams underfit streams. Manifestly underfit streams was a term applied to the case where a meandering stream occupies a more amply meandering valley (Figure 7.5A) or where the contemporary meanders follow a much more amply meandering course. The Osage type, named after a river in Missouri, U.S.A., was applied to the situation where a meandering valley lacks stream meanders but the spacing of the pool and riffle sequences does not correspond to valley meander wavelength. In this case, therefore, the present stream still imitates the meandering valley pattern but the pool and riffle sequences are much more closely spaced along the river channel than the meander wavelength would suggest. Such underfit streams have been described by Dury (1964) from a variety of areas including western Europe, where perhaps at least 50 per cent of the length of second and higher order streams are underfit; in Europe as far east as the Ukraine; in all major climatic regions of the U.S.A. including Alaska; and also in Australia.

The degree of underfitness of the present rivers can be indicated by comparing the meander wavelength of the valleys (M_{Ls}) with that of the streams (M_{Lv}). This ratio is generally of the order of 9 or 10 in lowland England, and it averages 5 in the U.S.A. but ranges from 10 near the former ice fronts in Wisconsin to about 3 in the Ozarks. The Osage type shows a ratio of meander wavelength to river width of approximately 40 which is substantially in excess of the usual value of this ratio, which is about 10. Analysis of the morphometric characteristics of the valley patterns and comparison with the present relations between river channel pattern and stream discharge (for example, Figure 5.5), in conjunction with evidence from sediments, allowed Dury to suggest how and when these meandering valleys and underfit streams had developed (Dury, 1964). The two problems which require solution are, firstly, the date of origin of the rivers which cut the deep channels and, secondly, the date by which shrinkage had reduced the river to an underfit form. In areas not covered by ice during the last glacial maximum the date of initiation of the valley meanders could range from early Pleistocene to late glacial. The last abandonment of the largest channels and the valley meanders generally occurred in the late Glacial, probably between 12,000 and 9000 years ago but such a reduction probably occurred on several occasions during the Pleistocene, although only the last phase is clearly evident. In some areas valley meanders may still be active and in Texas it has recently been argued (Tinkler, 1971) that the meandering valleys of south central Texas are still occupied by river discharge. Shrinkage usually occurred, however, and in the case of the Dorn valley, Oxfordshire (Figure 7.5C) pollen in the channel fill indicates that infilling had probably begun by 9000 B.P. An example is illustrated in Figure 7.5B. Explanations for the existence of underfit streams and meandering valleys include river capture, erosion by meltwater from ice or by water from glacial lakes, tidal action, and climatic change. Some examples of underfit streams could have been produced locally by river capture but Dury (1963) has shown that this explanation cannot account for the widespread distribution of underfit streams and that where it has been advocated as an explanation the evidence is often not entirely unequivocal. Similarly, although in some cases overspill from glacial lakes or glacial meltwater may have made a significant contribution, as in the case of some of the meandering valleys found in the Cotswolds, and although the explanation of tidal scour has recently been offered by Geyl (1968) to account for the character of underfit streams in larger valleys, these are of restricted application. Dury (1964) found that all such theories were of restricted areal application and could not be widely applied to explain all the characteristics and distribution of meandering valleys and their underfit streams. The hypothesis of climatic change is the one which remains to meet the requirements of a complete explanation.

Comparison of the meander wavelengths of valleys and of streams in relation to present channel-forming discharges indicated that as $M_L . \alpha Q_{bf}^{0.5}$ then a fivefold increase in meander wavelength (M_L) would perhaps require a bankfull discharge (Q_{bf}) 25 times greater than at present. However, when differences in channel slope, cross section, and velocity are incorporated into the estimates, Dury (1964) suggested that a bankfull discharge some 20 times greater than present was indicated, whereas where the ratio of valley meander wavelength to that of the streams was high (say 10) the former bankfull discharges may have been some 50 or 60 times greater than at present. Temperature change alone is insufficient to account for the greater runoff rates required to give bankfull discharges of this magnitude, especially as in some areas valley meanders may have been initiated during parts of the deglacial succession when air temperatures were rising. Greater rates of runoff would result from the presence of permafrost, which would eliminate infiltration, but this cannot supply a universally applicable explanation because manifestly underfit streams have been identified in the Gulf Coast of Texas and in Puerto Rico well beyond the possible influence of permafrost. Therefore, a change in total precipitation is most likely to be responsible and Dury visualised increases in precipitation by a factor of 1·5 to 2·0 which, in conjunction with temperature change (Table 7.4), are capable of increasing runoff by a factor of 5 to 10 times within a wide range of existing climates. This general suggestion of increased precipitation in early deglacial times is supported by the fluctuations of pluvial lakes and does not conflict with the reconstructions of world weather patterns or with analyses of evidence from deep sea sediments (Dury, 1964). In some areas an increase in storminess, associated with a slightly higher annual precipitation, could provide significantly higher runoff rates.

The remnants of former channel patterns, dating from contrasted climates in the Pleistocene, are therefore present in many areas. In addition, if the long profile of the river course is plotted, indications of several stages of down-cutting may be evident. The realisation that the long profile of the river is concave upwards may date back to the French engineer Surrell in 1848, the idea of base level as the level to which stream profiles are developed was originated by Powell in 1875, and these ideas were incorporated into the cycle of erosion by W. M. Davis in 1899 and assimilated into the interpretation of eustatic sea level changes by Baulig in 1935. Unfortunately, several concepts have become confused during this development and the idea of grade has been applied to the river profile and also to the stream. A graded stream was defined by Mackin (1948) as one which has 'slope delicately adjusted to provide, with available discharge and with prevailing channel characteristics, just the velocity required for transportation of the load supplied from the drainage basin'. The notion of the graded stream has been found difficult to apply because erosion in estuaries can occur below mean sea level, adjustments in the stream dynamics system can be accomplished by variables other than slope, and the sequence of recent evolution conditioned by the climatic changes of the Pleistocene necessarily provides a complex picture. The idea of the graded stream was therefore rejected as being unserviceable by Dury (1966) but this does not destroy the validity of the concept of a long profile which incorporates the remnants of several phases of development. Many long profiles consist of component curves and these are a response to either land uplift or sea level lowering which has occasioned the development of a new valley floor cut below the former one. At the junction of two former profiles a knickpoint may be preserved and this may sometimes be indicated by a waterfall (Plate 6) or by rapids. The development of a long profile in stages may also be accompanied by the comparable development of the cross profile which includes remnants of former valley floors as terraces or valley side benches. Other features of this 'rejuvenation' can be meanders which are incised; as ingrown forms if the meandering continued during deepening to give an asymmetrical cross profile, or as intrenched forms if the incised portion is symmetrical in cross profile and the meanders did not shift laterally during downcutting.

Also associated with valley development are remnants of former valley floors which may be preserved on the valley sides as river terraces. Terraces may be cut in bedrock or in alluvial deposits and they may therefore be erosional or depositional in origin, and some general sequences of development are indicated in Figure 7.4A. Furthermore, terraces on opposite sides of a valley may be paired or, particularly if they are developed when a meandering stream is cutting into its valley floor, they may be unpaired. The distribution of remnants of particular terraces may be employed to

Plate 23 Fluvial landforms
Braided river flanked by well-developed terraces flowing from the Southern Alps of New Zealand. Alluvial fans occur at the junction of the Alps with the terraces.

reconstruct the pattern and extent of the flood plain at previous stages (Figure 7.4B). Very often this information may be used in conjunction with a plot of the long profile of the river and with the height distribution of terrace fragments to facilitate the reconstruction of former valley long profiles (for example, Figure 7.11). A recurrent difficulty in terrace interpretation is that fluvial deposits and landforms may be associated with other deposits. Valley cross sections may, therefore, be very complex as illustrated by the view of the terraces of the Thames suggested in 1936 by King and Oakley (Figure 7.4). Although difficult to identify, a category of terraces, termed thalassostatic by Zeuner (1945), is represented by deposits which were formed by aggradation during the high sea levels of the Pleistocene and subsequent dissection left fragments of these deposits on valley sides.

7.3c   Drainage basin changes

The effects of changes of climate, of land level and of sea level have been reflected in the vertical development of valleys in stages during the Quaternary and it is inevitable that these controls should also have influenced the overall character of the drainage basin. The nature and extent of the drainage network is a characteristic particularly sensitive to change although increased dissection of the basin could also provide modified relief ratios of basins. Whereas expansion of the drainage network is most obvious at the short term scale (p. 369), at the intermediate scale contraction is more obvious because expansion may absorb and obscure traces of earlier more reduced networks. Contraction of drainage networks would be expected in many areas because as the drainage network has been shown to be related to climate as well as to other drainage basin characteristics, a change in climate such as is necessary to account for underfit streams should, at least in certain areas, have been accompanied by drainage network fluctuations.

The landform most indicative of drainage network change at the intermediate scale is the dry valley. A dry valley immediately presupposes a discrepancy between the more restricted extent of the present stream pattern and that of the valley pattern which it partly occupies. Definition of the stream network and of the valley network is not easy. The stream network cannot be defined simply according to the pattern of streamflow at a particular time because the stream network fluctuates during a particular year (pp. 86 and Figure 2.13) and also because such fluctuation is an expression of present climate in some morphogenetic systems. Thus in climates with a marked seasonal component seasonally dry valleys are a common phenomenon and are found in the Arctic, in arid and semi-arid, in temperate and tropical continental climates. In the subhumid warm temperate climate of part of Australia Jennings (1967) has noted that intermittently dry river beds are not restricted to limestone because all small creeks dry up for varying periods during the summer. The problem of seasonally dry valleys can be accommodated by considering the distribution of water courses including all definite channels to approximate to the present drainage network. However, in some areas water flow may in certain years occur beyond the limits of the stream channel net. In Britain it has been suggested (Kirkby and Chorley, 1967) that some short dry valleys may experience streamflow once in 200 years and in addition there are instances of flow being noted along the classic dry valleys of the English Chalk outcrops (for example, Anon, 1940).

Networks of valleys are equally difficult to define precisely because the valleys and depressions which contain no evidence of flowing water at the present time embrace a wide variety of morphological types. Dry valley is a term usually applied to clearly-defined valleys, especially on limestone outcrops (Plate 24), and on cuesta landscapes there is often a contrast between those of the dip slopes; which are longer, shallower and which grade gradually into the head of the dip slope; and those shorter, steep-sided features with a marked head that are found incised into the scarp slopes and often described in England by the term coombe. In addition, there are dry valleys which have a more subdued cross section, usually a convexo-concave form (for example, Figure 7.7B), which are distinct from the valleys at present occupied by streams and these features were reviewed by Schmitthenner (1925) and termed dells or dellen. Related features have subsequently been described by terms which imply mode of origin including corrasional valleys (Penck, 1924) and derasional valleys (Pecsi, 1964). Those forms with a flat floor have been termed cradle valleys (vallons en berceau) and Greenwood in 1877 regarded such shallow valleys at the head of present streams as an expectable phenomenon and described them as rain valleys.

Some of these terms indicate the areas in which dry valleys of several kinds have been described. Firstly, they were observed and discussed on areas of permeable rock, and particularly on limestones, but in Britain, although those on such limestone outcrops attracted most attention at first, it was later realised that they are present, although less extensively, on other rock outcrops as well (Gregory, 1966). Secondly, areas which experienced periglacial conditions during the Pleistocene have provided examples of dry valleys and they have been described extensively in western, central and eastern Europe and have often been referred to as dellen. Although less frequently described in the United States broad, rounded depressions in the Piedmont upland were referred to as dellen (Sharpe, 1941) and although many of these were originally streamless, many were dissected by gullies in 1941. Thirdly, in arid and semi-arid areas there are wadi-systems which may include systems which do not function during the occasional present storms and in the Kurkur oasis of Egypt Butzer (1965) mapped a valley system partly extending over limestone, which has a density of 2·3 and which is now a relict system (Figure 7.8). Fourthly, in areas of high relief some fossil valleys have been detected in Japan, in New Zealand where Cotton (1963) has recognised that the fine-textured relief may in part be a relict phenomenon, and in the Scandinavian mountains where the chute slopes offer a locally significant way in which valley-like forms can be produced which may not necessarily be a part of the present drainage network.

Although few studies have compared stream and valley networks to ascertain the morphometric properties of former networks, it is apparent that dry valleys are a widespread phenomenon. This conclusion is based upon the facts that there are valleys which do not have a stream channel at the present time, which have no instances of recorded water flow along their length, and which do not continue the lines of the present drainage network. Similar features, which are obviously fossil, have been recorded in sections in New Zealand (Cotton and Te Punga, 1956) and

Plate 24   Dry valleys
Above, two types of dry valley developed on sandstone in Devon described on page 395.
Below, dry valley on Carboniferous limestone in the southern Pennines.

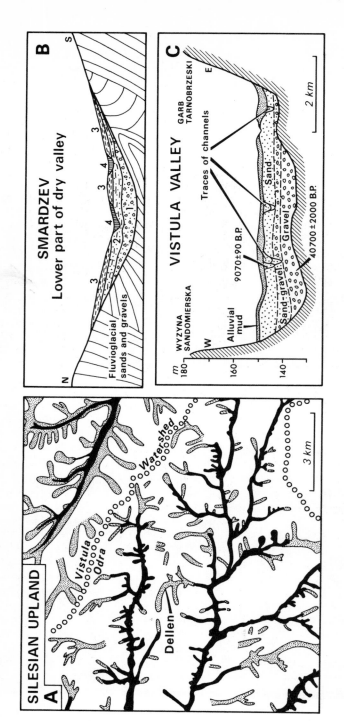

Figure 7.7 Some valley patterns and forms in Poland
In the Silesian Upland (A) the distribution of Holocene features (black) including young V-shaped valleys, flat-floored valleys and Holocene accumulation plains is complemented by periglacial dellen (stippled) after Starkel (1963). A cross-section through a dry valley is illustrated in B and the infilling deposits include (1) medium-grained sands with gravels of waxing Würm, (2) sands and silts of Würm climax, (3) rhythmically bedded sands with ferruginous striae, (4) silts and some stony horizons in present axes of dry valleys (after Dylik, 1961, Figure 8). The section across the Vistula valley is based upon Mycielska-Dowgiallo (1969) and shows the dateable sequence of deposits beneath the present valley floor and indicates the location of former river channels.

in Devon (Waters, 1966). The extent of dry valleys is difficult to describe quantitatively but on limestone outcrops valley densities are greatly in excess of stream densities and in the Dove basin (Figure 2.14) the valley density is 2.06 compared with the stream density of 1·36. On the chalk of Picardy in France a valley density of 0·74 contrasts with a stream density of 0·10 (Pinchemel, 1957). In south-east Devon in a drainage basin developed on sandstones, marls, and Greensand of 232 km² area the present stream density is 1·60 and the valley density is 2·67 (Gregory, 1971). Explanations for the presence of such discrepancies between networks of streams and valleys, and for the dry valleys which are responsible for this discrepancy, must be visualised in the context of the mechanics of the drainage basin and they must be set into scales of geomorphological time because some explanations are more relevant at one scale than another. Figure 7.9 attempts to achieve these two objectives and it is apparent that explanations for the existence of dry valleys can be envisaged in three main groups: non-fluvial, fluvial under a morphogenetic system similar to that of the present, and fluvial under a contrasted morphogenetic system.

Anomalous dry valleys can occur in a landscape as a result of a wide variety of non-fluvial causes including mass movement, which can leave depressions after removal of material especially by avalanches, nivation which can produce elongate hollows perhaps under snow patches as vallons de gelivation, and solution of calcareous rocks followed by collapse which could produce landforms reminiscent of valleys. More widely, development of surface depressions as alas may be typical of permafrost areas (Figure 7.8) and provide the basis for subsequent development of the valley and stream network; marine action may have been significant in some cases and Winslow (1966) has proposed that the dry valleys of the chalk of south-east England may have been initiated as submarine canyons during the Calabrian transgression; and in deglacierised areas a variety of glacial drainage channels has been distinguished, eroded by glacial meltwater, which may have remained independent of the present drainage network and therefore appear as dry valleys (for example, Bowen and Gregory, 1965). Such non-fluvial origins for dry valleys apply to specific and usually very localised areas and some of them are applicable to valley development in general and not merely to the valleys which are now dry. Therefore, more widely applicable explanations have been sought for the presence of dry valleys and these include those which envisage fluvial processes as being responsible under drainage basin conditions very similar to those of the present day.

That stream networks could have been more extensive in the past but under conditions similar to those of the present time appears incontrovertible in view of the fact that some dry valleys may experience occasional water flow and that the higher runoff rates of the past indicated by other evidence suggest the possibility of associated denser stream networks. Therefore, this group of explanations has focused upon the mechanism which could explain why stream networks were formerly more extensive and why they subsequently contracted. Two possibilities have been explored, namely fall in the level of the water table and a reduction in rainfall which could have caused a lowering of the water table. The existence of former spring lines at higher levels has stimulated the suggestion that some dry valleys, particularly those cut into escarpments, could have been produced by spring-sapping when water tables and the associated spring-lines were higher. In the case of southern England, Manley (1964) has argued that higher water tables could have been produced by seasons only slightly wetter than those of the present time. In the absence of a change of climate, water table lowering could have been achieved by sea level falling, by a rise in land level, or by a local change such as scarp recession (Fagg, 1923) or local downcutting whence incision of valleys would have led to a sympathetic lowering of the water table (Small, 1964). Such explanations obviously require consideration at the time scale of tens or hundreds of thousands of years (Figure 7.9). They are complemented by the suggestion that, in some areas over longer time periods, the gradual superimposition of a drainage system from a relatively impermeable to a more permeable rock, could have seen the inception of dry valleys either as a consequence of the contrast in lithology or as a consequence of the development of definite sub-surface water courses. In this manner it has been proposed that the dry valleys of the southern Pennines may have resulted from superimposition of the former drainage system from Millstone Grit on to the Carboniferous limestone beneath (Warwick, 1964).

In some areas it has appeared that fluvial processes under conditions similar to the present time are inadequate to account for either the character or the extent of dry valleys. Dry valleys have therefore been visualised in some areas as the product of past morphogenetic systems which involved either permanently frozen ground, different forms of mass movement combined with fluvial activity, higher seasonal runoff rates or some combination of these three causative factors. This type of explanation was introduced in the nineteenth century for the dry valleys of the Chalk areas of southern England (Reid, 1887). It was suggested that under periglacial conditions permafrost rendered infiltration impossible, and therefore that surface runoff was much greater and was sustained by a greatly expanded drainage network much of which was subsequently abandoned, as dry valleys, when the permafrost disappeared. This mechanism is undoubtedly important in explaining some dry valleys, but that it cannot account for all is indicated by the facts that permafrost did not occur in all areas where dry valleys have been described, and that in some areas mass movement by solifluction was much more significant than streamflow action and valley development during periglacial phases. Study of contemporary arctic areas shows that there are indeed numerous seasonally dry valleys in a network which is adjusted to spring flood discharges, that such valley networks can occur where there is little or no permafrost, and that melting of the permafrost can produce depressions or alases which subsequently can become the basis for valley development as a thermokarst phenomenon (Figure 7.8). One very well documented example of a dry valley, developed under periglacial morphogenetic conditions, is found on the scarp of the North Downs near Ashford in Kent (Figure 7.8). A detailed study (Kerney, Brown and Chandler, 1964) has shown that deposits spread out over the Gault Clay Vale were derived from the dry valley between 10,800 and 10,300 B.C. by niveofluvial processes without there necessarily having been any permafrost.

Figure 7.8 Examples of network changes
The development of alases by a thermokarst process (A) produces a pattern of depressions (stippled) in the Siberian Arctic (after Czudek and Demek, 1970), the valley pattern of the Kurkur oasis, Egypt (B) was shown to be largely a fossil phenomen by Butzer (1965), the Zone III deposits (shaded) beyond a scarp dry valley of the North Downs, Southern England (C) allowed Kerney, Brown and Chandler (1964) to date erosion of the dry valley between 10,800 and 10,300 B.C., on Triassic sandstones and marls in Devon, England (D) dry valleys (dashed lines) are frequent features (Gregory, 1966) and are illustrated in Plate 24, and the presence of a dry valley where sub-surface flow occurs in a humid tropical situation is shown in section E after S. B. St. Swan (1970).

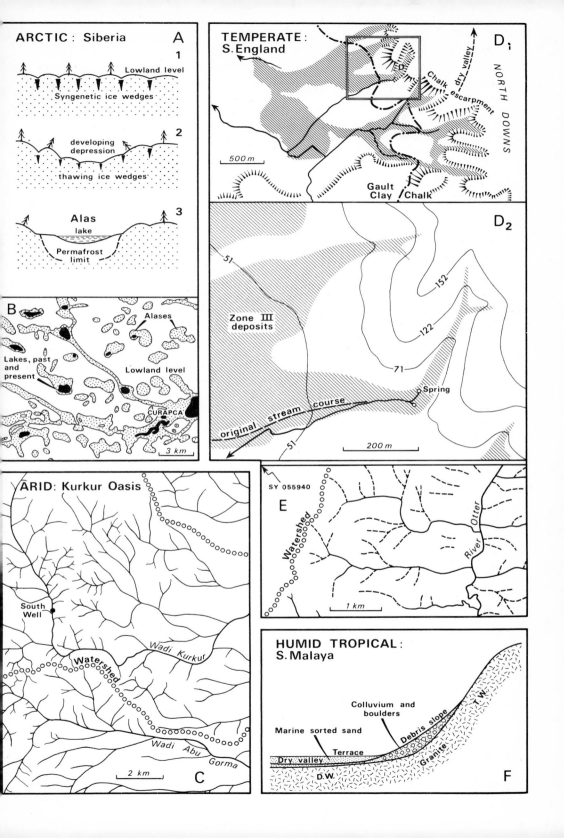

ARCTIC: Siberia A

1 — Lowland level
  Syngenetic ice wedges

2 — developing depression
  thawing ice wedges

3 — Alas
  lake
  Permafrost limit

B
Alases
Lakes, past and present
Lowland level
CURAPCA
3 km

ARID: Kurkur Oasis C
South Well
Watershed
Wadi Kurkur
Wadi Abu Gorma
2 km

TEMPERATE: S.England D₁
dry valley
Chalk escarpment
NORTH DOWNS
500 m
Gault Clay / Chalk

D₂
51
152
122
71
Zone III deposits
Spring
original stream course
51
200 m

E
SY 055940
Watershed
River Otter
1 km

HUMID TROPICAL: S.Malaya F
Colluvium and boulders
Marine sorted sand
Debris slope
T.W.
Terrace
Dry valley
D.W.
Granite

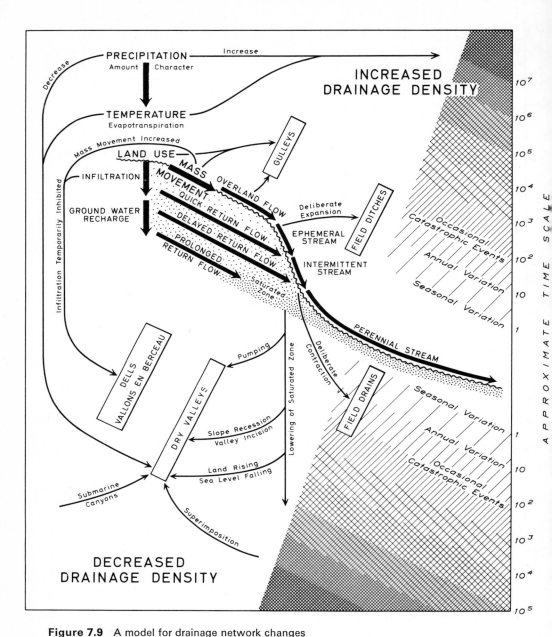

Figure 7.9 A model for drainage network changes
Based upon Gregory (1971) the diagram attempts to position the various explanations offered for drainage network changes according to the geomorphological time scale represented by shading and in relation to the function of the drainage basin shown in section. The types of water flow in the basin are based upon Jamieson and Amerman (1969).

Morphologically distinct from these large dry valleys are the dellen described especially in Poland (Figure 7.7) as including a variety of trough-like and cradle-like forms which often continue the lines of the present drainage network and which can be interpreted in relation to the deposits which occur in their floors (Figure 7.7A, B). These forms were developed by the combined action of a variety of mass movement processes and stream action during the periglacial morphogenetic system, when frost action was more significant than at the present time and when permafrost existed extensively.

These varied explanations are not necessarily alternatives but are often complementary explanations which may be required to account for the drainage network changes in a particular area. Thus in the Łodz upland Klatkowa (1965) has distinguished two forms; large dry valleys and vallons en berceau or dellen which occur on the sides of the larger dry valleys. Analysis of the deposits demonstrated that although the dellen were periglacial forms the larger dry valleys had functioned during interglacial type conditions as well and had subsequently become abandoned during the adjustment of the network to changes of climate and of basin characteristics. Similarly, in south-east Devon it has been demonstrated that there are two types of dry valley (Figure 7.8, Plate 24) which similarly reflect, firstly, adjustments of the drainage network during the Quaternary inspired by water table adjustments which produced large dry valleys, and, secondly, periglacial morphogenesis which produced a much smaller form of dry valley reminiscent of the dellen of Poland (Gregory, 1971). Thus a composite origin is indicated for the dry valleys of many areas and it is apparent that the valley networks of the present, like the stream net, reflect the total climatic and basin characteristics which the drainage basin has experienced over time. Therefore, in some areas where subsurface flow is an important phenomenon short dry valleys may appear above the lines of subsurface flow as noted in the humid tropics (St. C. Swan, 1970) (Figure 7.8), in parts of south-west Queensland, Australia (Arnett, 1971) and in parts of central Europe identified as suffosional forms.

During the Quaternary there is ample evidence for vertical changes in drainage basins and in valley dimensions testified by incised valleys as in the Silesian upland (Figure 7.7) and by river terraces on valley sides (Figure 7.4), and for lateral changes in the extent of the drainage network indicated by relict dry valleys. In addition there are numerous instances of changes of drainage pattern which involve shape of the network. Such changes have been accomplished through the course of gradual evolutionary change or through a more dramatic change which provided new basin characteristics as in the case of an area substantially modified by glaciation or by exposure from the sea.

Gradual evolutionary changes are manifested in the drainage pattern as it has progressively adjusted to the alteration of climate, of sea level and of land level or to the changing basin characteristics such as rock type which were exposed during the impact of these controls. Such gradual evolutionary change is illustrated by the lateral migration of the Arkansas river during the Quaternary (Figure 7.10). The sequence of changes was reconstructed from the terrace fragments (Sharps, 1969) and this demonstrated shifts in the position of the river during the gradual deepening of the valley which occurred during the Quaternary. East of La Junta the river apparently migrated southward during the Quaternary for as much as 14 km since early Nebraskan time. Lateral migration may have been occasioned by a greater discharge of

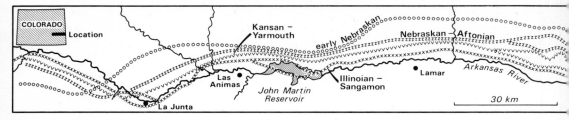

Figure 7.10   An example of change of drainage position
The successive positions of the Arkansas River were reconstructed by Sharps (1969).

unconsolidated sediment from one side of the valley than from the other. Therefore, east of La Junta, rivers from the north are long and flow over unconsolidated deposits; these rivers were therefore supplied with greater quantities of sediment than those from the south and this accumulated in fans which encouraged the southward migration of the river, undercutting the bedrock on the opposite side of the valley. On a longer term, changes in drainage pattern can be accomplished as rivers are adjusted to structural weaknesses in the rocks on the basin surface and this is one basis for the usually adopted classification of drainage patterns (Figure 2.5). Progressive development can take place in a drainage pattern and is frequently encouraged as rock contrasts are progressively exploited, as new rocks are exposed from beneath others, as sea level or climatic changes give advantages to particular streams, and as land movement or volcanic activity may determine the course of development of a pattern. The culmination of any of these influences can be shown in river capture, which leads to a modification of the drainage pattern and of the drainage basin form and is simply outlined and illustrated in Figure 7.11. Simulation studies applied to contemporary basin form and process problems can also be utilised in the study of river pattern changes. Howard (1971) has shown how many of the dimensionless properties of natural streams are well simulated by capture models rather than by random simulation models of stream development.

   During the Quaternary there are numerous examples of drainage patterns and, indeed, of drainage basins which have been fashioned largely or completely by nonfluvial processes and which thus provide examples of landform inheritance from contrasted landscape-forming conditions. Such changes are well documented in glacial geomorphology (for example, Flint, 1911, 227–43; Embleton and King, 1968, 286–93) but may profitably be illustrated by two small scale examples. On the

Figure 7.11   Illustrations of drainage pattern evolution
The drainage pattern (A2) of the Ontonagon Plain was shown by Hack (1965) to have been influenced by glacial grooves and has developed since Glacial Lake Duluth, evidenced by shorelines (A1) shrank from the plain. An anomalous drainage pattern in north-east Yorkshire is illustrated in B and derived from the sequence of glacial meltwater drainage during the Würm glaciation (Gregory, 1962). Many former valleys cut in bed rock (shaded) are infilled by glacial deposits (stippled) which have been dissected by two valleys to give twin parallel streams.
   On a longer time scale the general sequence of river capture in an area of belted rock outcrops is illustrated simply in C1 and an example of a drainage change in north-east Yorkshire evidenced by long profile, valley bench and col heights is depicted in C2.

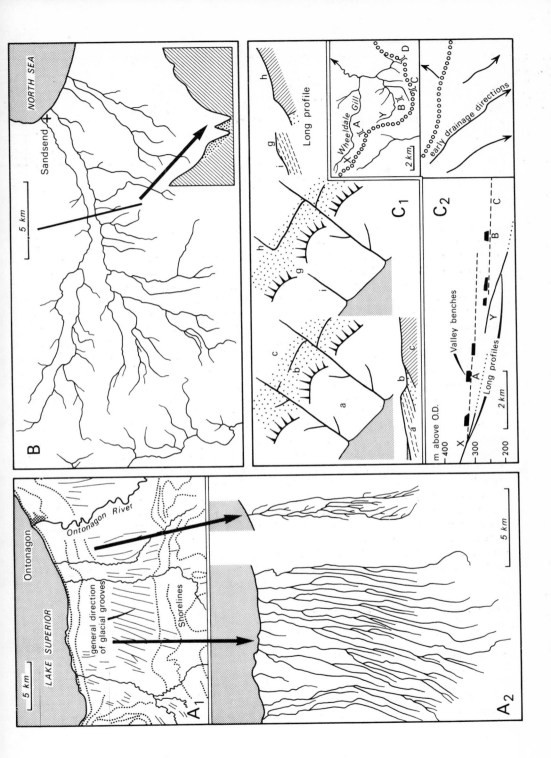

A₁

LAKE SUPERIOR

Ontonagon

Ontonagon River

general direction
of glacial grooves

Shorelines

5 km

A₂

5 km

B

NORTH SEA

Sandsend

5 km

C₁

h

g

i

c

b

a

b

a

c

Long profile

C₂

m above O.D.

Valley benches

Long profiles

X

300

400

200

A

Y

B

C

2 km

Wheeldale Gill

X

A

Y

B

D

C

early drainage directions

2 km

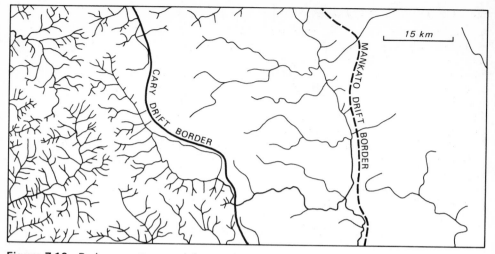

Figure 7.12   Drainage patterns and time
In this example based upon Ruhe (1952) the degree of drainage development in Iowa varies according to age of the drift deposits. Approximate ages are Mankato 13,000, Cary 15,000, Tazewell 17,000 B.P. and the drainage densities of these three ages were shown to be in the ratio 23 : 31 : 77 by Leopold, Wolman and Miller (1964).

south shore of Lake Superior the clay plain is founded on till, interbedded with fine-grained lacustrine sediments, and the plain is marked by glacial flutings parallel to the former direction of ice movement and averaging 120 m apart. As glacial Lake Duluth shrank from this plain (10,000–9500 B.P.) in stages marked by glacial lake shorelines, Hack (1965) has shown how a stream pattern developed progressively in the grooves and extended downslope towards the present lake (Figure 7.11). This sequence of development allowed the analysis of the progress of development of a pattern over a known time period and Hack (1965) concluded that the spacing of the stream junctions was random, that the depth of incision of the valleys is proportional to the size of the area drained, that there is little evidence to suggest that nickpoints migrate upstream, and that the valleys are cut to depths proportional to the available discharge. An example indicating a pattern also controlled by past development is illustrated in north east Yorkshire where, as a result of glacial deposition over the landscape and a subsequent development of a glacial drainage system, the stream pattern subsequently took the form of twin parallel streams where two streams flow parallel for distances of up to 8 km (Figure 7.11) separated by a narrow ridge of glacial deposits (Gregory, 1962). These examples are illustrative of the much larger adjustments and diversions which have occurred in drainage basin systems over time and these adjustments sometimes offer a valuable opportunity to examine the rate and nature of drainage development. Thus in Iowa, Ruhe (1952) was able to compare the drainage patterns developed on till of different ages (Figure 7.12) and Leopold, Wolman and Miller (1964) analysed this data to show how the rate of drainage development appears to be most rapid during the first 20,000 years but then subsequently decreases.

7.4   Towards the long term

Drainage basin development over time has been outlined above by a number of examples and these demonstrate how an understanding of the present can be invaluable as an aid to the understanding of the past. At the time scale of the pre-Quaternary, attention is focused upon the major elements of drainage basins. However, the major characteristics and outlines of drainage basins were very different from those of the present. Even so knowledge of present processes can provide the framework for the interpretation of the other lines of evidence; afforded morphologically by drainage alignments and by remnants of planation surfaces, and sedimentologically by sequences of deposits; to suggest the way in which the earliest outlines were fashioned and developed. The possible application of palaeohydrology to this end is indicated by the work of Menard (1961) and of Schumm (1968) in speculations of rates of sedimentation and erosion at various stages of the geological time scale (Table 7.5). A greater understanding of present drainage basin systems

Table 7.5   Past and present denudation rates (*after Menard, 1961*)
Rates based upon deposits in three geosynclinal areas

Area	Present denudation rate	Past denudation rate
Appalachians	0·8 cm/10^3 years	6·2 cm/10^3 years
Mississippi Basin	4·2	4·6
Himalayas	100	21
Rocky Mountains (Lower Cretaceous)		3
Rocky Mountains (Upper Cretaceous)		12–20

should in the future permit a greater understanding of river and drainage basin metamorphosis which has occurred in the past.

Selected reading

Some general considerations are included in:

s. a. schumm 1965: Quaternary Palaeohydrology. In h. e. wright and d. g. frey (eds.), *Quaternary of the United States*, Princeton, 783–94.

s. a. schumm 1969: River Metamorphosis, *Amer. Soc. Civil Engineers Proc.*, **95**, paper 6352, *J. Hydraulics Div.* **HY1**.

a. n. strahler 1956: The nature of induced erosion and aggradation. In w. l. thomas (ed.), *Man's role in changing the face of the earth*, Chicago, 621–38.

Some specific examples are afforded by:

g. h. dury 1964: Principles of underfit streams, *U.S. Geol. Survey, Prof. Paper*, **452**-A. 1964: Subsurface explorations and chronology of underfit streams, *ibid.*, **452**-B. 1965: Theoretical implications of underfit streams, *ibid.*, **452**-C.

k. j. gregory 1971: Drainage density changes in South West England. In k. j. gregory and w. l. d. ravenhill (eds.), *Exeter Essays in Geography*, Exeter, 33–53.

yi fu tan 1966: New Mexican gullies: A critical review and some recent observations, *Ann. Ass. Amer. Geog.*, **56**, 573–97.

8   The Prospect

We are now at the threshold of a new golden era in geomorphology. *A. L. Bloom, 1969*

For the drainage basin the prospect of a new golden era of geomorphology may, at first sight, seem unlikely in view of the number of unanswered questions which appear in the seven preceding chapters. However the development of methods for expressing the character of drainage basin form, for representing indices of drainage basin process, and for relating these two, can provide the basis for an examination of spatial and temporal variations of drainage basin form and process systems. Such development must rely heavily upon an adequate understanding of the underlying methods of measurement. Indeed, if this volume appears as a 'do-it-yourself' kit for drainage basin studies this may be justified by the convictions that firstly future developments may be facilitated by a greater awareness of the character of land form and of basin processes, and that secondly such an awareness is best achieved from the depth of understanding which can be gained from the methods and problems of measurement.

In the drainage basin the new golden era may be heralded by developments now beginning to emerge, facilitated by increasingly sophisticated measurement and analysis techniques. In the description of drainage basin characteristics the clearer identification of response units and the appreciation of their dynamic significance and, in the case of drainage basin processes, greater attention focused upon the nature and causes of detailed variations over time, indicate some of the ways in which developments are occurring. Hitherto form and process have tended to be studied separately at the level of the basin and of its components, but general principles are emerging from studies of the basin, of the network, of the channel pattern and of the hydraulic geometry of channels and these are increasingly visualised within a probability framework. Clearer understanding of the relations between, and the inter-dependence of, these hitherto separated scales of investigation may soon be achieved, and theory may be developed which perceives the drainage basin as a system containing component systems in which adjustments are made in time and space according to the principle of least work. The significance of such concepts can only be applied to geomorphological problems in the context of field investigations supplemented by laboratory and statistical analyses. Sufficient techniques for the measurement of form and process are already available and the future prospect for determining the most significant, for evaluating their relevance to landscape change, for mapping their variation, and for assessing their variation over time may open up the new era.

Is the geomorphologist the most appropriate person to solve these problems? Whereas methods of process measurement have been developed and continue to be developed within other disciplines, methods of land description are the province of the geographer who also wishes to know the nature of the form-process systems which maintain the land surface in a steady state and which determine the significance for human use. Concern for the character of land is a subject of increasing interest in the 1970s and the drainage basin is one functioning unit of the land surface to which the geomorphologist obviously directs his attention. However as realised long ago, and as emphasised by Horton in 1932, certain drainage basin characteristics cannot be considered independently from others; this fact has been confirmed and sustained by more recent work in which studies of dynamic topographic, soil, climatic and land use characteristics have tended to utilise similar methods and to realise obviously connected results. Therefore sections on rock type, on soil character and on land use have been included in the preceding chapters and their inclusion may have been such that the result does not appear to be geomorphology *sensu stricto*. The volume of information recently available increasingly dictates that boundary lines have to be drawn for the organisation and study of knowledge. It was felt that the drainage basin provides a much needed contemporary focus of attention and that the background necessary to its study was better served by a consideration of all physical basin characteristics and by an explanation of their significance in relation to basin processes, rather than by the alternative inclusion of slopes or of limestone geomorphology, both of which are the subjects of other recent volumes.

The result may be an incomplete geomorphological treatment of drainage basins, in that it lacks detailed treatment of certain themes such as slopes, but the result may be more reminiscent of a drainage basin physical geography. With increasing adoption of the systems approach perhaps we should turn our attention to the physical geography of the earth's surface and be not too constrained by the branches of geographical knowledge which have dominated the first half of this century. The alternative would have been to organise an approach to the drainage basin founded upon the established branches of geomorphology, namely weathering, slopes, and fluvial processes, and to rely upon separate treatments of rock, soil, vegetation and land use. This book has endeavoured to visualise the inter-relations of these themes and to introduce them where relevant, and has attempted to stress the similar problems and techniques which figure in the hitherto distinct branches. A physical geography approach certainly facilitates consideration of inter-relationships, it is in sympathy with the current interest in environment, and it may provide the means of 'realising the integration which our forefathers dreamed of' as Dury (1970) expressed the hope for a current trend towards integrated studies. If the result does not satisfy all geomorphologists at least a physical geography approach to the basin would have satisfied A. Tylor's 1875 (p. ix) request more completely than would a geomorphology of the basin which excluded specific consideration of rock, soil, vegetation, land use and hydrology.

The last decade has witnessed great concern over the extent of man's impact on environment. In physical geography this concern has been expressed by studies of the way in which the land surface has been changed directly and indirectly by man, and in geomorphology it has been registered by an increased interest in anthropogeomorphology. This movement has two main implications for geography. Firstly,

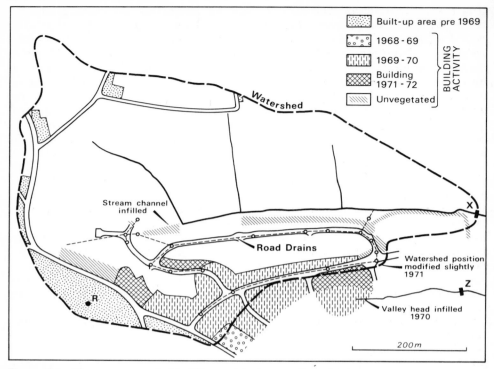

Figure 8.1 The measurement of building activity in a small catchment
This catchment, also illustrated in Figure 1.10, is being studied to assess the effect of building construction upon water and sediment yield in an area on the margin of Exeter, Devon. Streamflow and sediment are monitored continuously at X and precipitation is recorded autographically at R.

consideration of man-land relations have facilitated closer links between human and physical geography, so much so that physical geography may be confronted by a choice which could be resolved by adopting a path specifically towards man's effects and towards the significance of the earth's surface for man. The choice has been expressed by R. W. Kates in his view that 'physical geography is at a cross roads having suffered from the erosion of its domain by the many earth science specialities and the ageing of its competence and methods'. Competence and methods, according to the range recently employed by physical geographers, may no longer be in doubt. Erosion of its domain has necessarily occurred and could be rectified and adjusted by attention devoted to the study of physical geography of the form of the earth's surface as a basis for understanding its evaluation for human activity, and of the effect which man has had. The second implication, arising partly from the first, is that whereas study of present form-process systems has hitherto been the basis for assisting interpretation of past development, employing palaeohydrological approaches for example, opportunity is now afforded to include consideration of future consequences as well. This future dimension has increasingly been the concern of the human geographer (for example, Toyne, 1973) and it is available to a comparable extent to the physical geographer studying the drainage basin.

The magnitude of some of man's implications for the drainage basin are indicated in Table 8.1. Although the figures are by no means comprehensive they do indicate the extent of the impact of changes wrought by man and they underline the necessity for an understanding of the mechanism of drainage basin processes. The major changes in the drainage basin are those expressed in modifications of the basin character, particularly its land use cover; in changes of river courses arising from dam and reservoir construction, conservation practices and irrigation schemes; and in changes of water and sediment distribution by human use of drainage basin water derived from rivers and ground water sources. These changes of basin character, of the component subsystems of the basin, and of the processes operating have produced direct and indirect responses. The direct responses include changes in flood frequency, in the incidence of gullying and erosion, and in patterns of sediment production. Research has increasingly been devoted to such direct responses occasioned by changed surface processes. However, because these changes in process involve adjustments and changes of form within the basin, future research may see more attention directed to the significance of process changes for channel geometry, for channel patterns and for stream networks which may all incorporate indirect responses. Necessarily the geomorphologist is primarily concerned with the form of the land surface, with a clearer understanding of the detail of recent changes, and with their significance in relation to land geometry. This should afford opportunities for the application of geomorphological results to the problems of basin and land development.

Although geomorphological research can utilise a wide variety of methods for data acquisition and analysis it will remain essentially a field science because it is the field which produced the problems and which provides the data for their solution. The field problem provides scope for numerous laboratory experiments, and the detailed results from small catchments can provide one fruitful resource for extending our understanding of drainage basin processes. The small catchment employed to illustrate an approach to the drainage basin (Figure 1.10) is being studied over a number of years and some results have been included in Figures 4.5, 4.13. In Figure 8.1 the development of the catchment to 1972 is shown and the measurement of water and sediment at X, and of precipitation at R, have continued during the progress of building activity over this catchment. Hitherto studies have demonstrated increases of peak discharges of two to three times, of suspended sediment concentrations up to ten times, and of solute concentrations up to two times their former values. The subsequent changes of process, in relation to deliberate modification of the basin characteristics, are to be monitored and in addition the feed back effects on channel form and channel pattern are to be analysed.

Drainage basin geomorphology may have become increasingly concerned with small areas and with minute studies, but these provide some of the very necessary building blocks for a greater understanding. The greater understanding of the future will necessarily be attained with a variety of techniques because the techniques employed are determined by scale of study; it will be realised with more detailed investigation of the drainage basin components; and it will be facilitated by a more

Table 8.1   Examples of magnitude and cost of drainage basin changes (*after data in Guy, 1963; Glymph and Storey, 1960; Moore and Smith, 1968; Proc. Am. Soc. Civil Eng., 1969; Spraberry, 1965; Todd, 1971; Wolman, 1964*)

Drainage basin changes
Land use—England and Wales—70% agricultural land, 12% urbanised (16% by year 2000)
Drainage—England and Wales—3 km/km² ditch
Rivers—World Dams—1963: 341 high dams, 102 under construction, 237 planned
Reservoirs—U.S.A.—1963: 100 greater than 834 million m³ capacity
Water use—U.S.A.—1965: 594 lit/head/day (636 by year 2000)
Irrigation—World—198,770,460 hectares (1·9% surface); U.S.A.—1965: 41 million hectares (56 million by year 200)

Implications	*Cost*
Erosion and transport from rural cropland areas in the U.S.A. since settlement	Forced abandonment of crop production on 17,300,000 hectares. In the U.S.A. 150 million hectares are subject to serious erosion and 200 million hectares subject to erosion to a lesser extent. Rural erosion and sediment problems in the U.S.A. amount to over $1 billion per annum, $800 million of which results from erosion of cropland.
Increased flood hazard	Annual flood damage in the U.S.A. 1968: within catchment areas greater than 123,675 hectares $643 million; catchment areas less than this $1094 million ($1355 and $1548 by year 2000). Flood protection works in the U.S.A. 1966 greater than $3000 million.
Modified rivers and river regimes	Average loss of reservoir storage in the U.S.A. due to sediment deposition amounts to 0·2% each year. In an average year 344 million m³ of material are dredged from rivers and harbours in the U.S.A. Increased presence of sediment in the Potomac river is such that reduction of turbidity to an optimum level could produce an annual saving of $25,000 per annum in water-treatment costs for Washington, D.C.
Gully destruction of land in Iowa and Missouri	Capitalised value of land to society of $1220 per hectare. Average annual damage in Iowa due to gullying $2·26 per hectare.
Building activity and urbanisation	As a result of building construction in the area of metropolitan Washington, 10 tonnes of sediment are transported to the Potomac river for every person added to the city.

Plate 25   Effects of river flood
Damage occasioned when Cucamonga Canyon (*top left*) California flooded and flood water spread over fruit growing areas.

complete appreciation of physical principles and of the theoretical basis of drainage basin dynamics.

> And see the rivers how they run,
> Through woods and meadows, in shade and sun,
> Sometimes swift; sometimes slow,—
> Wave succeeding wave, they go
> A various journey to the deep,
> Like human life to endless sleep!   *John Dyer*

Further reading

Sedimentation Engineering, *Proceedings of the American Society of Civil Engineers* (1969), Chapter VI: Economic Aspects of Sedimentation. *J. Hydraulics Div.* **HY1**, 191–207.

K. J. GREGORY and D. E. WALLING (eds.), 1974: Fluvial Processes in Instrumented Watersheds: Studies of Small Watersheds in the British Isles, *Institute of British Geographers Special Publication* **6**.

References

ABRAHAMS, A. D. 1970: Towards a precise definition of drainage basin axis: Comment *Australian Geographical Studies* **8**, 84–6.

ABRAHAMS, A. D. 1970: An evolution of Melton's order-by-order growth analysis. *Australian Geographical Studies* **8**, 57–70.

ACKERMANN, W. C. 1966: *Guidelines for research on hydrology of small watersheds.* U.S. Dept. Interior O.W.R.R. Washington D.C. (26pp.)

ACKERS, P. and CHARLTON, F. G. 1970: The geometry of small meandering streams. *Proc. Inst. Civil Engineers, Paper* **73285**, 289–317.

AHNERT, F. 1970: Functional relationships between denudation, relief, and uplift in large mid-latitude drainage basins. *Amer. J. Sci.* **268**, 243–63.

AKROYD, T. N. W. 1958: *Laboratory testing in soil engineering.* Soil Mechanics Ltd., London. (233pp.)

ALBERTSON, M. L. and SIMONS, D. B. 1964: Fluid Mechanics. In V. T. CHOW (ed.), *Handbook of Applied Hydrology*, New York.

ALLEN, J. R. L. 1965: A review of the origin and characteristics of recent alluvial sediments. *Sedimentology* **5**, 89–191.

ALLEN, J. R. L. 1970: *Physical processes of sedimentation*, London. (248pp.)

AMERICAN GEOPHYSICAL UNION 1965: *Inventory of representative and experimental watershed studies conducted in the United States.* Amer. Geophys. Union.

AMERICAN SOCIETY OF CIVIL ENGINEERS 1962: Sediment transportation mechanics: erosion of sediment. Rept. Task Committee on preparation of Sedimentation Manual. *Proc. A.S.C.E., J.Hyd. Div.* **HY4**, 109–27.

AMERICAN SOCIETY OF CIVIL ENGINEERS 1969: Sediment measurement techniques: F Laboratory Procedures. Rept. Task Committee on preparation of Sedimentation Manual. *Proc. A.S.C.E., J.Hyd. Div.* **HY5**, 1515–42.

AMERICAN SOCIETY OF CIVIL ENGINEERS 1969: Sedimentation Engineering Chapter VI: Economic aspects of sedimentation. *Proc. A.S.C.E., J.Hyd. Div.* **HY1**, 191–207.

AMERICAN SOCIETY OF CIVIL ENGINEERS 1970: Sediment sources and sediment yields. Rept. Task Committee on preparation of Sedimentation Manual. *Proc. A.S.C.E., J.Hyd. Div.* **HY6**, 1283–329.

AMERICAN SOCIETY OF CIVIL ENGINEERS 1971: Sediment Transportation Mechanics H. Sediment Discharge Formulas. Rept. Task Committee on preparation of Sedimentation Manual. *Proc. A.S.C.E., J.Hyd. Div.* **HY4**, 523–67.

AMERICAN WATER WORKS ASSOCIATION 1966: *Standard methods for the examination of water and wastewater,* American Public Health Assoc., 12th edn.

AMERMAN, C. B. 1965: The use of unit source watershed data for runoff prediction. *Water Resources Res.* **1**, 499–507.

AMOROCHO, J. 1967: The nonlinear prediction problem in the study of the runoff cycle. *Water Resources Res.* **3**, 861–80.

AMOROCHO, J. and HART, W. E. 1964: A critique of current methods in hydrologic systems investigation. *Trans. Amer. Geophys. Union* **45**, 307–21.

AMOROCHO, J. and HART, W. E. 1965: The use of laboratory catchments in the study of hydrologic systems. *J. Hyd.*, 106–23.

ANDERSON, H. W. 1957: Relating sediment yield to watershed variables. *Trans. Amer. Geophys. Union* **38**, 921–4.

ANDERSON, H. W. and WALLIS, J. R. 1965: Some interpretations of sediment sources and causes, Pacific Coast basins in Oregon and California. *Proc. Fed. Inter-Agency Sedimentation Conf.*, *U.S.D.A. Misc. Pub.* **970**, 22–30.

ANDERSON, P. W. 1963: Variations in the chemical character of the Susquehanna River at Hamburg, Pennsylvania. *U.S. Geol. Survey Water Supply Paper* **1779B**.

ANON 1940: Occasional flow in a Chalk dry valley. *Nature* **466**.

ANTEVS, E. 1952: Arroyo-cutting and filling. *J. Geol.* **60**, 375–85.

ARNETT, R. R. 1971: Slope form and geomorphological process: an Australian example. *Inst. Brit. Geog.*, Special Publication **3**. Compiled by D. BRUNSDEN, 81–92.

ARRHENIUS, O. 1954: Chemical denudation in Sweden. *Tellus* **VI**, 326–41.

AULITZKY, H. 1967: Forest Hydrology Research in Austria. In W. E. SOPPER and H. W. LULL (eds.), *International Symposium on Forest Hydrology*, Oxford, 3–10.

AUSTRALIAN WATER RESOURCES COUNCIL 1969: The Representative Basin Concept in Australia. *Australian Water Resources Council Hydrological Series* **2**.

BAIRD, R. W. 1964: Sediment yields from Blackland Watersheds. *Trans. Amer. Soc. Agri. Eng.* **7**, 454–6.

BANNERMAN, R. B. W. 1966: River Flow Statistics. In R. B. THORN (ed.), *River Engineering and Water Conservation Works*, London, 156–82.

BARNES, B. S. 1939: The structure of discharge-recession curves. *Trans. Amer. Geophys. Union* **20**, 721–5.

BARNES, H. H. 1967: Roughness characteristics of natural channels. *U.S. Geol. Survey Water Supply Paper* **1849**.

BARNES, O. K. and COSTEL, G. 1957: A mobile infiltrometer. *J. Agron* **49**, 105–7.

BARNES, K. K. and FREVERT, R. K. 1954: A runoff sampler for large watersheds. *Agricultural Engineering* **35**, 84–90.

BARRY, R. G. 1969: The world hydrological cycle. In R. J. CHORLEY (ed.), *Water Earth and Man*, London, 11–29.

BARRY, R. G. 1969: Precipitation. In R. J. CHORLEY (ed.), *Water Earth and Man*, London, 114–29.

BARSBY, A. 1963: A wooden flume—the design and installation of a trapezoidal stream gauging flume of prefabricated wooden structure. *Water Research Assoc. Tech. Paper* **28**.

BAUER, L. and TILLE, W. 1967: Regional differentiations of the suspended sediment transport in Thuringia and their relation to soil erosion. *Internat. Assoc. Sci. Hyd. Pub.* **75**, 367–77.

BAULIG, H. 1935: *The changing sea level.* Institute of British Geographers. (46pp.)

BAVER, L. D. 1933: Some factors affecting erosion. *Agricultural Engineering*, **14**, 51–2.

BAVER, L. D. 1948: *Soil Physics*, New York. (398pp.)

BAVER, L. D. 1956: *Soil Physics*, 3rd edn., New York.

BEARDMORE, N. 1862: *Manual of Hydrology*, London.

BEATY, C. B. 1963: Origin of alluvial fans, White Mountains, California and Nevada. *Ann. Ass. Amer. Geog.* **53**, 516–35.

BEATY, C. B. 1970: Age and estimated rate of accumulation of an alluvial fan, White Mountains, California, U.S.A. *Amer. J. Sci.* **268**, 50–77.

BEDEUS, K. and IVICSICS, L. 1964: Observation of the noise of bedload. *Internat. Assoc. Sci. Hyd. Pub.* **65**, 384–90.

BEER, C. E. and JOHNSON, H. P. 1965: Factors related to gully growth in the deep loess area of western Iowa. *U.S. Dept. of Agriculture, Misc. Pub.* **970**, 37–43.

BELL, J. P. and MCCULLOCH, J. S. G. 1966: Soil moisture estimation by the neutron scattering method in Britain. *J. Hyd.* **4**, 254–63.

BENECKE, P. 1967: Investigation into the behaviour of precipitation water in soils by means of ^{131}I. *Isotope and Radiation Techniques in Soil Physics and Irrigation Studies*, Internat. Atomic Energy Agency Vienna, 227–37.

BENEDICT, J. B. 1970: Downslope soil movement in a Colorado Alpine Region: Rates, processes and climatic significance. *Arctic and Alpine Res.* **2**, 165–226.

BENEDICT, P. C. *et al.* 1955: Total sediment load measured in turbulence flume. *Trans. A.S.C.E.* **120**, 457–89.

BENNETT, H. H. 1939: *Soil Conservation*, London.

BERRY, L. and RUXTON, B. P. 1960: The evolution of the Hong Kong harbour basin. *Zeitschrift für Geomorphologie* **4**, 97–115.

BERRY, L. 1970: Some erosional features due to piping and sub-surface wash with special reference to the Sudan. *Geografiska Annaler* **52A**, 113–19.

BEVERAGE, J. P. and CULBERTSON, J. K. 1964: Hyperconcentrations of suspended sediment. *Proc. A.S.C.E. J. Hyd. Div.* **HY6**, 117–28.

BINNIE, A. R. 1892: On mean or average rainfall and the fluctuations to which it is subject. *Proc. Inst. Civ. Eng.* **109**, 89–172.

BISWAS, A. K. and FLEMING, G. 1966: Floods in Scotland: Magnitude and Frequency. *Water and Water Engineering*, 246–52.

BLACK, P. E. 1970: Runoff from watershed models. *Water Resources Res.* **6**, 465–77.

BLAKE, D. H. and OLLIER, C. D. 1971: Alluvial plains of the Fly River, Papua. *Zeitschrift für Geomorphologie* **12**, 1–17.

BLISSENBACH, E. 1954: Geology of alluvial fans in semi-arid regions. *Geol. Soc. Amer. Bull.* **65**, 175–90.

BLOCKER, W. and BOWER, D. 1963: Filter for sampling sediment in small streams. *J. Soil and Water Conservation* **18**, 222.

BLONG, R. J. 1970: The development of discontinuous gullies in a pumice catchment. *Amer. J. Sci.* **268**, 369–83.

BLOOM, A. L. 1969: *The surface of the earth*. New Jersey. (152 pp.)

BLUME, H. P. *et al.* 1967: Tritium tagging of soil moisture: the water balance of forest soils. *Isotope and Radiation Techniques in Soil Physics and Irrigation Studies*, Internat. Atomic Energy Agency Vienna, 315–30.

BOCHKOV, A. P. 1959: The forest and river runoff. *Internat. Assoc. Sci. Hyd. Pub.* **48**, 174–81.

BORMANN, F. H. and LIKENS, G. E. 1969: The watershed-ecosystem concept and studies of nutrient cycles. In G. M. VAN DYNE (ed.), *The ecosystem concept in natural resource management*, New York, 49–76.

BOUGHTON, W. C. 1968: Hydrological studies of changes in land use. *Soil and Water* **4**, 19–23.

BOWDEN, K. L. and WALLIS, J. R. 1964: Effect of stream-ordering technique on Horton's laws of drainage composition. *Geol. Soc. Amer. Bull.* **75**, 767–74.

BOWDEN, K. L. and WALLIS, J. R. 1966: Watershed shape as a hydrologic parameter. *Proc. First Annual Meeting, Amer. Water Resources Assoc.*

BOWEN, D. Q. and GREGORY, K. J. 1965: A glacial drainage system near Fishguard, Pembrokeshire. *Proc. Geol. Ass., Lond.* **76**, 275–82.

BOWMAN, D. H. and KING, K. M. 1965: Convenient cadmium-metal standard for checking neutron, soil moisture probes. *Soil Sci. Soc. Amer. Proc.* **25**, 339–42.

BOYCE, R. B. and CLARK, W. A. V. 1964: The concept of shape in geography. *Geog. Review* **54**, 561–72.

BOYER, M. C. 1957: A correlation of the characteristics of great storms. *Trans. Amer. Geophys. Union* **38**, 233–8.

BRICE, J. C. 1960: Index for description of channel braiding. *Geol. Soc. Amer. Bull.*, (Abst) **71**, 1833.

BRICE, J. C. 1964: Channel patterns and terraces of the Loup River in Nebraska. *U.S. Geol. Survey Prof. Paper* **422–D**, 1–41.

BRICE, J. C. 1966: Erosion and deposition in the loess-mantled Great Plains, Medicine Creek drainage basin, Nebraska. *U.S. Geol. Survey Prof. Paper* **352–H**, 255–339.

BRIGGS, W. et al. 1969: *Erosion on Wisconsin Roadsides. A report to Wisconsin Citizens*. Wisconsin Chapter, Soil Conservation Society of America.

BRITISH STANDARDS INSTITUTION 1965: *B.S. 3680. Methods of measurement of liquid flow in open channels*, London.

BRITISH STANDARDS INSTITUTION 1967: *B.S. 1377. Methods of testing soils for Civil Engineering purposes*, London (c.f. *B.S. 3406* and *B.S. 1796*).

BROOKS, R. H. and COREY, A. T. 1966: Properties of porous media affecting fluid flow. *Proc. A.S.C.E. J. Irrig. Drainage Div.* **IR2**, 61–88.

BROSCOE, A. J. 1959: Quantitative analysis of longitudinal stream profiles of small watersheds. *Office of Naval Research, Geography Branch, Project NR 389–42, Technical Report* **18**.

BROWN, C. B. 1950: Effects of Soil Conservation. In P. D. TRASK (ed.), *Applied Sedimentation*, New York, 380–406.

BROWN, E. H. 1969: Jointing, aspect and orientation of scarp-face dry valleys, near Ivinghoe, Bucks. *Trans. Inst. Brit. Geog.* **48**, 61–73.

BROWN, E. H. 1970: Man shapes the earth. *Geog. J.* **136**, 74–84.

BROWN, D. 1954: Methods of surveying and measuring vegetation. *Commonwealth Bureau of Pastures and Field Crops Bulletin* **42**, London.

BROWN, G. W. and KRYGIER, J. T. 1967: Changing water temperatures in small mountain streams. *J. Soil and Water Conservation*, 242–4.

BROWN, H. E. et al. 1970: A system for measuring total sediment yield from small watersheds. *Water Resources Res.* **6**, 818–26.

BROWN, J. A. H. 1972: Hydrologic effects of a bushfire in a catchment in south-eastern New South Wales. *J. Hyd.* **15**, 77–96.

BRUNE, G. M. 1951: Sediment records in midwestern United States. *Internat. Assoc. Sci. Hyd. Pub.* **33**, 29–38.

BRUNE, G. M. 1953: Trap efficiency of reservoirs. *Trans. Amer. Geophys. Union.* **34**, 407–18.

BRUSH, L. M. 1961: Drainage basins, channels, and flow characteristics of selected streams in central Pennsylvania. *U.S. Geol. Survey Prof. Paper* **282F**.

BRYAN, K. 1928: Historic evidence on changes in the channel of Rio Puerto, a tributary of the Rio Grande in New Mexico. *J. Geol.* **36**, 265–82.

BRYAN, R. B. 1968: The development, use and efficiency of indices of soil erodibility. *Geoderma* **2**(1), 5–25.

BRYAN, R. B. 1969: The relative erodibility of soils developed in the Peak District of Derbyshire. *Geografiska Annaler* **51A**, 145–59.

BUCKHAM, A. F. and COCKFIELD, W. E. 1950: Gullies formed by sinking of the ground. *Amer. J. Sci.* **248**, 137–41.

BUDEL, J. 1948: Das System der klimatischen Geomorphologie (Beiträge zur Geomorphologie der klimazonen und Vorzeitklimate V). *Verhandlungen Deutscher Geographertag, München* **27**, 65–100.

BUDEL, J. 1963: Klima-genetische geomorphologie. *Geog. Rundschau* **7**, 269–86.

BUDEL, J. 1969: Das system der klima-genetischen geomorphologie. *Erdkunde* **23**, 165–82.

BULL, W. B. 1963: Geomorphology of segmented alluvial fans in Western Fresno County, California. *U.S. Geol. Survey Prof. Paper* **352E**.

BULL, W. B. 1964: Alluvial fans and near surface subsidence in Western Fresno County, California. *U.S. Geol. Survey Prof. Paper* **437A**.

BUNGE, W. 1962: Theoretical Geography. *Lund Studies in Geography, Series C, General and Mathematical Geography* **1**.

BUNTING, B. T. 1961: The role of seepage moisture in soil formation, slope development and stream initiation. *Amer. J. Sci.* **259**, 503–18.

BUNTING, B. T. 1964: Slope development and soil formation on some British sandstones. *Geog. J.* **130**, 73–9.

BURDON, D. J. 1966: The largest karst spring. *J. Hyd.* **4**, 104.

BUSBY, M. W. 1964: Relation of annual runoff to meteorological factors. *U.S. Geol. Survey Prof. Paper,* **501C**.

BUTLER, S. S. 1957: *Eng. Hyd.,* New Jersey, 227–9.

BUTZER, K. W. 1965: Desert landforms at the Kurkur oasis, Egypt. *Ann. Ass. Amer. Geogr.* **55**, 578–91.

BUTZER, K. W. 1971: Fine alluvial fills in the Orange and Vaal basins of South Africa. *Proc. Ass. Am. Geog.* **3**, 41–8.

CAILLEUX, A. 1947: L'indice d'émoussé: définition et première application. *Comptes Rendus sommaires de la Société geologique de France,* 165–7.

CALKINS, D. and DUNNE, T. 1970: A salt trace method for measuring channel velocities in small mountain streams. *J. Hyd.* **11**, 379–92.

CAMPBELL, A. P. 1962: The use of suspended sediment measurements to assist in the study of catchment condition. *Hyd. and Land Management, S.C.R.C.C., N.Z., Pub.,* 102–3.

CAMPBELL, A. P. 1966: Measurement of movement of an earthflow. *Soil and Water* **2** (3) 23–4.

CAMPBELL, A. P. and CADDIE, G. H. 1963: Erosion rate coefficients for some New Zealand catchments. *J. Hyd. (N.Z.)* **2**, 22–3.

CAMPBELL, D. A. 1955: Down to the sea in slips. *N.Z. Soil Cons. and Rivers Control Council Bulletin* **5**, Wellington.

CAMPBELL, F. B. and BAUDER, H. A. 1940: A rating-curve method for determining silt-discharge of streams. *Trans. Amer. Geophys. Union* **21**, 603–7.

CANADIAN NATIONAL I.H.D. COMMITTEE 1966: Guide lines for research basin studies. *Proc. of the National Workshop Seminar on Research Basin Studies.* (44pp.)

CAPPER, P. L. and CASSIE, W. F. 1969: *The mechanics of engineering soils*, London. (309pp.)

CAREY, W. C. 1969: Formation of flood plain lands. *Proc. A.S.C.E. J. Hyd. Div.* **HY3**, 981–94.

CARLSTON, C. W. 1963: Drainage density and streamflow. *U.S. Geol. Survey Prof. Paper* **422C**.

CARLSTON, C. W. 1965: The relation of free meander geometry to stream discharge and its geomorphic implications. *Amer. J. Sci.* **263**, 864–85.

CARLSTON, C. W. and LANGBEIN, W. B. 1960: Rapid approximation of drainage density: line intersection method. *U.S. Geol. Survey Water Resources Div. Bulletin* **11**.

CARSON, M. A. 1971: *The mechanics of erosion*, London. (174pp.)

CARSON, M. A. and KIRKBY, M. J. 1971: *Hillslope Form and Process*, Cambridge. (475pp.)

CARSON, M. A. and SUTTON, E. A. 1971: The hydrologic response of the Eaton River Basin, Quebec. *Canadian J. of Earth Sci.* **8**, 102–15.

CARTER, R. W. 1961: Magnitude and frequency of floods in suburban areas. *U.S. Geol. Survey Prof. Paper* **424B**, 9–11.

CASTLE, G. H. 1965: Determination of reservoir deposits by reconnaissance methods. *Proc. Fed. Inter-Agency Sedimentation Conf., U.S.D.A. Misc. Pub.* **970**, 857–67.

CHANG, J. and OKIMOTO, G. 1970: Global Water Balance according to the Penman approach. *Geographical Analysis*, 55–67.

CHEBOTAREV, N. P. 1966: Theory of stream runoff. *Israel Program for Scientific translations*, Jerusalem. (464pp.)

CHERY, D. L. 1966: Design and test of a physical watershed model. *J. Hyd.* **4**, 224–35.

CHIDLEY, T. R. E. and PIKE, J. G. 1970: A generalised computer program for the solution of the Penman equation for evapotranspiration. *J. Hyd.* **10**, 75–89.

CHISHOLM, M. 1967: General systems theory and geography. *Trans. Inst. Brit. Geog.* **42**, 42–52.

CHORLEY, R. J. 1957: Climate and morphometry. *J. Geol.* **65**, 628–68.

CHORLEY, R. J. 1959: The geomorphic significance of some Oxford Soils. *Amer. J. Sci.* **257**, 503–15.

CHORLEY, R. J. 1962: Geomorphology and general systems theory. *U.S. Geol. Survey Prof. Paper* **500-B**. (10pp.)

CHORLEY, R. J. 1966: The application of statistical methods to geomorphology, In G. H. DURY ed., *Essays in Geomorphology*, London, 275–387.

CHORLEY, R. J. 1967: Models in Geomorphology In R. J. CHORLEY and P. HAGGETT (eds.), *Models in Geography*, London, 59–96.

CHORLEY, R. J. 1969: R. J. CHORLEY (ed.), *Water, Earth and Man*, London, Methuen. (588pp.)

CHORLEY, R. J. 1969: The drainage basin as the fundamental geomorphic unit. In R. J. CHORLEY (ed.), *Water, Earth and Man*, London, 77–100.

CHORLEY, R. J. 1971: The role and relations of physical geography. *Progress in Geography* **2**, 87–109.

CHORLEY, R. J., MALM, D. E. G. and POGORZELSKI, H. A. 1957: A new standard for estimating basin shape. *Amer. J. Sci.* **255**, 138–41.

CHORLEY, R. J. and MORGAN, M. A. 1962: Comparison of morphometric features, Unaka Mountains Tennessee and North Carolina, and Dartmoor, England. *Geol. Soc. Amer. Bull.* **73**, 17–34.

CHORLEY, R. J. and HAGGETT, P. (eds.) 1967: *Models in Geography*, London. (816pp.)

CHORLEY, R. J. and KENNEDY, B. A. 1971: *Physical Geography: A systems approach*, London. (370pp.)

CHOW, V. T. 1959: *Open-channel Hydraulics*, New York.

CHOW, V. T. (ed.) 1964: *Handbook of Applied Hydrology; a compendium of water-resources technology*, New York.

CHOW, V. T. 1964: Runoff, In V. T. CHOW (ed.), *Handbook of Applied Hydrology*, New York, 14-1–14-54.

CHOW, V. T. 1964: Statistical and probability analysis of hydrologic data Part 1. Frequency Analysis, In V. T. CHOW (ed.), *Handbook of Applied Hydrology*, New York, 8-1–8-42.

CHOW, V. T. 1967: Laboratory study of watershed hydrology. *Proc. Internat. Hydrology Symp.* **1967**. *Fort Collins Colorado, U.S.A.* **1**, 194–202.

CHURCH, J. E. 1935: Principles of snow surveying as applied to forecasting streamflow. *J. Agr. Res.* **51** (2), 97–130.

CHURCH, M. and KELLERHALS, R. 1970: *Stream gauging techniques for remote areas using portable equipment, Technical Bulletin* **25**, Inland Waters Branch; Dept. of Energy, Mines and Resources.

CLARIDGE, G. G. C. 1970: Studies in element balances in a small catchment at Taita, New Zealand. *Internat. Assoc. Sci. Hyd. Pub.* **96**, 523–40.

CLARKE, F. W. 1924: The data of geochemistry. *U.S. Geol. Survey Bull.* **770**. (841 pp.)

CLEAVES, E., GODFREY, A. E. and BRICKER, O. P. 1970: Geochemical balance of a small watershed and its geomorphic implications. *Geol. Soc. Amer. Bull.* **81**, 3015–32.

COHEN, O. P. and TADMORE, H. H. 1966: A comparison of neutron moderation and gravimetric sampling for soil-moisture determination with emphasis on the cost factor. *Agricultural Meteorology* **3**, 97–102.

COHEN, P. *et al.*, 1968: An atlas of Long Island's Water Resources. *New York Water Res. Committee Bull.* **62**.

COLBY, B. R. 1956: Relationship of sediment discharge to streamflow. *U.S. Geol. Survey Open File Report.* (170pp.)

COLBY, B. R. and HUBBELL, D. W. 1961: Simplified methods for computing total sediment discharge with the modified Einstein procedure. *U.S. Geol. Survey Water Supply Paper* **1593**.

COLE, G. 1966: An application of the regional analysis of flood flows. *River Flood Hydrology*, Inst. Civ. Engrs., London, 39–57.

COLLIER, C. R. *et al.*, 1964: Influences of strip mining on the hydrologic environment of parts of Beaver Creek Basin Kentucky, 1955–9. *U.S. Geol. Survey Prof. Paper* **427B**. (83pp.)

COLLIER, C. R. *et al.* 1970: Influences of strip mining on the hydrological environment of parts of Beaver Creek Basin, Kentucky, 1955–66. *U.S. Geol. Survey Prof. Paper* **427C**. (80pp.)

COMER, G. H. and ZIMMERMAN, R. C. 1969: Low flow and basin characteristics of two streams in northern Vermont. *J. Hyd.* **7**, 98–108.

CONWAY, E. J. 1942: Mean geochemical data in relation to oceanic evolution. *Royal Irish Acad. Proc.* **B48**, 119–59.

COPELAND, O. L. 1965: Land use and ecological factors in relation to sediment yields. *Proc. Fed. Inter-Agency Sedimentation Conf.*, *U.S.D.A. Misc. Pub.* **970**, 72–84.

CORBEL, J. 1964: L'érosion terrestre, étude quantitative (Méthodes-techniques-résultats). *Annales de Geographie* **73**, 385–412.

CORBETT, D. 1943: Stream Gauging Procedure. *U.S. Geol. Survey Water Supply Paper* **588**.

CORBETT, E. S. 1967: Measurement and estimation of precipitation on experimental watersheds, In W. E. SOPPER and H. W. LULL (eds.), *International Symposium on Forest Hydrology*, Oxford.

CORDERY, I. 1967: Synthetic unitgraphs for small catchments in eastern New South Wales. *Inst. of Eng., Australia, Hyd. Symp. Brisbane*, 49–59.

COTTON, C. A. 1963: Development of fine-textured landscape relief in temperate pluvial climates. *N.Z. J. Geol. and Geophys.* **6**, 528–33.

COTTON, C. A. 1964: The control of drainage density. *N.Z. J. Geol. and Geophys.* **7**, 348–52.

COTTON, C. A. and TE PUNGA, M. 1956: Fossil gullies in the Wellington landscape. *N.Z. Geog.* **11**, 72–5.

COURT, A. 1961: Area-depth rainfall formulas. *J. Geophys. Res.* **66**, 1823–31.

COURT, A. 1962: Measures of streamflow timing. *J. Geophys. Res.* **67**, 4335–9.

COUTTS, J. R. H. *et al.* 1968: Use of radioactive 59Fe for tracing soil particle movement. Pt. 1. Field studies of splash erosion. *J. Soil Sci.* **19**, 311–24.

CRAMPTON, C. B. 1969: The chronology of certain terraced river deposits in the north east Wales area. *Zeitschrift für Geomörphologie* **13**, 245–59.

CRAWFORD, N. H. and LINSLEY, R. K. 1966: *The Stanford Watershed Model Mk. IV., Tech. Rept.* **39**, Dept. Civ. Eng., Stanford University.

CRICKMORE, M. J. and LEAN, G. M. 1962: The measurement of sand transport by the time integration method with radioactive tracers. *Proc. Royal. Soc. A.* **270**, (1340), 27–47.

CROUSE, R. P., CORBETT, E. S. and SEEGRIST, D. W. 1966: Methods of measuring and analyzing rainfall interception by grass. *Internat. Assoc. Sci. Hyd. Bull.* **11**, 110–20.

CULLING, W. E. H. 1960: Analytical theory of erosion. *J. Geol.* **68**, 336–44.

CUMBERLAND, K. B. 1944: *Soil erosion in New Zealand, a geographic reconnaissance*, Christchurch, New Zealand. (228pp.)

CZUDEK, T. and DEMEK, J. 1970: Thermokarst in Siberia and its influence on the development of lowland relief. *Quaternary Research* **1** (1), 103–20.

DALRYMPLE, T. 1960: Flood frequency analyses. *U.S. Geol. Survey Water Supply Paper* **1543A**.

DANIEL, J. F. 1970: Channel movement of meandering Indiana streams. *U.S. Geol. Survey Open File Rep.*

DANIELS, R. B. 1966: Physiographic history and the soils, entrenched stream systems and gullies, Harrison Co., Iowa. *U.S. Dept. Ag. Tech. Bull.* **1348**, 51–83.

DAVIES, G. L. 1968: *The earth in decay: a history of British geomorphology 1578–1878*, London. (390pp.)

DAVIS, G. 1967: Hydrology of contour strip mines in the Appalachian Region of the United States. *Int. Union of Forestry Res. Orgs. 14th Cong. Section* **01-02-11**, 420–43.

DAVIS. S. N. and DEWIEST, R. J. M. 1966: *Hydrogeology*, New York. (463pp.)

DAVIS, W. M. 1899: The geographical cycle. *Geog. J.* **14**, 481–504.

DAVIS, W. M. 1900: Physical Geography in the High School. *School Review* **8**, 328–39. Reprinted in D. W. JOHNSON (ed.), *Geographical Essays*, New York (1954), 129–45.

DE COURSEY, D. G. 1971: A stochastic approach to watershed modelling. *Nordic Hyd.* **2**, 186–216.

DENEVAN, W. M. 1967: Livestock numbers in nineteenth century New Mexico, and the problem of gullying in the south west. *Ann. Ass. Amer. Geog.* **57**, 691–703.

DE PLOEY, J. 1967: Erosion pluvial au Congo occidental. *Isotopes in Hydrology*, I.A.E.A. Vienna, 291–300.

DETWYLER, T. R. 1971: *Man's impact on environment*, New York. (731pp.)

DE VRIES, J. and KING, K. M. 1961: Note on the volume of influence of a neutron surface moisture probe. *Canadian J. Soil Sci.* **41**, 253–7.

DE WIEST, R. J. M. 1965: *Geohydrology*, New York. (366pp.)

DIACONU, C. 1971: Probleme ale scurgerii de aluviuni a riurilor Romaniei. *Studii de Hidrologie XXXI*, Bucarest Institute de Meteorologie si Hidrologie.

DISEKER, E. G. and RICHARDSON, E. C. 1962: Erosion rates and control methods on highway cuts. *Trans. Amer. Soc. Agr. Eng.* **5**, 153–5.

DISKIN, M. H. 1967: A Laplace transform proof of the theories of moments for the instantaneous unit hydrograph. *Water Resources Res.* **3**, 385–8.

DOBROUMOV, B. M. 1967: Evaluation of the effect of mining the iron ore deposits of the Kursk Magnetic Anomaly on the flow regime of the Oskolets River. *Transactions State Hyd. Inst. (Trudy GGI)* **319**, 206–23.

DOKUCHAYEV, V. V. 1899: *Towards a theory of Natural Zones. Horizontal and Vertical Soil Zones*, St. Petersburg.

DOOGE, J. C. I. 1968: The hydrologic cycle as a closed system. *Internat. Assoc. Sci. Hyd. Bull.* **13**, 58–68.

DOORNKAMP, J. C. and KING, C. A. M. 1971: *Numerical Analysis in Geomorphology*, London, Arnold. (372pp.)

DORAN, F. J. 1942: High water gaging. *Civil Eng.* **12**, 103–4.

DORTINGNAC, E. J. and BEATTIE, B. 1965: Using representative watersheds to manage forest and range lands for improved water yield. *Internat. Assoc. Sci. Hyd. Pub.* **66**, 480–88.

DOTY, R. 1970: A portable automatic water sampler. *Water Resources Res.* **6**, 1787–88.

DOUGLAS, I. 1963: Field methods for hardness determinations of cave and river waters. *Cave Res. Gp. Gt. Brit. Newsletter* **88**, 3–6.

DOUGLAS, I. 1967: Man, vegetation and the sediment yield of rivers. *Nature* **215**, 925–8.

DOUGLAS, I. 1968: Field methods of water hardness determination. *British Geomorph. Research Group Tech. Bull.* **1**.

DOUGLAS, I. 1968: The effects of precipitation chemistry and catchment area lithology on the quality of river water in selected catchments in Eastern Australia. *Earth Sci. J.* **2**, 126–44.

DOUGLAS, I. 1968: Erosion in the Sungei Gombak catchment, Selangor, Malaya. *J. Tropical Geog.* **26**, 1–16.

DOUGLAS, I. 1969: Sediment yields from forested and agricultural lands in J. A. TAYLOR (ed.), *Univ. College of Wales, Aberystwyth Memo.* **12**, E1–22.

DOUGLAS, I. 1971: Comments on the determination of fluvial sediment discharge. *Australian Geog. Studies* **9**, 172–6.

E. MYCIELSKA-DOWGIALLO 1969: Proba rekonstrukcji warunkow paleohydro-dynamicznych rzeki na podstawie badan sedymentologicznych w dolinie wisly pod Tarnobrzegiem. *Przeglad Geograficzny* **41**(3), 409–29.

DOWNING, R. A., ALLENDER, R. and BRIDGE, L. R. 1970: The hydrogeology of the Trent River Basin. *Water Supply Paper of the Institute of Geological Sciences, Hydro-geological Rpt.* **5**, Natural Environment Research Council. (104pp.)

DRAGOUN, F. J. 1962: Rainfall energy as related to sediment yield. *J. Geophys. Res.* **67**, 1495–501.

DRAGOVICH, D. 1966: Gullying in the Mount Lofty Ranges. *Australian J. Sci.* **29**, 80–81.

DREW, D. P., NEWSON, M. D. and SMITH, D. I. 1968: Mendip Karst Hydrology Research Project. Phase Three. *Wessex Cave Club Occasional Pub. Series* **2** (2).

DREW, D. P. and SMITH, D. I. 1969: Techniques for the tracing of subterranean drainage. *British Geomorph. Res. Group Tech. Bull.* **2**.

DROST, H. 1966: Precalibrated fibreglass flow measurement structures for experi-mental basins. *J. Hyd. (N.Z.)* **5**, 20–24.

DUNNE, T. and BLACK, R. D. 1970: An experimental investigation of runoff production in permeable soils. *Water Resources Res.* **6**, 478–90.

DUNNE, T. and BLACK, R. D. 1970: Partial area contributions to storm runoff in a small New England watershed. *Water Resources Res.* **6**, 1296–311.

DURUM, W. H. 1953: Relationship of the mineral constituents in solution to stream flow, Saline River near Russell, Kansas. *Trans. Amer. Geophys. Union* **34**, 435–42.

DURUM, W. H., HEIDEL, S. G. and TISON, L. J. 1960: World-wide runoff of dissolved solids. *Internat. Assoc. Sci. Hyd. Pub.* **51**, 618–28.

DURY, G. H. 1963: Underfit streams in relation to capture: A reassessment of the ideas of W. M. Davis. *Trans. Inst. Brit. Geog.* **32**, 83–94.

DURY, G. H. 1964: Principles of underfit streams. *U.S. Geol. Survey Prof. Paper* **452A**. (67pp.)

DURY, G. H. 1964: Subsurface explorations and chronology of underfit streams. *U.S. Geol. Survey Prof. Paper* **452B**. (56pp.)

DURY, G. H. 1965: Theoretical implications of underfit streams. *U.S. Geol. Survey Prof. Paper* **452C**. (43pp.)

DURY, G. H. 1966: The concept of grade. In G. H. DURY (ed.), *Essays in Geomorphology*, London, 211–34.

DURY, G. H. 1967: Climatic change as a geographical backdrop. *Australian Geographer* **10**(4), 231–42.

DURY, G. H. 1969: Relation of morphometry to runoff frequency. In R. J. CHORLEY (ed.), *Water, Earth and Man*, London, 419–30.

DURY, G. H. 1970: Merely from nervousness. *Area* **2**(4), 29–32.

DURY, G. H. (ed.) 1970: *Rivers and River Terraces*, London. (283pp.)

DUVDEVANI, S. 1947: An optical method of dew estimation. *Q.J. Royal Met. Soc.* **73**, 282–96.

DYLIK, J. 1961: The Łodz Region. *Guide Book of Excursion* **C**, INQUA VIth Congress.

EAGLESON, P. S. 1969: Potential of physical models for achieving better understanding and evaluation of watershed changes. In W. L. MOORE and C. W. MORGAN (eds.), *Effects of watershed changes on streamflow. Water Resources Symposium* **2**, University of Texas, 12–25.

EDMONDS, D. T., PAINTER, R. B. and ASHLEY, G. D. 1970: A semi-quantitative hydrological classification of soils in north-east England. *J. Soil Sci.* **21**(2), 256–64.

EDWARDS, K. A. and RODDA, J. C. 1970: A preliminary study of the water balance of a small clay catchment. *J. Hyd. (N.Z.)* **9**, 202–18.

EGNER, H. and ERIKSSON, E. 1955: Current data on the chemical composition of air and precipitation. *Tellus* **VII**, 134–9.

EINSTEIN, H. A. 1950: The bed-load function for sediment transportation in open channel flows. *U.S. Dept. Ag. SCS Tech. Bull.* **1026**.

EINSTEIN, H. A. 1964: River Sedimentation. In V. T. CHOW (ed.), *Handbook of Applied Hyd.*, 17-35–17-67.

ELLISON, W. D. 1945: Some effects of raindrops and surface-flow on soil erosion and infiltration. *Trans. Amer. Geophys. Union* **26**, 415–29.

ELLISON, W. D. 1950: Splash erosion in pictures. *J. Soil and Water Conservation* **5**, 71–3.

ELRICK, D. E. and LAWSON, D. W. 1969: Tracer techniques in hydrology. *Instrumentation and Observation Techniques. Proceedings of Hydrology Symp.* **7**, National Research Council of Canada, 155–77.

EMBLETON, C. and KING, C. A. M. 1968: *Glacial and periglacial geomorphology*, London. (608pp.)

EMMETT, W. W. 1965: The vigil network: methods of measurement and a sample of data collected. *Internat. Assoc. Sci. Hyd. Pub.* **66**, 89–106.

EMMETT, W. W. and LEOPOLD, L. B. 1965: Downstream pattern of riverbed scour and fill. *Proc. Fed. Inter-Agency Sedimentation Conf. U.S.D.A. Misc. Pub.* **970**, 399–409.

EMMETT, W. W. and LEOPOLD, L. B. 1967: On the observation of soil movement in excavated pits. *Revue de Geomorph. Dynamique* **17**, 157.

ENGLAND, C. B. and HOLTAN, H. N. 1969: Geomorphic grouping of soils in watershed engineering. *J. Hyd.* **7**, 217–25.

ENGLAND, C. B. and ONSTAD, C. A. 1968: Isolation and characterisation of hydrologic response units within agricultural watersheds. *Water Resources Res.* **4**(1), 73–7.

EVERETT, K. R. 1962: Quantitative measurement of soil movement. *Geol. Soc. Amer. Special Paper* **73**, 147–8.

EYLES, R. J. 1966: Stream representation on Malayan maps. *J. Tropical Geog.* **22**, 1–9.

EYLES, R. J. 1968: Stream net ratios in west Malaysia. *Geol. Soc. Amer. Bull.* **79**, 701–12.

EYLES, R. J. 1971: A classification of west Malaysian drainage basins. *Ann. Ass. Amer. Geog.* **61**, 460–67.

FAGG, C. C. 1923: The recession of the Chalk escarpment and the development of the dry chalk valley. *Proc. Croydon Nat. Hist. and Scientific Soc.* **9**, 93–112.

FAHNESTOCK, R. K. 1963: Morphology and hydrology of a glacial stream—White River, Mount Rainier, Washington. *U.S. Geol. Survey Prof. Paper*, **422A**. (70pp.)

FANIRAN, A. 1969: The index of drainage intensity a provisional new drainage factor. *Australian J. Sci.* **31**(9), 328–30.

F.A.O., 1965: Soil erosion by water, some measures for its control on cultivated lands. *F.A.O. Agricultural Dev. Paper* **81**.

FELS, E. 1965: Nochmals: Anthropogene Geomorphologie. *Petermanns Geographische Mitteilungen* **109**, 9–15.

F.I.A.S.P., 1961: The single-stage sampler for suspended sediment. Federal *Inter-Agency Sedimentation Project Report* **13**.

F.I.A.S.P., 1962: Investigation of a pumping sampler with alternate suspended sediment handling systems. *Federal Inter-Agency Sedimentation Project Report* **Q**.

F.I.A.S.P., 1963: Determination of fluvial sediment discharge. *Federal Inter-Agency Sedimentation Project Report* **14**.

FINLAYSON, G. D. 1969: Telemetry of meaningful suspended sediment data in the Umgeni catchment. In *Progress in River Engineering*, South African Institution of Civil Engineers, **B17–B20**.

FISHMAN, M. J. and DOWNS, S. C. 1966: Methods for analysis of selected metals in water by atomic absorption. *U.S. Geol. Survey Water Supply Paper* **1540C**.

FLEMING, G. 1967: The computer as a tool in sediment transport research. *Internat. Assoc. Sci. Hyd. Bull.*, 45–54 (12pp.)

FLEMING, G. 1968: The Stanford Sediment Model: I Translation. *Internat. Assoc. Sci. Hyd. Bull.* **13**, 108–25.

FLEMING, G. 1969: *The Clyde Basin: hydrology and sediment transport.* Unpublished Ph.D. thesis, Strathclyde University, Glasgow.

FLEMING, G. 1969: Suspended solids monitoring: A comparison between three instruments. *Water and Water Engineering*, 377–82.

FLEMING, G. 1969: Design curves for suspended load estimation. *Proc. Inst. Civ. Eng.* **43**, 1–9.

FLEMING, G. 1970: Simulation of streamflow in Scotland. *Internat. Assoc. Sci. Hyd. Bull.* **15**, 53–9.

FLEMING, N. C. 1964: Form and function of sedimentary particles. *J. Sedimentary Petrology* **35**, 381–90.

FLINT, R. F. 1971: *Glacial and Pleistocene Geol.*, New York. (892pp.)

FLORKOWSKI, T. and CAMERON, J. F. 1966: A simple radioisotope X-ray transmission gauge for measuring suspended sediment concentrations in rivers. In *Radioisotope Instruments in Industry and Geophysics*, Int. Atomic Energy Agency, Vienna, 395.

FOSTER, E. E. 1949: *Rainfall and Runoff*, New York. (487 pp.)

FOURNIER, F. 1960: *Climat et érosion: la relation entre l'érosion du sol par l'eau les précipitations atmospheriques*, Paris. (201pp.)

FOURNIER, F. 1960: Debit solide des cours d'eau. Essai d'estimation de la perte en terre subie par l'ensemble du globe terrestre. *Internat. Assoc. Sci. Hyd. Pub.* **53**, 19–22.

FOURNIER, F. 1969: Transports solides effectués par les cours d'eau. *Internat. Assoc. Sci. Hyd. Bull.* **14**, 7–47.

FREDIKSEN, R. L. 1969: A battery powered proportional stream water sampler. *Water Resources Research* **5**, 1410–13.

FREDIKSEN, R. L. 1970: Erosion and sedimentation following road construction and timber harvest on unstable soils in three small western Oregon watersheds. *U.S.D.A. Forest Service Res. Paper* **PNW104**. (15pp.)

FREE, G. R., BROWNING, G. M. and MUSGRAVE, G. W. 1940: Relative infiltration and related physical characteristics of certain soils. *U.S. Dept. Agr. Tech. Bull.* **729**.

FREISE, F. 1936: Das binnenklima van Urwaldern in subtropischen Brasilien. *Petermanns Mitt.* **82**, 301–7.

FRIEDMAN, G. M. 1967: Dynamic processes and statistical parameters compared for size frequency distribution of beach and river sands. *J. Sedimentary Petrology* **37**, 327–54.

GALFI, J. and PALOS, M. 1970: Use of seismic refraction measurements for ground water prospecting. *Internat. Assoc. Sci. Hyd.* **15**, 41–6.

GALON, R. 1964: Hydrological research for the needs of the regional economy. *Problems of Applied Geography II, Geographica Polonica* **3**, 239–50.

GARDINER, V. 1971: A drainage density map of Dartmoor. *Trans. Devon. Assoc.* **103**, 167–80.

GEIB, H. V. 1933: A new type of installation for measuring soil and water losses from control plots. *J. Amer. Soc. Agronomy* **24**, 429–40.

GERASIMOV, I. P. 1967: Sovetskaya fizicheskaya geografiya i ee nov'e Koustructivn'e napral'eniya. *Geograficky Casopis* **19**, 257–62.

GERLACH, T. 1967: Hillslope troughs for measuring sediment movement. *Revue de Geomorph. Dynamique* **17**, 173.

GERMAN, R. 1963: Taldichte und Flussdichte in Südwestdentschland. Ein Beitrag zur klimabedingten Oberflachenformung. *Berichte zur Deutschen Laudeskunde* **1**, 12–32.

GEYL, W. F. 1968: Tidal stream action and sea level change as one cause of valley meanders and underfit streams. *Australian Geographical Studies* **6**, 24–42.

GILBERT, G. K. 1887: *Report on the Geology of the Henry Mountains*, Washington. (160pp.)

GILBERT, G. K. 1914: The transportation of debris by running water. *U.S. Geol. Survey Prof. Paper* **86**, (263pp.)

GIUSTI, E. V. 1962: A relation between floods and drought flows in the Piedmont Province in Virginia. *U.S. Geol. Survey Prof. Paper* **450C**, 128–9.

GIUSTI, E. V. and SCHNEIDER, W. J. 1962: Comparison of drainage on topographic maps of the Piedmont province. *U.S. Geol. Survey Prof. Paper* **450E**, E118–19.

GIUSTI, E. V. and SCHNEIDER, W. J. 1965: The distribution of branches in river networks. *U.S. Geol. Surv. Prof. Paper* **422G**.

GLEASON, G. B. 1952: Consumptive use of water, municipal and industrial areas. *Trans. Am. Soc. Agr. Engineers* **117**, 1004–9.

GLUSHKOV, V. G. 1933: The Geographical-Hydrologic Method. *Bulletin State Hydrologic Institute*, (I3v. G.G.I.), 57–8.

GLYMPH, L. M. 1954: Studies of sediment yields from watersheds. *Internat. Assoc. Sci. Hyd. Pub.* **36**, 178–91.

GLYMPH, L. M. and STOREY, H. N. 1967: Sediment—its consequences and control. In N. C. BRADY (ed.), Agriculture and the quality of our environment. *American Assoc. Adv. Sci. Pub.* **85**, 205–20.

GLYMPH, L. M. and HOLTAN, H. N. 1969: Land treatment in agricultural watershed hydrology research. In W. L. MOORE and C. W. MORGAN (eds.), *Effects of watershed change on streamflow*, Texas, 44–68.

GOLTERMAN, H. L. and CLYMO, R. S. (eds.) 1969: Methods for chemical analysis of fresh waters. *International Biological Programme Handbook* **8**.

GOLF, I. W. 1967: Contribution concerning flow rates of rivers transporting drain waters of open cast mines. *Internat. Assoc. Sci. Hyd. Pub.* **76**, 306–16.

GORHAM, E. 1955: On the acidity and salinity of rain. *Geochim. Cosmochim. Acta.* **7**, 231–9.

GORHAM, E. 1961: Factors influencing supply of major ions to inland waters with special reference to the atmosphere. *Geol. Soc. Amer. Bull.* **72**, 795–840.

GOSH, A. K. and SCHEIDEGGER, A. E. 1970: Dependence of stream link lengths and drainage areas on stream order. *Water Resources Res.* **6**, 336–40.

GOTTSCHALK, L. C. 1946: Silting of stock ponds in land utilisation area SD-LU-2, Pierre, South Dakota. *U.S. Soil Conserv. Ser. Spec. Rept.* **9**.

GOTTSCHALK, L. C. 1962: Effects of watershed protection measures on reduction of erosion and sediment damages in the United States. *Internat. Assoc. Sci. Hyd. Pub.* **59**, 426–47.

GOTTSCHALK, L. C. 1964: Reservoir Sedimentation. In V.T. CHOW (ed.), *Handbook of Applied Hydrology*, 17-1–17-34.

GRAF, W. H. 1971: *Hydraulics of sediment transport.* New York. (544pp.)

GRAY, D. M. 1962: Derivation of hydrographs for small watersheds from measurable physical characteristics. *Agricultural and Home Economics Experimental Station, Iowa State University of Science and Technology Res. Bull.* **506**, 514–570.

GRAY, D. M. 1965: Physiographic characteristics and the runoff pattern. In *Research Watersheds Proceedings of Hydrology Symposium* **4**, National Research Council of Canada, 147–64.

GREEN, M. J. 1970: Effects of exposure on the catch of rain gauges. *J. Hyd. (N.Z.)* **9**, 55–71.

GREENWOOD, G. 1877: *River Terraces*, London. (247pp.)

GREGORY, K. J. 1962: The deglaciation of eastern Eskdale, Yorkshire. *Proc. Yorks. Geol. Soc.* **33**, 363–80.

GREGORY, K. J. 1966: Dry valleys and the composition of the drainage net. *J. Hyd.* **4**, 327–40.

GREGORY, K. J. 1968: The composition of the drainage net: Morphometric analysis of maps. British Geomorphological Research Group Occasional Paper **4**, 9–11.

GREGORY, K. J. 1971: Drainage density changes in south west England. K. J. GREGORY and W. L. D. RAVENHILL (eds.), *Exeter Essays in Geography*, Exeter, 33–53.

GREGORY, K. J. and BROWN, E. H. 1966: Data processing and the study of land form. *Zeitschrift für Geomorphologie* N.F.I.O., 237–63.

GREGORY, K. J. and WALLING, D. E. 1968: The variation of drainage density within a catchment. *Bull. Int. Assoc. Sci. Hyd.* **13**, 61–8.

GREGORY, K. J. and WALLING, D. E. 1971: Field measurements in the drainage basin. *Geography* **56**, 277–92.

GREIG-SMITH, P. 1964: *Quantitative Plant Ecology*, London. (198pp.)

GRINDLEY, J. and SINGLETON, F. 1967: The routine estimation of soil moisture deficits. *Internat. Assoc. Sci. Hyd. Pub.* **85**, 811–20.

GUNNERSON, C. G. 1967: Streamflow and quality in the Columbia River basin. *Proc. A.S.C.E., J. San. Eng. Div.* **39**, 1–16.

GUY, H. P. 1964: An analysis of some storm-period variables affecting stream sediment transport. *U.S. Geol. Survey Professional Paper* **462E**.

GUY, H. P. 1965: Residential construction and sedimentation at Kensington, Md.

Proc. Federal Inter-Agency Sedimentation Conference, U.S.D.A. Misc. Pub. **970**, 30–37.

GUY, H. P. 1969: Laboratory theory and methods for sediment analysis. Laboratory analysis: *Techniques of Water-Resource Investigations of the United States Geological Survey.* **C1**(5), Washington. (58pp.)

GUY, H. P. 1970: Fluvial sediment concepts. *Techniques of Water-Resource Investigations of the United States Geological Survey* **C1**(3), Washington. (55pp.)

GUY, H. P. and NORMAN, V. W. 1970: Field methods for measurement of fluvial sediment. *Techniques of Water-Resource Investigations of the United States Geological Survey* **C2**(3).

GUY, H. P., SIMONS, D. B. and RICHARDSON, E. V. 1966: Summary of alluvial channel data from flume experiments 1956–61. *U.S. Geol. Survey Professional Paper* **462I**.

GUY, L. T. 1942: Unique method of gauging high water. *Civil Engineering* **12**, 397.

HACK, J. T. 1957: Studies of longitudinal stream profiles in Virginia and Maryland. *U.S. Geol. Surv. Professional Paper* **294B**. (53pp.)

HACK, J. T. 1960: Interpretation of erosional topography in humid temperate regions. *Amer. J. Sci.* **258**, 80–97.

HACK, J. T. 1960: Relation of solution features to chemical character of water in the Shenandoah Valley, Virginia. *U.S. Geol. Survey Professional Paper* **400B**, 387–90.

HACK, J. T. 1965: Geomorphology of the Shenandoah Valley Virginia and West Virginia and origin of the residual ore deposits. *U.S. Geol. Survey Professional Paper* **484**.

HACK, J. T. 1965: Postglacial drainage evolution and stream geometry in the Ontonagon area, Michigan. *U.S. Geol. Surv. Professional Paper* **504B**, B1–40.

HACK, J. T. 1966: Interpretation of Cumberland escarpment and Highland rim, South-central Tennessee and north east Alabama. *U.S. Geol. Survey Professional Paper* **524C**.

HADLEY, R. F. 1960: Recent sedimentation and erosional history of Fivemile Creek, Fremont County, Wyoming. *U.S. Geol. Survey Professional Paper* **352A**, 1–16.

HADLEY, R. F. 1967: On the use of holes filled with colored grains. *Revue de Geomorph. Dynamique* **17**, 158–9.

HAGGETT, P. 1961: Land use and sediment yield in an old plantation tract of the Serra do Mar, Brazil. *Geog. J.* **127**, 50–62.

HAGGETT, P. and CHORLEY, R. J. 1969: *Network Analysis in Geography*, London. (348pp.)

HALEVY, E., MOSER, H., ZELLHOFER, O. and ZUBER, A. 1967: Borehole dilution techniques: A critical review. In *Isotopes in Hydrology*. Internat. Atomic Energy Agency, Vienna, 531.

HALL, D. G. 1967: The pattern of sediment movement in the River Tyne. *Internat. Assoc. Sci. Hyd. Pub.* **75**, 117–42.

HALL, D. G. 1967: The assessment of water resources in Devon, England using limited hydrometric data. *Internat. Assoc. Sc. Hyd. Pub.* **76**, 110–20.

HALL, D. G., PRAIN, A. F. and HOER, J. J. 1968: Automatic instrumentation for hydrometric data observation. *Water and Water Engineering* **72**, 51–7.

HALL, F. R. 1970: Dissolved solids-discharge relationships: 1 Mixing models. *Water Resources Res.* **6**, 845–50.

HALL, F. R. 1971: Dissolved solids-discharge relationships: 2 Application to field data. *Water Resources Res.* **7**, 591–601.

HAMILTON, E. L. 1954: Rainfall sampling in rugged terrain. *U.S. Dept. Agriculture Tech. Bull.* **1096**.

HAMMOND, E. 1964: Analysis of properties in landform geography: An application to broad-scale land form mapping. *Ann. Ass. Amer. Geog.* **54**, 11–18.

HANWELL, J. D. and NEWSON, M. D. 1970: The great storms and floods of July 1968 on Mendip. *Wessex Cave Club Occasional Publication* **1**(2). (72pp.)

HARMESON, R. H. and LARSON, T. E. 1969: Quality of surface water in Illinois 1956–66. *Illinois State Water Survey Bulletin* **54**.

HARRIS, D. R. and VITA-FINZI, C. 1968: Kokkinopilos—A Greek badland. *Geog. J.* **134**, 537–45.

HARRISON, A. J. M. 1966: Plynlimon experimental catchments model investigation of a structure for flow measurement in steep streams. *Hydraulics Research Station Rept.* **EX335**.

HARROLD, L. L. and KRIMGOLD, D. B. 1943: Devices for measuring rates and amounts of runoff employed in soil conservation research. *U.S. Soil Conservation Service Tech. Paper* **51**. (42pp.)

HARTSHORNE, R. 1959: *Perspective on the Nature of Geography*. Assoc. Amer. Geographers. (201pp.)

HARVEY, A. M. 1969: Channel capacity and the adjustment of streams to hydrologic regime. *J. Hyd.* **8**, 82–98.

HAYNES, C. V. 1968: Geochronology of late-Quaternary alluvium. In R. B. MORRISON and H. E. WRIGHT (eds.), Means of Correlation of Quaternary Successions. *Proc. VII Inqua Congress* **8**, Utah, 591–631.

HEIDEL, S. G. 1956: The progressive lag of sediment concentration with flood waves. *Trans. Amer. Geophys. Union* **37**, 56–66.

HEINEMANN, H. G. 1962: Using the gamma probe to determine the volume-weight of reservoir sediment. *Internat. Assoc. Sci. Hyd. Pub.* **59**, 411–23.

HEM, J. D. 1970: Study and interpretation of the chemical characteristics of natural water. *U.S. Geol. Survey Water Supply Paper* **1373** (2nd edn.).

HENDERSON, F. M. 1966: *Open Channel Flow*, London.

HENDRICKSON, G. E. and KRIEGER, R. A. 1960: Relationship of chemical quality of water to stream discharge in Kentucky. *Geochemical cycles: Internat. Geol. Cong. 21st, Rept., pt.* **1**, 66–75.

HENDRICKSON, G. E. and KRIEGER, R. A. 1964: Geochemistry of natural waters of the Blue Grass Region Kentucky. *U.S. Geol. Survey Water Supply Paper* **1700**.

HERBERTSON, J. G. 1969: A critical review of conventional bed load formulae. *J. Hyd.* **8**, 1–26.

HERDMAN, E. R. 1970: Mean annual runoff as related to channel geometry of selected streams in California. *U.S. Geol. Survey Water Supply Paper* **1999E**.

HERMSMEIER, L. et al. 1963: *Construction and operation of a 16-unit rainulator*. U.S. Dept. Agriculture ARS, 41–62.

HERSHFIELD, D. M. 1961: Rainfall frequency atlas of the United States for durations from 30 minutes to 24 hours and return periods from 1 to 100 years. *U.S. Weather Bureau Tech. Rept.* **40**.

HERSHFIELD, D. M. and WILSON, W. T. 1958: Generalizing of rainfall-intensity-frequency data. I.U.G.G. *Assemblee generale de Toronto, 1957* **1**, 499–506.

HEWLETT, J. D. 1961: Soil moisture as a source of base flow from steep mountain watersheds. *U.S. Dept. Agriculture Forest Service, S.E. Forest Expt. Sta. Paper* **132**.

HEWLETT, J. D. 1967: A hydrologic response map for the state of Georgia. *Water Resources Bull.* **13**, 4–20.

HEWLETT, J. D. and HIBBERT, A. R. 1961: Increases in water yield after several types of forest cutting. *Internat. Assoc. Sci. Hyd. Bull.* **5–17**.

HEWLETT, J. D. and HIBBERT, A. R. 1967: Factors affecting the response of small watersheds to precipitation in humid areas. In W. E. SOPPER and H. W. LULL (eds.), *International Symposium on Forest Hydrology*, Oxford, 275–90.

HEWLETT, J. D., LULL, H. W. and REINHART, K. G. 1969: In Defense of Experimental Watersheds. *Water Resources Res.* **5**, 306–16.

HIBBERT, A. R. and CUNNINGHAM, G. B. 1967: Streamflow data processing opportunities and application. In W. E. SOPPER and H. W. LULL (eds.), *International Symposium on Forest Hydrology*, Oxford, 725–36.

HICKOK, R. B., KEPPEL, R. Y. and RAFFERTY, B. R. 1959: Hydrograph synthesis for small aridland watershed. *Agricultural Engineering* **40**, 608–11, 615.

HILLS, R. C. 1971: Lateral flow under cylinder infiltrometers—a graphical correction procedure. *J. Hyd.* **13**, 153–62.

HILLS, R. C. 1971: The influence of land management and soil characteristics on infiltration and the occurrence of overland flow. *J. Hyd.* **13**, 163–81.

HIRANANDANI, M. G. and CHITALE, S. V. 1960: *Stream Gauging*, Central Water and Power Research Station, Poona, India.

HJULSTROM, F. 1935: Studies of the morphological activity of rivers as illustrated by the River Fyris. *Bull. Geol. Inst. Univ. Uppsala* **25**, 221–527.

HOLEMAN, J. N. 1968: The sediment yield of major rivers of the world. *Water Resources Res.* **4**, 737–47.

HOLMES, C. D. 1964: Equilibrium in humid-climate physiographic processes. *Amer. J. Sci.* **262**, 436–45.

HOLMES, J. W., TAYLOR, S. A. and RICHARDS, S. J. 1967: Measurement of Soil Water. In H. R. HAGAN *et al.* (eds.), *Irrigation of Agricultural Lands, Agronomy* **11**(15).

HOLSTENER-JORGENSEN, H. 1967: Influences of forest management and drainage on ground-water fluctuations. In W. E. SOPPER and H. W. LULL (eds.), *International Symposium on Forest Hydrology*, Oxford, 325–34.

HORNBECK, J. W., PIERCE, R. S. and FEDERER, C. A. 1970: Streamflow changes after forest clearing in New England. *Water Resources Res.* **6**, 1124–31.

HORNER, W. W. and LLOYD, C. L. 1940: Infiltration-capacity values as determined from a study of an eighteen-month record at Edwardsville Illinois. *Trans. Amer. Geophys. Union* **21**, 522–41.

HORTON, R. E. 1924: Discussion of 'The distribution of intense rainfall and some other factors in the design of storm-water drains'. *Proc. A.S.C.E.* **50**, 660–67.

HORTON, R. E. 1932: Drainage basin characteristics. *Trans. Amer. Geophys. Union* **13**, 350–61.

HORTON, R. E. 1933: The role of infiltration in the hydrologic cycle. *Trans. Amer. Geophys. Union* **14**, 446–60.

HORTON, R. E. 1945: Erosional development of streams and their drainage basins:

hydrophysical approach to quantitative morphology. *Geol. Soc. Amer. Bull.* **56**, 275–370.

HOWARD, A. D. 1965: Geomorphological systems—equilibrium and dynamics. *Amer. J. Sci.* **263**, 303–12.

HOWARD, A. D. 1967: Drainage analysis in geologic interpretation: a Summation. *Bull. Amer. Assoc. Petroleum Geologists* **51**, 2246–59.

HOWARD, A. D. 1971: Simulation model of stream capture. *Geol. Soc. Amer. Bull.* **82**, 1355–76.

HOWARD, A. D. 1971: Simulation of stream networks by headward growth and branching. *Geog. Analysis* **3**, 29–50.

HOWARD, A. D., KEETCH, M. E. and LINWOOD VINCENT, C. 1970: Topological and geometrical properties of braided streams. *Water Resources Res.* **6**, 1674–88.

HOWE, G. M., SLAYMAKER, H. O. and HARDING, D. M. 1966: Flood Hazard in Mid-Wales. *Nature* **212**, 584–5.

HOWE, G. M., SLAYMAKER, H. O. and HARDING, D. M. 1967: Some aspects of the flood hydrology of the upper catchments of the Severn and Wye. *Trans. Inst. Brit. Geog.* **41**, 33–58.

HUBBELL, D. W. 1964: Apparatus and techniques for measuring bedload. *U.S. Geol. Survey Water Supply Paper* **1748**.

HUBBELL, D. W. and SAYRE, W. W. 1965: Application of Radioactive tracers in the study of sediment movement. *Proc. Federal Inter-Agency Sedimentation Conference 1963 U.S.D.A. Misc. Pub.* **970**, 569–78.

HUDSON, N. 1971: *Soil Conservation*, London. (320pp.)

HUFF, F. A. and STOUT, G. E. 1952: Area-depth studies for thunderstorm rainfall in Illinois. *Trans. Amer. Geophys. Union* **33**, 495–8.

HUTCHINSON, J. N. 1970: A coastal mudflow on the London Clay Cliffs at Beltinge, North Kent. *Geotechnique* **20**, 412–38.

HUTTON, J. 1795: *Theory of the earth*, Edinburgh. (2 vols.)

INGLIS, C. C. 1949: The behaviour and control of rivers and canals. *Res. Pub., Poona* **13**, (India). (2 vols.)

INSTITUTE OF HYDROLOGY, 1967: *Record of Res. 1967*, Natural Environment Research Council, Wallingford.

INSTITUTE OF HYDROLOGY, 1968: *Record of Res. 1968*, Natural Environment Research Council, Wallingford.

INSTITUTE OF HYDROLOGY, 1971: *Record of Research 1970–71*, Natural Environment Research Council, Wallingford.

INSTITUTE OF HYDROLOGY, 1971–2: *Record of Res. 1971–2*, Natural Environment Research Council, Wallingford.

INSTITUTION OF WATER ENGINEERS, 1960: *Approved methods for the physical and chemical examination of water*, London.

INTERNATIONAL ASSOCIATION OF SCIENTIFIC HYDROLOGY, 1965: *Symposium of Quebec Internat. Assoc. Sci. Hyd. Pubs.* **67**, **68**.

INTERNATIONAL ASSOCIATION OF SCIENTIFIC HYDROLOGY, 1965: *Symposium of Budapest Internat. Assoc. Sci. Hyd. Pub.* **66**.

INTERNATIONAL ASSOCIATION OF SCIENTIFIC HYDROLOGY, 1969: *Symposium of Leningrad Internat. Assoc. Sci. Hyd. Pubs.* **84**, **85**.

INTERNATIONAL ATOMIC ENERGY AGENCY, 1968: *Guidebook on nuclear techniques in Hydrology: Tech. Rept.* **91**, I.A.E.A., Vienna.

JACKSON, W. H. 1964: An investigation into silt in suspension in the River Humber. *The Dock and Harbour Authority* **45** (526).

JAMES, L. D. 1965: Using a digital computer to estimate the effects of urban development on flood peaks. *Water Resources Res.* **1**, 223–34.

JAMIESON, D. G. and AMERMAN, C. R. 1969: Quick return subsurface flow. *J. Hyd.* **8**, 122–36.

JAROCKI, W. 1957: *A study of sediment* (Bodanie Rumowiska), Wydawnictwo Morskie, Poland.

JAWORSKA, M. 1968: Erozja chemiczna i denudacja zlewni rzek Wieprza i Pilicy. *Prace Panstwowego Instytutu Hydrologiczno-Meteorologicznego* **95**, 29–47.

JENNINGS, E. G. and MONTEITH, I. L. 1954: A sensitive recording dew balance. *J. Royal Met. Soc.* **80**, 344.

JENNINGS, J. N. 1965: Man as a geological agent. *Australian J. Sci.* **28**, 150–56.

JENNINGS, J. N. 1967: Some karst areas in Australia. In J. N. JENNINGS AND J. A. MABBUTT (eds.), *Landform studies from Australia and New Guinea*, 256–92.

JENNINGS, J. N. 1971: *Karst*, Cambridge. (253pp.)

JEPPSON, R. W., ASHCROFT, G. L., HUBER, A. L., SKOGERBOE, G. B. and BAGLEY, J. M. 1968: *Hydrologic Atlas of Utah*, **PRWG 35–1**, Utah Water Research Laboratory, Utah State University.

JOHNSON, D. W. 1931: *Stream sculpture on the Atlantic slope*, New York. (142 pp.)

JOHNSTONE, D. and CROSS, W. P. 1949: *Elements of Applied Hydrology*, New York. (275pp.)

JOHNSON, E. A. and DILS, R. E. 1956: Outline for compiling precipitation, runoff and groundwater data from small watersheds. *U.S. Dept. Agriculture, Forest Service, Technical Note* **34**.

JOHNSON, E. A. G. 1966: Land drainage in England and Wales. In R. B. THORNE (ed.), *River Engineering and Water Conservation Works*, London, 29–46.

JOHNSON, F. A. 1971: A note on river water sampling and testing. *Water and Water Engineering* **75**, 59–61.

JOHNSON, J. W. 1943: Distribution graphs of suspended matter concentration. *Trans. A.S.C.E.* **69**, 941–56.

JOHNSON, N. M., LIKENS, G. E., BORMANN, F. H., FISHER, D. W. and PIERCE, R. S. 1969: A working model for the variation in stream water chemistry at the Hubbard Brook Experimental Forest, New Hampshire. *Water Resources Res.* **5**, 1353–63.

JOHNSON, P. 1970: Calculation of the instantaneous unit hydrograph using Laplace transforms. *J. Hyd. (N.Z.)* **9**, 307–22.

JOHNSON, P. and MUIR, T. C. 1969: Acoustic detection of sediment movement. *J. Hydraulic Res.* **7**, 519–40.

JONES, A. 1971: Soil piping and stream channel initiation. *Water Resources Res.* **7**, 602–10.

JORDAN, C. F. *et al.*, 1971: Size analysis of silt and clay by hydrophotometer. *J. Sedimentary Petrology* **41**(2), 489–96.

JUDSON, S. and ANDREWS, G. W. 1955: Pattern and form of some valleys in the Driftless Area, Wisconsin. *J. Geol.* **63**, 328–36.

JUDSON, S. and RITTER, D. F. 1964: Rates of regional denudation in the United States. *J. Geophys. Res.* **69**, 3395–401.

WIT-JOZWIK, K. 1968: Przyktady map hydrograficznych z Potudniowej Polski. *Przeglad Geograficzny* **40**(2), 271–83.

JUNGE, C. E. and WERBY, R. T. 1958: The concentration of chloride, sodium, potassium, calcium and sulphate in rain water over the United States. *J. Meteorol.* **15**, 417–25.

KAZO, B. and KLIMES-SZMIK, A. 1962: A method of artificial sprinkling for the investigation of the process of erosion. *Internat. Assoc. Sci. Hyd. Pub.* **59**, 52–61.

KELLER, R. 1968: The role of geography within the International Hydrological Decade. *Fluss regime und Wasserhaushalt 1. Bericht der IGU—Commission on the International Hydrological Decade. Freiburger Geografische Hefte* **6**, 7–14.

KELWAY, P. S. and HERBERT, S. I. 1969: Short term rainfall analysis. *Weather* **24**, 342–54.

KERNEY, M. J., BROWN, E. H. and CHANDLER, T. J. 1964: The late-glacial and post-glacial history of the Chalk escarpment near Brook, Kent. *Phil. Trans. Roy. Soc. Series* **B248**, 135–204.

KIDSON, C. and CARR, A. P. 1962: Marking beach material for tracing experiments. *Proc. A.S.C.E. J. Hyd. Div.* **88**, 43–60.

KILPATRICK, F. A. and BARNES, H. H. 1964: Channel geometry of Piedmont streams as related to frequency of floods. *U.S. Geol. Surv. Prof. Paper* **422E**.

KINDSVATER, C. E. and CARTER, R. W. 1959: Discharge characteristics of rectangular thinplate weirs. *Trans. A.S.C.E.* **124**, 772–802.

KING, C. A. M. 1960: *Techniques in Geomorphology*, London. (342pp.)

KING, C. A. M. 1970: Feedback relationships in geomorphology. *Geografiska Annaler* **52A**, 147–59.

KING, H. W. 1954: *Handbook of Hydraulics*, New York.

KING, L. M. 1967: Soil moisture—Instrumentation, measurement and general principles of network design. In *Soil Moisture, Hydrology Symposium* **6**, Canada National Research Council, 269–314.

KING, N. J. 1968: Restoration of gullied valley floors in semi-arid regions. *XXII Int. Geol. Congress* **12**, 187–195.

KING, W. B. R. and OAKLEY, K. P. 1936: The Pleistocene succession in the lower part of the Thames valley. *Proceedings Prehistoric Society* **2**, 52–76.

KIRKBY, M. J. 1967: Measurement and theory of soil creep. *J. Geol.* **75**, 359–78.

KIRKBY, M. J. 1969: Infiltration, throughflow, and overland flow. In R. J. CHORLEY (ed.), *Water, Earth and Man*, London, 215–28.

KIRKBY, M. J. 1969: Erosion by water on hillslopes. In R. J. CHORLEY (ed.), *Water, Earth and Man*, London, 229–38.

KIRKBY, M. J. and CHORLEY, R. J. 1967: Throughflow, overland flow and erosion. *Internat. Assoc. Sci. Hyd. Bull.* **12**, 5–21.

KLATKOWA, H. 1965: Bowl-shaped basins and dry valleys in the region of Łodz. *Acta Geographica Lodziensia* **3**.

KLIMASZEWSKI, M. 1956: The detailed Hydrographical map of Poland. *Przeglad Geograficzny* **XXVIII**, 41–7.

KNEDLHANS, S. 1971: Mechanical Sampler for determining the water quality of ephemeral streams. *Water Resources Research* **7**, 728–30.

KOCH, E. 1970: The effects of urbanisation in the quality of selected streams in southern Nassau County, Long Island, New York. *U.S. Geol. Survey Professional Paper* **700C**, 189–92.

KOHLER, V. D. 1954: Quoted by L. M. Glymph, Studies of sediment yields from Watersheds, *Internat. Assoc. Sci. Hyd. Pub.* **36**, 188.

KOHNKE, H. and BERTRAND, A. R. 1959: *Soil Conservation*, New York. (298pp.)

KRAMMES, J. S. 1965: Seasonal debris movement from mountainside slopes in Southern California. In *Proc. Federal Inter-Agency Sedimentation Conference: U.S.D.A. Misc. Pub.* **970**, 85–8.

KRIGSTRÖM, A. 1962: Geomorphological studies of sandur plains and their braided rivers in Iceland. *Geografiska Annaler* **44**, 328–46.

KRIVENTSOV, M. I. 1961: Precipitation in the area of the Proletarskoye Reservoir on the Zapadnyy Manych River. *Gidrokhimicheskiye Materialy* **31**.

KRUMBEIN, W. C. 1941: The effect of abrasion on the size and shape and roundness of rock fragments. *J. Geol.* **49**, 482–520.

KRUMBEIN, W. C. and GRAYBILL, F. A. 1965: *An introduction to Statistical models in Geology*, New York. (475pp.)

KUNKLE, G. R. 1962: The baseflow—duration curve, a technique for the study of groundwater discharge from a drainage basin. *J. Geophys. Res.* **67**, 1543–54.

KUNKLE, G. R. 1965: Computation of ground-water discharge to streams during floods, or to individual reaches during baseflow, by use of specific conductance. *U.S. Geol. Survey Professional Paper* **525D**, 207–10.

KUTAL, Z. 1971: *Geology of Recent Sediments*, London. (490pp.)

KUZNIK, I. A. 1954: Runoff from different agricultural lands and some preliminary considerations on the modern stream-flow magnitude in the Transvolga Region. *Meteorologia and Hidrologia* **2**.

LAMPLUGH, G. W. 1914: Taming of streams. *Geog. J.* **43**, 651–6.

LANE, E. W. 1955: The importance of fluvial morphology in hydraulic engineering. *Proc. A.S.C.E., J. Hyd. Div.* **81**(745), 1–17.

LANE, E. W. and KALINSKE, A. A. 1941: Engineering calculations of suspended sediment. *Trans. Amer. Geophys. Union* **22**, 603–7.

LANE, E. W. and LEI, K. 1950: Streamflow variability. *Trans. A.S.C.E.* **115**, 1084–134.

LANG, S. M. and RHODEHAMEL, E. C. 1962: Movement of ground water beneath the bed of the Mullica River in the Wharton Tract, Southern New Jersey. *U.S. Geol. Survey Prof. Paper* **450B**, 90–92.

LANGBEIN, W. B. 1940: Some channel storage and unit hydrograph studies. *Trans. Am. Geophys. Union* **21**, 620–27.

LANGBEIN, W. B. 1947: Topographic characteristics of drainage basins. *U.S. Geol. Survey Water Supply Paper* **968C**, 125–57.

LANGBEIN, W. B. 1964: Geometry of river channels. *Proc. A.S.C.E., J. Hyd. Div.* **HY2**, 90, 301–12.

LANGBEIN, W. B. et al. 1949: Annual Runoff in the United States. *U.S. Geol. Survey Circ.* **52**. (14pp.)

LANGBEIN, W. B. and DAWDY, D. R. 1963: Some general comments on the occurrence of dissolved solids in the surface waters of the United States. *U.S. Geol. Survey unpublished Report* referred to in L. B. LEOPOLD, M. G. WOLMAN and J. P. MILLER (eds.), *Fluvial Processes in Geomorphology*, New York, 77–8.

LANGBEIN, W. B. and LEOPOLD, L. B. 1964: Quasi-equilibrium states in channel morphology. *Amer. J. Sci.* **262**, 782–94.

LANGBEIN, W. B. and LEOPOLD, L. B. 1966: River meanders—theory of minimum variance. *U.S. Geol. Surv. Prof. Paper* **422H**.

LANGBEIN, W. B. and LEOPOLD, L. B. 1968: River channel bars and dunes—theory of kinematic waves. *U.S. Geol. Surv. Prof. Paper* **122L**, L1–20.

LANGBEIN, W. B. and SCHUMM, S. A. 1958: Yield of sediment in relation to mean annual precipitation. *Trans. Amer. Geophys. Union* **39**, 1076–84.

LAPWORTH, C. F., GLASSPOOLE, J. and LLOYD, D. 1948: Report on standard methods of measurement of evaporation. *J. Inst. Water Engineers* **2**, 257–66.

LAUDER, T. L. 1830: *An account of the great floods of August 1829 in the Province of Moray and the adjoining districts*, Edinburgh.

LAW, F. 1956: The effect of afforestation upon the yield of water catchment areas. *J. Brit. Waterworks Assoc.*, 489–94.

LAW, F. 1957: Measurement of rainfall, interception and evaporation losses in a plantation of Sitka spruce trees. *Internat. Assoc. Sci. Hyd. Pub.* **44**, 397–411.

LEDGER, D. C. 1964: Some hydrological characteristics of West African rivers. *Trans. Inst. Brit. Geog.* **35**, 73–90.

LEES, G. 1964: A new method for determining the angularity of particles. *Sedimentology* **3**, 2–21.

LELIAVSKY, S. 1955: *An introduction to fluvial hydraulics*, London. (257pp.)

LEOPOLD, L. B. 1951: Rainfall frequency: An aspect of climatic variation. *Trans. Amer. Geophys. Union* **32**.

LEOPOLD, L. B. 1962: The vigil network. *Internat. Assoc. Sci. Hyd. Bull.* **7**, 5–9.

LEOPOLD, L. B. 1962: Rivers. *Amer. Scientist* **50**, 511–37.

LEOPOLD, L. B. 1968: Hydrology for urban land planning—a guidebook on the Hydrologic Effects of Urban Land Use. *U.S. Geol. Survey Circular* **554**.

LEOPOLD, L. B. 1969: The rapids and pools—Grand Canyon. In *The Colorado River Region and John Wesley Powell. U.S. Geol. Survey Prof. Paper* **669**, 131–45.

LEOPOLD, L. B. 1970: An improved method for size distribution of stream bed gravel. *Water Resources Res.* **6**, 1357–66.

LEOPOLD, L. B. and EMMETT, W. W. 1965: Vigil network sites: a sample of data for permanent filing. *Internat. Assoc. Sci. Hyd. Bull.* **10**, 12–21.

LEOPOLD, L. B., EMMETT, W. W. and MYRICK, R. W. 1966: Channel and hillslope processes in a semi-arid area, New Mexico. *U.S. Geol. Survey Professional Paper* **352G**.

LEOPOLD, L. B. and LANGBEIN, W. B. 1962: The concept of entropy in landscape evolution. *U.S. Geol. Survey Professional Paper* **500A**. (20pp.)

LEOPOLD, L. B. and LANGBEIN, W. B. 1963: Association and indeterminacy in geomorphology. In C. C. ALBRITTON (ed.), *The fabric of Geology*, New York, 184–92.

LEOPOLD, L. B. and MADDOCK, T. 1953: The hydraulic geometry of stream channels and some physiographic implications. *U.S. Geol. Survey Prof. Paper* **252**. (56pp.)

LEOPOLD, L. B. and O'BRIEN MARCHAND, M. 1968: On the quantitative inventory of riverscape. *Water Resources Res.* **4**(4), 709–17.

LEOPOLD, L. B. and MILLER, J. P. 1954: A postglacial chronology for some alluvial valleys in Wyoming. *U.S. Geol. Survey Water Supply Paper* **1261**, 1–90.

LEOPOLD, L. B. and MILLER, J. P. 1956: Ephemeral streams—hydraulic factors and their relation to the drainage net. *U.S. Geol. Survey Prof. Paper* **282A**. (37pp.)

LEOPOLD, L. B. and SKIBITZKE, H. E. 1967: Observations on unmeasured rivers. *Geografiska Annaler* **49A**, 247–55.

LEOPOLD, L. B. and WOLMAN, M. G. 1957: River channel patterns—braided, meandering, and straight. *U.S. Geol. Survey Prof. Paper* **282B**, 39–85.

LEOPOLD, L. B. and WOLMAN, M. G. 1960: River Meanders. *Geol. Soc. Amer. Bull.* **71**, 769–94.

LEOPOLD, L. B., WOLMAN, M. G. and MILLER, J. P. 1964: *Fluvial Processes in Geomorphology*, San Francisco. (522pp.)

LEWIN, J. 1970: A note on stream ordering. *Area* **2**, 32–5.

LEWIS, A. D. 1921: Silt observations of the River Tigris. *Minutes Proc. Inst. Civ. Eng.* **212**, 393–9.

LEYTON, L., REYNOLDS, E. R. C. and THOMPSON, F. B. 1965: Water relations of trees and forests. In *Rept. on Forest Res. 1965* (Forestry Commission), 119–23.

LIAKOPOULOS, A. C. 1965: Theoretical solution of the unsteady unsaturated flow problem in soils. *Internat. Assoc. Sci. Hyd. Bull.* **10**, 58–69.

LIGHT, P. 1947: Area-depth relations. Storm profiles *Thunderstorm Rainfall, Hydrometeorol Rept.* **5**, U.S. Weather Bureau, 264–73.

LIKENS, G. E. 1967: The calcium, magnesium, potassium and sodium budgets for a small forested ecosystem. *Ecology* **48**, 772–85.

LINSLEY, R. K., KOHLER, M. A. and PAULHUS, J. L. H. 1949: *Applied Hydrology*, New York. (689pp.)

LINTON, D. L. 1957: The everlasting hills. *Advancement of Science* **14**(54), 58–67.

LIVINGSTONE, D. A. 1963: Chemical composition of rivers and lakes. *U.S. Geol. Survey Prof. Paper* **440G**. (64pp.)

LJUNGGREN, P. and SUNDBORG, A. 1968: Some aspects of fluvial sediments and fluvial morphology II A study of some heavy mineral deposits in the valley of the river Lule Älv. *Geografiska Annaler* **50A**, 121–35.

LOHMAN, S. W. 1972: Ground-Water Hydraulics. *U.S. Geol. Survey Prof. Paper* **708**.

LUBOWE, J. K. 1964: Stream junction angles in the dendritic drainage pattern. *Amer. J. Sci.* **262**, 325–39.

LUCKMAN, B. H. 1971: The role of snow avalanches in the evolution of alpine talus slopes. In *Slopes, form and process* comp. D. Brunsden, *Institute of British Geographers Special Pub.* **3**, 93–110.

LUSBY, G. C. 1970: Hydrologic and biotic effects of grazing versus nongrazing near Grand Junction, Colorado. *U.S. Geol. Survey Prof. Paper* **700B**, 232–6.

LUTTIG, G. 1962: The shape of pebbles in the continental, fluviatile and marine facies. *Internat. Assoc. Sci. Hyd. Pub.* **59**, 253–8.

LVOVITCH, M. I. 1958: Streamflow formation factors. *Internat. Assoc. Sci. Hyd. Pub.* **45**, 122–32.

LVOVITCH, M. I. 1970: World water balance (general report). *Internat. Assoc. Sci. Hyd. Pub.* **93**, 401–15.

LYELL, C. 1830: *Principles of Geology*.

MACKENZIE, F. T. and GARRELS, R. M. 1966: Chemical mass balance between rivers and oceans. *Amer. J. Sci.* **264**, 507–25.

MACKIN, J. H. 1948: Concept of the graded river. *Geol. Soc. Amer. Bull.* **59**, 463–512.

MADEREY, L. E. 1971: *Balance hidrologico de la cuenca del Rio Tizar, durante el periodo 1967–68.* Universidad Nacional Autonuma de Mexico, Tesis.

MANDEVILLE, A. N. and RODDA, J. C. 1970: A contribution to the objective assessment of areal rainfall amounts. *J. Hyd. (N.Z.)* **9**, 281–91.

MANER, S. B. 1958: Factors affecting sediment delivery rates in the Red Hills physiographic area. *Trans. Amer. Geophys. Union* **39**, 669–75.

MANLEY, G. 1955: A climatological survey of the retreat of the Laurentide ice sheet. *Amer. J. Sci.* **243**, 256–73.

MANLEY, G. 1964: The evolution of the climatic environment. In J. WREFORD WATSON and J. B. SISSONS (eds.), *The British Isles*, London, 152–76.

MANN, J. C. 1970: Randomness in nature. *Geol. Soc. Amer. Bull.* **81**, 95–104.

MAROSI, S. and SZILARD, J. 1964: Landscape evaluation as an applied discipline of geography. In M. PECSI (ed.), *Applied Geography in Hungary*, Budapest, 20–35.

MARSH, G. P. 1864: *Man and nature; or, Physical geography as modified by human action*, New York.

MATHER, P. and DOORNKAMP, J. C. 1970: Multivariate analysis in geography, with particular reference to drainage basin morphometry. *Trans. Inst. Brit. Geog.* **51**, 163–87.

MATVEYEV, A. A. and BASHMAKOVA, O. I. 1967: Chemical composition of atmospheric precipitation in some regions of the U.S.S.R. *Soviet Hyd.* 480–91.

MAXEY, G. B. 1964: Hydrostratigraphic units. *J. Hyd.* **2**, 124–9.

MAXWELL, J. C. 1960: Quantitative geomorphology of the San Dimas Experimental Forest, California. *Office of Naval Research, Geography Branch, Project NR 389–042: Technical Report* **19**.

MCCOY, R. M. 1969: Drainage networks with K-band radar imagery. *Geog. Rev.* **59**, 493–512.

MCCOY, R. M. 1971: Rapid Measurement of Drainage Density. *Geol. Soc. Amer. Bull.* **82**, 757–62.

MCHENRY, J. R. 1964: A two-probe nuclear device for determining the density of sediments. *Internat. Assoc. Sci. Hyd. Pub.* **65**, 189–202.

MCPHERSON, H. J. 1971: Dissolved, suspended and bedload movement patterns in Two O'clock Creek, Rocky Mountains, Canada, Summer 1969, *J. Hyd.* **12**, 221–33.

MCQUEEN, I. S. 1963: Development of a hand portable rainfall-simulator infiltrometer. *U.S. Geol. Survey Circular* **482**.

MEAD, W. R. 1969: The course of geographical knowledge. In R. U. COOKE and J. H. JOHNSON (eds.), *Trends in Geography*, Oxford, 3–12.

MEADE, R. H. 1969: Errors in using modern stream-load data to estimate natural rates of denudation. *Geol. Soc. Amer. Bull.* **80**, 1265–74.

MEINZER, O. E. 1923: Outline of ground-water hydrology. *U.S. Geol. Survey Water Supply Paper* **494**.

MEINZER, O. E. (ed.) 1942: *Hydrology*, New York. (712pp.)

MELTON, F. A. 1936: An empirical classification of flood plain streams. *Geographical Review* **26**, 593–609.

MELTON, M. A. 1957: An analysis of the relations among elements of climate, surface properties and geomorphology. *Office of Naval Research, Geography Branch, Project NR 389–042: Technical Report* **11**, Columbia University.

MELTON, M. A. 1958: Geometric properties of mature drainage systems and their representation in an E4 phase space. *J. Geol.* **66**, 25–54.

MELTON, M. A. 1958: Correlation structure of morphometric properties of drainage systems and their controlling agents. *J. Geol.* **66**, 442–60.

MELTON, M. A. 1959: A derivation of Strahler's channel-ordering system. *J. Geol.* **67**, 345–6.

MENARD, H. W. 1961: Some rates of erosion. *J. Geol.* **69**, 154–61.

MEYBOOM, P. *et al.* 1966: Patterns of groundwater flow in seven discharge areas in Saskatchewan and Manitoba. *Geol. Survey Canada Bull.* **147**.

MIDDLETON, H. E. 1930: Properties of soils which influence soil erosion. *U.S. Dept. Agr. Tech. Bull.* **178** (16pp.)

MIKULSKI, Z. 1963: *Zarys Hydrografii Polski*, Pantwowe Naukowe, Poland.

MILLER, C. R. 1951: *Analysis of flow duration sediment rating curve method of computing sediment yield.* U.S. Bureau Reclamation, Denver, Colorado.

MILLER, J. P. 1961: Solutes in small streams draining single rock types, Sangre de Cristo Range, New Mexico. *U.S. Geol. Survey Water Supply Paper* **1535F**.

MILLER, J. P. and LEOPOLD, L. B. 1963: Simple measurements of morphological changes in river channels and hill slopes. In *Changes of climate, UNESCO Arid Zone Research Series* **XX**, 421–7.

MILLER, V. C. 1953: A quantitative geomorphic study of drainage basin characteristics in the Clinch mountain area: Va. and Tenn. *Office Naval Research Project NR 389–042, Tech. Rept.* **3**, Columbia University.

MILTON, L. E. 1965: Quantitative expression of drainage net patterns. *Australian J. Sci.* **27**, 238–40.

MILTON, L. E. 1966: The geomorphic irrelevance of some drainage net laws. *Australian Geographical Studies* **4**, 89–95.

MIRCEA, M. D. 1970: Estimation de l'influence des facteurs d'erosion in *Proceedings International Water Erosion Symposium, Praha*, Vol. **II**, 43–58.

MOORE, W. R. and SMITH, C. E. 1968: Erosion control in relation to Watershed management. *Proc. A.S.C.E. J. Irrig. Dr. Div.* **94**, IR3, 321–31.

MOORE, W. L., CLABORN, B. J. and COSKUN, E. 1969: Application of continuous accounting techniques to evaluate the effects of small structures on Mukewater Creek, Texas. In W. L. MOORE and C. W. MORGAN (eds.), *Effects of watershed changes on streamflow, Water Resources Symposium* **2**, University of Texas, 79–99.

MORE, R. 1967: Hydrological models and geography. In R. J. CHORLEY and P. HAGGETT (eds.), *Models in Geography*, London, 145–85.

MORGAN, J. P. (ed.) 1970: Deltaic Sedimentation Modern and Ancient. *Society of Economic Paleontologists and Mineralogists, Special Publication* **15**, Tulsa, Oklahoma. (312pp.)

MORGAN, R. P. C. 1970: Climatic geomorphology: its scope and future. *Geographica* **6**, 26–35.

MORISAWA, M. E. 1957: Accuracy of determination of stream lengths from topographic maps. *Trans. Amer. Geophys. Union* **38**, 86–8.

MORISAWA, M. E. 1962: Quantitative geomorphology of some watersheds in the Appalachian plateau. *Geol. Soc. Amer. Bull.* **73**, 1025–46.

MORISAWA, M. E. 1963: Distribution of stream-flow direction in drainage patterns. *J. Geol.* **71**, 528–9.

MORISAWA, M. E. 1964: Development of drainage systems on an upraised lake floor. *Amer. J. Sci.* **262**, 340–54.

MORISAWA, M. E. 1967: Relation of discharge and stream length in eastern United States. *Proc. Int. Hyd. Symp.*, Fort Collins, Colorado, 173–6.

MORISAWA, M. E. 1968: *Streams: Their dynamics and morphology*, New York. (175pp.)

MORRIS, D. A. and JOHNSON, A. I. 1967: Summary of hydrologic and physical properties of rock and soil materials, as analyzed by the Hydrologic Laboratory of the U.S. Geological Survey 1948–60. *U.S. Geol. Survey Water Supply Paper* **1839D**.

MORRISSEY, W. B. 1970: Field testing of the IAEA Portable Radioisotope sediment-concentration gauge. *N.Z. Ministry of Works, Water and Soil Division. Misc. Hydrological Pub.* **5**.

MUELLER, J. E. 1968: An introduction to the hydraulic and topographic sinuosity indexes. *Ann. Ass. Amer. Geog.* **58**, 371–85.

MÜHLHOFER, L. 1933: Untersuchungen über Schwebstaff-und Geschiebeführung des Inns nächst Kirchbichl. *Wasserwirtschaft*, Heft 1–6.

MUIR, T. C. 1970: Bedload discharge of the River Tyne, England. *Internat. Assoc. Sci. Hyd. Bull.* **15**, 35–9.

MUNDORFF, J. C. 1968: *Fluvial Sediment in Utah, 1905–65 A Data Compilation Information Bulletin* **20**, State of Utah, Dept. of Natural Resources, Division of Water Rights.

MURAVEYSKIY, S. D. 1956: *The process of runoff as a geographical factor*, Izb. Sochineniya, Acad. Sci. USSR Press, Moscow.

MURPHREE, C. E., BOLTON, G. C., MCHENRY, J. R. and PARSONS, D. A. 1968: Field test of an X-ray sediment concentration gage. *Proc. A.S.C.E. J. Hyd. Div.* **94**, 515–28.

MUSGRAVE, G. W. 1947: The quantitative evaluation of factors in water erosion, a first approximation. *J. Soil and Water Conservation* **2**, 133–8.

MUSGRAVE, G. W. and HOLTAN, H. N. 1964: Infiltration. In V. T. CHOW (ed.), *Handbook of Applied Hydrology*, New York, 12–1–12–30.

MUSSER, J. J. 1965: Acid mine drainage in the Appalachian Region (Sheet 9). In *Water Resources of the Appalachian Region, Pennsylvania to Alabama, U.S. Geol. Survey Hyd. Inv. Atlas* **HA198**.

MUSTONEN, S. E. 1967: Effects of climatologic and basin characteristics on annual runoff. *Water Resources Res.* **3**, 123–30.

NAGEL, J. F. 1956: Fog precipitation on Table Mountain. *Q.J. Royal Met. Soc.* **82**, 452–60.

NASH, J. E. 1966: Applied Flood Hydrology. In R. B. THORN (ed.), *River Engineering and Water Conservation Works*, London, 63–110.

NASH, J. E. and SHAW, B. L. 1966: Flood Frequency as a function of catchment characteristics. In *River Flood Hydrology*. Institution of Civil Engineers, London, 115–36.

N.E.D.E.C.O. 1959: *River studies and recommendations on improvement of Niger and Benue*, Amsterdam.

NEGEV, M. 1967: *A sediment model on a digital computer Rept.* **76**. Stanford University, California.

NEWBOULD, P. J. 1967: Methods for estimating the primary production of forests. *International Biological Programme Handbook* **2**, Oxford. (62pp.)

NEWBURY, R. W., CHERRY, J. A. and COX, R. A. 1969: Groundwater-streamflow systems in the Wilson Creek Experimental Watershed, Manitoba. *Canadian J. Earth Sci.* **6**, 613–23.

NEW ZEALAND MINISTRY OF WORKS, 1966: *Hydrology Annual* **14**, Wellington.

NEW ZEALAND MINISTRY OF WORKS, 1968: *Annual hydrological research report for Moutere* **1**, Wellington.

NEW ZEALAND MINISTRY OF WORKS, 1970: Representative basins of New Zealand. *Water and Soil Division, Miscellaneous Hydrological Pub.* **7**, Wellington.

NEW ZEALAND MINISTRY OF WORKS, 1970: *Annual hydrological research report for Otutira* **2**, Wellington.

NEW ZEALAND MINISTRY OF WORKS, 1970: *Annual hydrological research report for Puketurua* **2**, Wellington.

NEWSON, M. D. 1971: The role of abrasion in cavern development. *Trans. Cave Research Group G.B.* **B**, 101–7.

NEWSON, M. D. 1971: A model of subterranean limestone erosion in the British Isles based on hydrology. *Trans. Inst. Brit. Geog.* **54**, 55–70.

NING, CHIEN, 1961: The braided stream of the lower Yellow River. *Scientia Sinica* **10**, 734–54.

NIXON, M. 1959: A study of the bank-full discharges of rivers in England and Wales. *Inst. Civil Engineers Proc.* Paper **6322**, 157–74.

NOBLE, E. L. 1965: Sediment reduction through watershed rehabilitation in *Proc. Federal Inter-Agency Sedimentation Conference, U.S.D.A. Misc. Pub.* **970**, 114–23.

NOVAK, P. 1957: Bedload meters—Development of a new type and determination of their efficiency with the aid of scale models. *Int. Assoc. Hydraulic Res. 7th gen. mtg. Lisbon* (Transl.) Vol. **1**, A9–1.

OFOMATA, G. E. K. 1967: Some observations on relief and erosion in eastern Nigeria. *Revue de Geomorphologie Dynamique* **17**, 21–9.

OLLIER, C. D. 1968: Open systems and dynamic equilibrium in geomorphology. *Australian Geographical Studies* **6**, 167–70.

OLLIER, C. D. 1969: *Weathering*. Geomorphology Texts. General Editor, K. M. CLAYTON Edinburgh. (304pp.)

O'LOUGHLIN, C. L. 1969: Streambed investigations in a small mountain catchment. *New Zealand J. Geol. Geophys.* **12**, 684–706.

OLTMAN, R. E. 1968: Reconnaissance investigations of the discharge and water quality of the Amazon River. *U.S. Geol. Survey Circular* **552**. (16pp.)

ONGLEY, E. D. 1968: Towards a precise definition of drainage basin axis. *Australian Geographical Studies* **6**, 84–8.

ONGLEY, E. D. 1970: Drainage basin axial and shape parameters from mouent measures. *Canadian Geographer* **14**, 38–44.

ORLOVA, V. V. 1966: *Hydrometry*, Hydrometisdat, Leningrad.

ORSBORN, J. F. 1970: Drainage density in drift-covered basins. *Proc. A.S.C.E., J. Hyd. Div.* **HY1**, 183–92.

OSBORN, H. B. and LANE, L. 1969: Precipitation-runoff relations for very small semi-arid rangeland watersheds. *Water Resources Research* **5**, 419–25.

OSBORNE, B. 1953: Field measurements of soil splash to evaluate ground cover. *J. Soil and Water Conservation* **8**, 225–60, 266.

OUMA, J. P. B. M. 1967: Fluviatile morphogenesis of roundness: The Hacking River, New South Wales, Australia. *Internat. Assoc. Sci. Hyd. Pub.* **75**, 319–44.

OWENS, M. and EDWARDS, R. W. 1964: A chemical survey of some English Rivers. *Proc. Soc. Water Treatment Examination* **13**, 145–52.

PARDÉ, M. 1933: *Fleuves et Rivières*, Paris. (224pp.)

PARDÉ, M. 1955: *Fleuves et Rivières*, Paris, 3rd edn. (224pp.)

PARDÉ, M. 1960: Le régime des rivières en Nouvelle-Zélande. *Revue de Geographie Alpine* **48**, 383–429.

PARDÉ, M. 1961: Sur la puissance des crues en diverses parties du monde. *Geographica Saragossa.* (293pp.)

PARIZEK, R. R. and LANE, B. E. 1970: Soil water sampling using Pan and Deep Pressure-Vacuum lysimeters. *J. Hyd.* **11**, 1–21.

PARSONS, D. A. 1955: Coshocton—Type Runoff samplers. *U.S. Agricultural Res. Service Pub.* ARS **41–2**.

PASSARGE, S. 1926: Geomorphologie der Klimazonen oder Geomorphologie der Landschaftgürtel. *Petemanns Mitteilungen* **72**, 173–5.

PASSEGA, R. 1964: Grain size representation by CM patterns as a geological tool. *J. Sedimentary Petrology* **34**, 830–47.

PASTERNAK, K. 1963: Sklad chemiczny wody rzek i potokow o zlewniach zbudowanych z roznych skal i gleb. *Acta Hydrobiol.* **10**, 1–25.

PAVELKO, I. M. and TARASOV, M. N. 1967: Hydrochemical maps of the rivers of Kazakhstan and their use for the rapid forecasting of the mineralization and ion composition of waters of prospective reservoirs. *Soviet Hydrology*, 495–507.

PECK, A. J. and RABBIDGE, R. M. 1966: Soil water potential: direct measurement by a new technique. *Science* **151**, 1385–6.

PECSI, M. 1964: The role of derasion in modelling the earth's surface. M. PECSI (ed.), *Ten years of physicogeographic research in Hungary*, Budapest, 40–47.

PELTIER, L. C. 1950: The geographic cycle in periglacial regions as it is related to climatic geomorphology. *Ann. Ass. Amer. Geog.* **40**, 214–36.

PELTIER, L. C. 1962: Area sampling for terrain analysis. *Professional Geographer* **14**, 24–8.

PENCK, W. 1924: *Morphological analysis of landforms*, trans. H. CZECH and K. C. BOSWELL, 1953, 111–18.

PENMAN, H. L. 1949: The dependence of transpiration on weather and soil conditions. *J. Soil Sci.* **1**, 74–89.

PENMAN, H. L. 1963: Vegetation and Hydrology. *Technical Communication* **53**. Commonwealth Bureau of Soils Harpenden, Commonwealth Agricultural Bureau.

PETERSON, H. V. 1950: The problem of gullying in western valleys. In P. D. TRASK (ed.), *Applied Sedimentation*, New York, 412–36.

PHILIP, J. R. 1957: The theory of infiltration. *Soil Sci.* **83**, 345–57, 435–48, 84, 163–77, 257–64, 329–39; 85, 278–86, 333–6.

PIERCE, R. S. *et al.* 1970: Effect of elimination of vegetation on stream water quantity and quality. *Internat. Assoc. Sci. Hyd. Pub.* **96**, 311–28.

PIEST, R. F. 1965: The role of the large storm as a sediment contribution. *Proc. Fed. Int-Agency, Sed. Conf. U.S.D.A. Misc. Pub.* **970**, 97–108.

PILGRIM, D. H. 1966: Radioactive tracing of storm runoff on a small catchment: 1 experimental technique. *J. Hyd.* **4**, 289–305.

PINCHEMEL, P. H. 1957: Densités de drainage et densités des vallées. *Tijdschrift van het Koninklijk Nederlandsch Aardrijks-Kundig Genootschap*, Amsterdam, 373–6.

PINDER, G. F. and JONES, J. F. 1969: Determination of the groundwater component of peak discharge from the chemistry of total runoff. *Water Resources Res.* 5, 438–45.

PITTMAN, E. D. and OVENSHINE, T. 1968: Pebble Morphology in the Merced River (California). *Sedimentary Geology* 2, 125–40.

PLAYFAIR, J. 1802: *Illustrations of the Huttonian theory of the earth*, Edinburgh.

PLUHOWSKI, E. J. and KANTROWITZ, I. H. 1962: Source of groundwater runoff at Champlin Creek, Long Island, New York. *U.S. Geol. Survey Prof. Paper* **450B**, 95–7.

POLEMAN, J. N. B. and PAPE, J. C. 1967: The river landscapes in the Netherlands. *Internat. Assoc. Sci. Hyd. Pub.* **75**, 7–13.

POTTER, W. D. 1961: Peak rates of runoff from small watersheds. *U.S. Dept. Commerce, Bureau of Public Roads, Hydraulic Design Series* **2**. (35pp.)

POTTER, W. B., STOVICEK, F. K. and WOO, D. C. 1968: Flood frequency and channel cross section of a small natural stream. *Internat. Assoc. Sci. Hyd. Bull.* **13**, 66–76.

PRETL, J. 1970: The possibility of applying the Wischmeier-Smith's Relation in estimating the soil loss caused by water erosion in Czechoslovak conditions. *Proc. International Water Erosion Symposium* Vol. **III**, 83–96, Praha.

PRIOR, D. B. and STEPHENS, N. 1971: A method of monitoring mudflow movements. *Engineering Geology* 5, 239–46.

PRIOR, D. B., STEPHENS, N. and DOUGLAS, G. R. 1971: Some examples of mudflow and rockfall activity in north-east Ireland. In *Slopes, form and process*, comp. D. BRUNSDEN, *Institute of British Geographers Special Pub.* **3**, 129–40.

RAGAN, R. M. 1967: An experimental investigation of partial area contributions. *Internat. Assoc. Sci. Hyd. Pub.* **76**, 241–9.

RAISZ, E. and HENRY, J. 1937: An average slope map of southern New England. *Geographical Review* **27**, 467–72.

RAINWATER, F. H. and THATCHER, L. L. 1960: Methods for collection and analysis of water samples. *U.S. Geol. Survey Water Supply Paper* **1454**.

RAINWATER, F. H. 1962: Stream composition of the conterminous United States. *U.S. Geol. Survey Hyd. Inv. Atlas* **HA61**.

RAKHMANOV, V. V. 1962: *Role of forests in water conservation*, Gaslesbumizdat Moskva (Israel Program for Scientific Translations, 1966).

RANGO, A. 1970: Possible effects of precipitation modification on stream channel geometry and sediment yield. *Water Resources Res.* **6**, 1765–70.

RAPP, A. 1960: Recent development of mountain slopes in Karkevagge and surroundings, Northern Scandinavia. *Geografiska Annaler* **42**, 73–200.

RAPP, A. 1967: On the measurements of solifluction movements. *Revue de Geomorph. Dynamique* **17**, 162–3.

RATHBUN, R. E. and NORDIN, C. F. 1971: Tracer studies of sediment transport processes. *Proc. A.S.C.E. J. Hyd. Div.* **HY9**, 1305–16.

REID, C. 1887: On the origin of dry Chalk valleys and of Coombe rock. *Quart. J. Geol. Soc. Lond.* **43**, 364–73.

REYNOLDS, E. R. C. 1966: The percolation of rainwater through soil demonstrated by fluorescent dyes. *J. Soil Sci.* **17**, 127–32.

REYNOLDS, E. R. C. and LEYTON, L. 1967: Research data for forest policy: the purpose, methods and progress of forest hydrology. *Proc. 9th Brit. Commonwealth Forestry Conf. Commonwealth Forestry Inst., Univ. of Oxford.* (16pp.)

REYNOLDS, R. C. 1971: Analysis of Alpine waters by Ion Electrode Methods. *Water Resources Res.* **7**, 1333–7.

REYNOLDS, S. G. 1970: The gravimetric method of soil moisture determination Pts. 1, 2, and 3. *J. Hyd.* **11**, 258–300.

RICHARDS, L. A. 1931: Capillary conduction of liquids through porous mediums. *Physics* **1**, 318–33.

RIGGS, H. C. 1965: Effect of land use on the low flow of streams in Rappahannock County, Virginia. *U.S. Geol. Survey Prof. Paper* **525C**, C196–8.

RIGGS, H. C. 1967: Some statistical tools in hydrology. *U.S. Geol. Surv. Tech. Water-Resources Inv.*, Book **4**, Chap. A1.

RILEY, J. P., CHADWICK, D. G. and ISRAELSON, E. K. 1967: Application of an electronic analog computer for the simulation of hydrologic events on a southwest watershed. *Utah State Univ. Water Res. Laboratory Rept.* **PRWG 38–1**. (53pp.)

RILEY, J. P. and NARAYANA, V. V. D. 1969: Modelling the runoff characteristics of an urban watershed by means of an analog computer. In W. L. MOORE and C. W. MORGAN (eds.), *Effects of watershed changes on streamflow, Water Resources Symposium* **2**, University of Texas, 183–200.

RIVIÈRE, A. and VILLE, P. 1967: Sur l'utilisation d'une indice morphologique nouveau dans la representation d'une formation détritique grossiere. *Comptes Rendus, Acad. Sci. Paris, Ser. D.*, **265**, 1369–72.

ROAD RESEARCH LABORATORY 1968: The design of urban sewer systems. *Road Research Technical Paper* **55**.

ROBERTS, M. C. and KLINGEMAN, P. C. 1970: The influence of landform and precipitation parameters on flood hydrographs. *J. Hyd.* **11**, 393–411.

ROCHE, M. 1963: *Hydrologie de Surface.* Gauthier-Villars, Paris.

RODDA, J. C. 1967: The rainfall measurement problem. *Internat. Assoc. Sci. Hyd. Pub.* **78**, 215–31.

RODDA, J. C. 1967: A countrywide study of intense rainfall for the United Kingdom. *J. Hyd.* **5**, 58–69.

RODDA, J. C. 1969: The assessment of precipitation. In R. J. CHORLEY (ed.), *Water, Earth and Man*, London, 130–34.

RODDA, J. C. 1969: On more realistic rainfall measurements and their significance for agriculture. In *The Role of Water in Agriculture Aberystwyth Symposia in Agricultural Meteorology, Symposium* **XII**, A1–15.

RODDA, J. C. 1969: The flood hydrograph. In R. J. CHORLEY (ed.), *Water, Earth and Man*, London, 405–18.

RODDA, J. C. 1969: The significance of characteristics of basin rainfall and morphology in a study of floods in the United Kingdom. *UNESCO Symposium on Floods and their compilation. Internat. Assoc. Sci. Hyd.—UNESCO—WMO*, Vol. **2**, 834–45.

RODDA, J. C. 1970: Rainfall excesses in the United Kingdom. *Trans. Inst. Brit. Geog.* **49**, 49–60.

ROEHL, J. W. 1962: Sediment source areas, delivery ratios and influencing morphological factors. *Internat. Assoc. Sci. Hyd. Pub.* **59**, 202–13.

ROYSE, C. F. 1968: Recognition of fluvial environments by particle-size characteristics. *J. Sedimentary Petrology* **38**(4), 1171–8.

RUDBERG, S. 1967: On the use of test pillars. *Revue de Geomorph. Dynamique* **17**, 164–5.

RUHE, R. V. 1952: Topographic discontinuities of the Des Moines lake. *Amer. J. Sci.* **250**, 46–56.

RUSSELL, R. J. 1949: Geographical geomorphology. *Ann. Ass. Amer. Geog.* **39**, 1–11.

RUSSELL, R. J. 1954: Alluvial morphology of Anatolian rivers. *Ann. Ass. Amer. Geog.* **44**, 363–91.

RUSSELL, R. J. 1967: *River plains and sea coast*, Los Angeles. (173pp.)

RYAN, A. P. 1966: Radar estimation of rainfall, *J. Hyd.* (*N.Z.*) **5**, 100–110.

SANTOS, A. 1966: Statistical study of the flood flows of two Brazilian Rivers. In *River Flood Hydrology*, Institute of Civil Engineers, 59–64.

SATTERLUND, D. R. and ESCHNER, A. R. 1965: Land use, snow and streamflow regimen in central New York. *Water Resources Res.* **1**, 397–405.

SAVIGEAR, R. A. G. 1965: A technique of morphological mapping. *Ann. Ass. Amer. Geog.* **55**, 514–38.

SAWYER, R. M. 1963: Effect of urbanisation on storm discharge and groundwater recharge in Massau County, New York. *U.S. Geol. Survey Prof. Paper* **475C**, 185–7.

SCHEIDEGGER, A. E. 1965: The algebra of stream-order numbers. *U.S. Geol. Survey Prof. Paper* **525B**, B187–9.

SCHEIDEGGER, A. E. 1966: Stochastic branching processes and the law of stream orders. *Water Resources Res.* **2**, 199–203.

SCHEIDEGGER, A. E. 1969: Geomorphology. In *The Progress of Hydrology* **2**, U.S. Contribution to the International Hydrological Decade, University of Illinois, 777–87.

SCHEIDEGGER, A. E. and LANGBEIN, W. B. 1966: Probability concepts in geomorphology *U.S. Geol. Survey Professional Paper* **500C**. (14pp.)

SCHENCK, H. 1963: Simulation of the evolution of drainage basin networks with a digital computer. *J. of Geophys. Res.* **68**, 5739–45.

SCHICK, A. P. 1965: The effects of lineative factors on stream courses in homogeneous bedrock. *Internat. Assoc. Sci. Hyd. Bull.* **10**, 5–11.

SCHICK, A. P. 1967: Suspended sampler, *Revue de Geomorph. Dynamique* **17**, 181–2.

SCHICK, A. P. 1967: On the construction of troughs. *Revue de Geomorph. Dynamique* **17**, 170–72.

SCHMITTHENNER, H. 1925: Die enstehung der dellen und ihre morphologische bedetung. *Zeitschrift für Geomorphologie* I, 3–28.

SCHNEIDER, W. J. 1957: Relation of geology to streamflow in the Upper Little Miami Basin. *Ohio J. of Sci.* **57**, 11–14.

SCHNEIDER, W. J. 1965: Floods in the Appalachian Region (Sheet 6). In *Water Resources of the Appalachian Region, Pennsylvania to Alabama*, U.S.G.S. *Hyd. Inv. Atlas* **HA198**.

SCHNEIDER, W. J. 1965: Areal variability of low flows in a basin of diverse geologic units. *Water Resources Res.* 1, 4, 509–15.

SCHNEIDER, W. J. and FRIEL, E. A. 1965: Low flows in the Region (Sheet 5). In *Water Resources of the Appalachian Region, Pennsylvania to Alabama*, U.S.G.S. *Hyd. Inv. Atlas* **HA198**.

SCHUMM, S. A. 1954: The relation of drainage basin relief to sediment loss. *Internat. Assoc. Sci. Hyd. Pub.* **36**, 216–19.

SCHUMM, S. A. 1956: The evolution of drainage systems and slopes in badlands at Perth Amboy, New Jersey. *Geol. Soc. Amer. Bull.* **67**, 597–646.

SCHUMM, S. A. 1960: The shape of alluvial channels in relation to sediment type. *U.S. Geol. Survey Prof. Paper* **352B**, 17–30.

SCHUMM, S. A. 1961: Dimensions of some stable alluvial channels. *U.S. Geol. Survey Prof. Paper* **424B**, 26–7.

SCHUMM, S. A. 1961: The effect of sediment characteristics on erosion and deposition in ephemeral stream channels. *U.S. Geol. Survey Prof. Paper* **352C**, 31–70.

SCHUMM, S. A. 1963: Sinuosity of alluvial rivers on the Great Plains. *Geol. Soc. Amer. Bull.* **74**, 1089–100.

SCHUMM, S. A. 1963: The disparity between present rates of denudation and orogeny. *U.S. Geol. Survey Prof. Paper* **454H**. (13pp.)

SCHUMM, S. A. 1963: A tentative classification of river channels. *U.S. Geol. Survey Circular* **477**. (10pp.)

SCHUMM, S. A. 1965: Quaternary Palaeohydrology. In H. E. WRIGHT and D. G. FREY, (eds.), *The Quaternary of the United States*, Princeton, 783–94.

SCHUMM, S. A. 1967: Palaeohydrology: Application of modern hydrologic data to problems of the ancient past. *Internat. Assoc. Sci. Hyd. Pub.* 185–93.

SCHUMM, S. A. 1968: River adjustment to altered hydrologic regimen—Murrumbidgee River and palaeochannels, Australia. *U.S. Geol. Survey Prof. Paper* **598**. (65pp.)

SCHUMM, S. A. 1968: Speculations concerning palaeohydrologic controls of terrestrial sedimentation. *Geol. Soc. Amer. Bull.* **79**, 1573–88.

SCHUMM, S. A. 1969: River Metamorphosis. *Proc. A.S.C.E. J. Hyd. Div.*, **HY1**, 6352, 255–73.

SCHUMM, S. A. and HADLEY, R. F. 1957: Arroyos and the semi-arid cycle of erosion. *Amer. J. Sci.* **255**, 161–74.

SCHUMM, S. A. and KHAN, H. R. 1971: Experimental study of channel patterns. *Nature* **233**, 407–9.

SCHUMM, S. A. and LICHTY, R. W. 1963: Channel widening and flood plain construction along Cimarron river in south-western Kansas. *U.S. Geol. Survey Prof. Paper* **352D**, 71–88.

SCHUMM, S. A. and LICHTY, R. W. 1965: Time, space and causality in geomorphology. *Amer. J. Sci.* **263**, 110–19.

SEARCY, J. K. 1959: Flow Duration Curves. *U.S. Geol. Survey Water Supply Paper* **1542A**, 1–33.

SELBY, M. J. 1968: Cones for measuring soil creep. *J. Hyd. (N.Z.)* **7**, 136–7.

SELBY, M. J. 1968: Morphometry of drainage basins in areas of pumice lithology. *Proc. fifth New Zealand Geog. Conference, New Zealand Geog. Soc.*, 169–74.

SELBY, M. J. 1970: A flume for studying the relative erodibility of soils and sediments. *Earth Sci. J.* **4**, 32–5.

SELBY, M. J. 1970: Design of a hand-portable rainfall-simulating infiltrometer with results from the Otutira Catchment. *J. Hyd. (N.Z.)* **9**, 117–32.

SHAHJAHAN, M. 1970: Factors controlling the geometry of fluvial meanders. *Internat. Assoc. Sci. Hyd. Bull.* **15**, 13–24.

SHAMBLIN, O. H. 1965: Reservoir sedimentation survey methods in the U.S. Army

Engineer District, Vicksburg, Miss. *Proc. Federal Inter-Agency Sedimentation Conference 1963 U.S.D.A. Misc. Pub.* **970**, 810–18.

SHAROV, D. G. 1965: On the coefficient of underground flow on the territory of the European part of the U.S.S.R. *Trans. State Hyd. Inst. (Trudy GGI)* **122**, 217–25.

SHARP, A. L. and HOLTAN, H. N. 1940: A graphical method of analysis of sprinkled-plot hydrographs. *Trans. Amer. Geophys. Union* **21**, 558–70.

SHARPE, C. F. S. 1941: Geomorphic aspects of normal and accelerated erosion Symposium on Dynamics of Land Erosion. *Trans. Amer. Geophys. Union* **22**, 236–40.

SHARPS, J. A. 1969: Lateral migrations of the Arkansas River during the Quaternary–Fowler, Colorado, to the Colorado–Kansas state line. *U.S. Geol. Survey Prof. Paper* **650C**, C66–70.

SHEPHERDSON, I. 1965: Using Raydist for sedimentation surveys on larger reservoirs. *Proc. Federal Inter-Agency Sedimentation Conference U.S.D.A. Misc. Pub.* **970**, 869–79.

SHERLOCK, R. L. 1922: *Man as a geological agent*, London. (372pp.)

SHERMAN, L. K. 1932: Stream flow from rainfall by the unit-graph method. *Engineering News Record* **108**, 501–5.

SHEVCHENKO, M. A. 1962: Effect of various methods of tillage on the reduction of snowmelt runoff from sloping land. *Soviet Hydrology*, 27–33.

SHOWN, L. M. 1970: Evaluation of a method for estimating sediment yield. *U.S. Geol. Survey Prof. Paper* **700B**, 245–9.

SHREVE, R. L. 1966: Statistical law of stream numbers. *J. Geol.* **74**, 17–37.

SHREVE, R. L. 1967: Infinite topologically random channel networks. *J. Geol.* **75**, 178–86.

SHULITS, S. 1967: Quantitative formulation of stream and watershed morphology. Two decades of search in the U.S.A. *Internat. Assoc. Sci. Hyd. Pub.* **75**, *Symposium on River Morphology* 201–8.

SHUTER, E. and JOHNSON, A. I. 1961: Evaluation of equipment for measurement of water levels in wells of small diameter. *U.S. Geol. Survey Circular* **453**.

SHYKIND, E. B. 1956: Quantitative studies in geomorphology: subaerial and submarine erosional environments. *Geol. Soc. Amer. Bull.* **67**, 1733–4.

SIMONS, D. B. and RICHARDSON, E. V. 1962: The effect of bed roughness on depth–discharge relations in alluvial channels. *U.S. Geol. Survey Water-Supply Paper* **1498E**. (26pp.)

SIMONS, D. B. and RICHARDSON, E. V. 1966: Resistance to flow in alluvial channels. *U.S. Geol. Survey Prof. Paper* **422J**. (61pp.)

SKAKALSKIY, B. G. 1966: Basic geographical and hydrochemical characteristics of the local runoff of natural zones in the European territory of the U.S.S.R. *Trans. State Hyd. Inst. (Trudy GGI)* **137**, 125–80.

SKARZYNSKA, K. 1965: Paleohydrological Research in the Territory of Ancient Poland I and II. *Bulletin de l'Academie Polonaise des Sciences Serie des sci. geol. et geog.* **13**(3), 237–47.

SKIBITZKE, H. E. 1963: The use of analogue computers for studies on groundwater hydrology. *J. Inst. Water Engineers* **17**, 216–30.

SLIVITZKY, M. S. and HENDLER, M. 1965: Watershed research as a basis for water resources development. In *Research Watersheds, Fourth Canadian Hydrol. Symp. Canadian Natl. Res. Council*, 289–94.

SMALL, R. J. 1964: Geomorphology. In F. J. MONKHOUSE (ed.), *A survey of Southampton and its region*, Southampton, 37–50.

SMALLEY, I. J. and VITA-FINZI, C. 1969: The concept of 'system' in the earth sciences. *Geol. Soc. Amer. Bull.* **80**, 1591–4.

SMART, J. S. 1967: A comment on Horton's law of stream numbers. *Water Resources Res.* **3**, 773–6.

SMART, J. S. 1969: Topographical properties of channel networks. *Geol. Soc. Amer. Bull.* **80**, 1757–74.

SMART, J. S. 1972: Channel networks. In *Advances in Hydroscience* **8**, 305–46, New York.

SMART, J. S. and SURKAN, A. J. 1967: The relation between mainstream length and area in drainage basins. *Water Resources Res.* **3**, 963–74.

SMART, J. S., SURKAN, A. J. and CONSIDINE, J. P. 1967: Digital simulation of channel networks. *Internat. Assoc. Sci. Hyd. Pub.* **75**, 87–98.

SMART, J. S. and MORUZZI, V. L. 1971: Computer simulation of Clinch Mountain drainage networks. *J. Geol.* **79**, 572–84.

SMITH, C. T. 1969: The drainage basin as an historical basis for human activity. In R. J. CHORLEY (ed.), *Water, Earth and Man*, London, 101–10.

SMITH, K. 1969: The baseflow contribution to runoff in two upland catchments. *Water and Water Engineering*, 18–20.

SMITH, K. and LAVIS, M. E. 1969: A prefabricated Flat-Vee weir for experimental catchments. *J. Hyd.* **8**, 217–26.

SMITH, K. G. 1950: Standards for grading texture of erosional topography. *Amer. J. Sci.* **248**, 655–68.

SMITH, K. G. 1958: Erosional processes and landforms in Badlands National Monument, South Dakota. *Geol. Soc. Amer. Bull.* **69**, 975–1008.

SMITH, N. D. 1970: Braided stream depositional environment—comparison of the Platte river with some Silurian clastic rocks, north central Appalachians. *Geol. Soc. Amer. Bull.* **81**, 2993–3013.

SNEED, E. D. and FOLK, R. L. 1958: Pebbles in the Lower Colorado river. A study in particle morphogenesis. *J. Geol.* **66**, 114–50.

SNYDER, W. M. 1971: The parametric approach to watershed modelling. *Nordic Hydrology* **2**, 167–85.

SOKOLOV, A. A. 1969: Interrelationship between geomorphological characteristics of a drainage basin and the stream. *Soviet Hydrology*, Selected Papers **1**, 16–22.

SOLOVYEV, N. Y. 1967: Improvement and test of an instrument for recording coarse sediments. *Trans. of the State Hydrologic Inst. (Trudy GGI)* **141**, 58–78.

SOONS, J. M. 1967: Erosion by needle ice in the Southern Alps, New Zealand. In H. E. WRIGHT and W. H. OSBURN (eds.), *Arctic and Alpine environments*, 217–27.

SOONS, J. M. and RAINER, J. N. 1968: Micro-climate and erosion processes in the southern Alps, New Zealand. *Geografiska Annaler* **50A**, 1–15.

SOPPER, W. E. and LULL, H. W. 1965: Streamflow characteristics of physiographic units in the Northeast. *Water Resources Res.* **1**, 115–24.

SOPPER, W. E. and LULL, H. W. 1967: An evaluation of streamflow timing for small watersheds in N.E. United States. *14th Int. Forest Res. Organ. Kongr. Paper* Sect. **11**, 281–97.

SOPPER, W. E. and LULL, H. W. 1967: In W. E. SOPPER and H. W. LULL (eds.), *International Symposium on Forest Hydrology*, Oxford.

SPEIGHT, J. G. 1965: Meander spectra of the Angabunga river. *J. Hyd.* **3**, 1–15.

SPEIGHT, J. G. 1969: Parametric description of land form. In G. A. STEWART (ed.), *Land Evaluation*, 239–50.

SPRABERRY, J. A. 1965: Summary of reservoir sediment deposition surveys made in the United States through 1960. *U.S. Dept. Agr. Misc. Pub.* **964**. (61pp.)

STALL, J. B. and BARTELLI, L. J. 1959: Correlation of reservoir sedimentation and watershed factors, Springfield Plain, Illinois. *Illinois State Water Surv. Div. Rept. Invest.* **37**.

STALL, J. B. and YU-SI FOK 1967: Discharge as related to stream system morphology. *Symposium on River Morphology. Internat. Assoc. Sci. Hyd. Pub.* **75**, 224–35.

STALL, J. B. and YANG, C. T. 1970: Hydraulic geometry of 12 selected stream systems of the United States. *University of Illinois Water Resources Center, Res. Rept.* **32**. (73pp.)

STANKOWSKI, S. J. 1972: Population density as an indirect indicator of urban and suburban land-surface modifications. *U.S. Geol. Survey Prof. Paper* **800B**, B219–24.

STARKEL, L. 1963: The significance of the Holocene for the moulding of the relief of southern Poland. Geomorphological Section, *Report of the VIth International Congress on the Quaternary Warsaw 1961*, Vol. **III**, 341–4, Lodz.

STEHLIK, O. 1967: On methods of measuring sheet wash and rill erosion. *Revue de Geomorph. Dynamique* **17**, 176.

STEPHENSON, G. R. and ENGLAND, C. B. 1969: Digitized physical data of a rangeland watershed. *J. Hyd.* **8**, 442–50.

STEVENS, J. C. 1942: Device for measuring static heads. *Civil Engineering* **12**, 461.

STEWART, G. (ed.) 1969: *Landscape Evaluation*, Australia.

STICHLING, W. 1969: Instrumentation and Techniques in Sediment Surveying. In *Instrumentation and Observation Techniques, Proc. Hyd. Symp. No.* **7**, National Research Council of Canada, 81–139.

STODDART, D. R. 1965: Geography and the ecological approach. The ecosystem as a geographic principle and method. *Geography* **50**, 242–51.

STODDART, D. R. 1969: World erosion and sedimentation. In R. J. CHORLEY (ed.), *Water, Earth and Man*, London, 43–64.

STRAHLER, A. N. 1950: Equilibrium theory of erosional slopes approached by frequency distribution analysis. *Amer. J. Sci.* **248**, 673–96, 800–14.

STRAHLER, A. N. 1952: Hypsometric (area-altitude) analysis of erosional topography. *Geol. Soc. Amer. Bull.* **63**, 1117–42.

STRAHLER, A. N. 1952: Dynamic basis of geomorphology. *Geol. Soc. Amer. Bull.* **63**, 923–38.

STRAHLER, A. N. 1956: The nature of induced erosion and aggradation. In W. L. THOMAS (ed.), *Man's role in changing the face of the earth*, Chicago, 621–38.

STRAHLER, A. N. 1956: Quantitative slope analysis. *Geol. Soc. Amer. Bull.* **67**, 571–96.

STRAHLER, A. N. 1957: Quantitative analysis of watershed geomorphology. *Trans. Amer. Geophys. Union* **38**, 913–20.

STRAHLER, A. N. 1958: Dimensional analysis applied to fluvially eroded landforms. *Geol. Soc. Amer. Bull.* **69**, 279–300.

STRAHLER, A. N. 1964: Quantitative geomorphology of drainage basins and channel networks. In V. T. CHOW (ed.), *Handbook of Applied Hydrology*, 4–39—4–76.

STRAHLER, A. N. and KOONS, D. 1960: Objective and quantitative field methods of terrain analysis. *Final Report of Project NR 387-021 Office of Naval Research Geography Branch*, Columbia University, New York. (51pp.)

STRAKHOV, N. M. 1967: *Principles of Lithogenesis* **1**, trans. J. P. FITZSIMMONS, S. I. TOMKIEFF, and J. E. HEMINGWAY, New York. (245pp.)

STRIFFLER, W. D. 1964: Sediment streamflow, and land use relationships in northern Lower Michigan. *U.S. Forest Service Res. Paper* **LS16**. (12pp.)

STRIFFLER, W. D. 1965: Suspended sediment concentrations in a Michigan trout stream as related to watershed characteristics. *Proc. Fed. Inter-agency sedimentation Conf.*, *U.S.D.A. Misc. Pub.* **970**, 144–50.

STRINGFIELD, V. T. and LEGRAND, H. E. 1969: Hydrology of carbonate rock terrains—a review with special reference to the United States. *J. Hyd.* **8**, 349–417.

STRUTHERS, P. H. 1961: 180,000 strip mine acres: Ohio's largest chemical works. *Ohio Farm and Home Res.* **46**. (4pp.)

STRUZER, L. R., NECHAYEV, I. N. and BOGDANOVA, E. G. 1965: Systematic errors of measurements of atmospheric precipitation. *Soviet Hydrology*, 500–504.

SUNDBORG, A. 1956: The river Klarälven, a study of fluvial processes. *Geografiska Annaler* **38**, 127–316.

SUNDBORG, A. 1967: Some aspects on fluvial sediments and fluvial morphology: 1 General views and graphic methods. *Geografiska Annaler* **49A**, 333–43.

SUNLEY, C. S. 1969: Precision in terminology: an example from fluvial morphology. *Geographica Polonica* **18**, 183–97.

SURRELL, A. L. 1841: *Études sur les torrents des Hautes Alpes*, Paris.

ST. C. SWAN, S. B. 1970: Piedmont slope studies in a humid tropical region, Johar, southern Malaya. *Zeitschrift für Geomorphologie Suppl.* **10**, 30–39.

SWANSON, N. P. *et al.* 1965: Rotating boom rainfall simulator. *Trans. Amer. Soc. Agricultural Engineers* **8**, 71–2.

SWANSON, R. H. 1970: Sampling for direct transpiration estimates. *J. Hyd.* (*N.Z.*) **9**, 72–7.

SWEETING, M. M. 1972: *Karst Landforms*, London.

SWISS FEDERAL AUTHORITY 1939: Untersuchungen in der Natur über Bettbildung, Geschiebe-und Schwebest-offührung. Mitteilung des Antes fur Wasser wirtschaft **33**, Bern.

SYLVESTER, R. O. and SEABLOOM, R. W. 1963: Quality and significance of irrigation return flow. *Proc. A.S.C.E. J. Irrig. and Drainage Div.* **89**, IR3, 1–27.

TAKEDA, S. 1967: Movement of subsurface flow from forest area and mechanisms that affect soil erosion. *Int. Union of Forestry Res. Orgs. 14th Cong.*, 298–318.

TAMM, C. O. 1953: Growth, yield and nutrition in carpets of forest moss (Hylocomium splendens). *Skogstorskn Inst. Stockholm Medd.* **43**. (140pp.)

TANNER, W. F. 1960: Helicoidal flow, a possible cause of meandering. *J. Geophys. Res.* **65**, 993–5.

TANNER, W. F. 1968: Rivers—meandering and braiding. In R. W. FAIRBRIDGE (ed.), *Encyclopedia of Geomorphology*, New York.

TANNER, W. F. 1971: The river profile. *J. Geol.* **79**, 482–92.

TAYLOR, A. B. and SCHWARZ, H. E. 1952: Unit-hydrograph lag and peak flow related to basin characteristics. *Trans. Amer. Geophys. Union* **32**, 235–46.

TAYLOR, C. H. 1967: Relations between geomorphology and streamflow in selected New Zealand river catchments. *J. Hyd. (N.Z.)* **6**, 106–12.

TENNESSEE VALLEY AUTHORITY 1965: Area-stream factor correlation. *Internat. Assoc. Sci. Hyd. Bull.* **10**, 22–37.

THOMAS, D. M. and BENSON, M. A. 1970: Generalisation of streamflow characteristics from drainage basin characteristics. *U.S. Geol. Survey Water Supply Paper* **1975.**

THOMAS, W. L. 1956: *Man's role in changing the face of the earth*, Chicago. (1193pp.)

THOMPSON, J. R. 1964: Quantitative effect of watershed variables on rate of gully-head advancement. *Trans. Amer. Agr. Engineers* **7**(1), 54–5.

THOMPSON, J. R. 1970: Soil erosion in the Detroit Metropolitan area. *J. Soil and Water Conservation*, 8–10.

THORNES, J. B. 1970: The hydraulic geometry of stream channels in the Xingu-Araguaia headwaters. *Geog. J.* **136**, 376–82.

THORNTHWAITE, C. W. 1948: An approach towards a rational classification of climate. *Geog. Rev.* **38**, 55–94.

THORPE, G. B. 1964: A new suspended solids recorder. *Industrial Electronics*, 415–18.

TINKER, J. 1971: The forest: Umbrella or Sponge. *New Scientist and Sci. J.* **51**, 608–9.

TINKLER, K. J. 1971: Active valley meanders in south central Texas and their wider implications. *Geol. Soc. Amer. Bull.* **81**, 1873–99.

TINLIN, R. M. 1969: A passive direct electric analog of a watershed. *Dept. of Watershed Management, Univ. Arizona Prof. Paper.* (22pp.)

TINLIN, R. M. and THAMES, J. L. 1969: A passive direct electric analog of a watershed. Abstract in *Trans. Am. Geophys. Un.* **50**, 609.

TINLIN, R. M. and THAMES, J. L. 1971: *The analysis and application of a digitally simulated electronic watershed analog*, Dept. Watershed Management University, Arizona.

TODD, D. K. 1959: *Ground Water Hydrology*, New York.

TODD, D. K. 1970: *The water encyclopedia*, New York. (559pp.)

TODD, O. J. and ELIASSEN, C. E. 1938: The Yellow River Problem. *Proc. A.S.C.E.* **64**.

TOEBES, C. and STRANG, D. D. 1964: On recession curves. *J. Hyd. (N.Z.)* **3**, 2–14.

TOEBES, C. and OURYVAEV, V. (eds.) 1970: *Representative and experimental basins. An international guide for research and practice*, UNESCO.

TOLER, L. G. 1965: Relation between chemical quality and water discharge in Spring Creek, Southwestern Georgia. *U.S. Geol. Survey Prof. Paper* **525C**, 209–13.

TOYNE, P. 1973: *Organisation, location and behaviour. An introduction to Economic Geography*, London.

TRAINER, F. W. 1969: Drainage density as an indicator of baseflow in part of the Potomac river. *U.S. Geol. Survey Prof. Paper* **650C**, C177-83.

TRICART, J. 1953: Geomorphologie dynamique de la Steppe Russe. *Revue de Geomorphologie Dynamique* **4**, 1–32.

TRICART, J. 1957: Application du concept de zonalité à la geomorphologie. *Tijdschrift van het Koninklijk Nederlandsch Aardrijkskundig Geomootschap*, Amsterdam, 422–34.

TRICART, J. 1966: Schéma des mécanismes de causalité en géomorphologie. *Annales de Geographie* **403**, 322–6.

TRICART, J. and CAILLEUX, A. 1965: *Introduction à la géomorphologie climatique*, Paris.

TRICART, J. and SHAEFFER, R. 1950: L'indice d'émoussé des galets. Moyen d'étude des systèmes d'erosion. *Revue de Geomorphologie Dynamique* **1**, 151–79.

TRICART, J. and VOGT, H. 1967: Quelques aspects du transport des alluvions grossières et du façonnement des lits fluviaux. *Geografiska Annaler* **49A**, 351–66.

TRIMBLE, S. W. 1970: The Alcovy River Swamps: The result of culturally accelerated sedimentation. *Bull. Georgia Academy of Sci.* **28**, 131–41.

TROEH, F. R. 1965: Landform equations fitted to contour maps. *Amer. J. Sci.* **263**, 616–27.

TROSKELANSKI, A. T. 1960: *Hydrometry*, New York.

TUCKFIELD, C. G. 1964: Gully erosion in the New Forest, Hampshire. *Amer. J. Sci.* **262**, 795–807.

TYLOR, A. 1875: On the action and the formation of rivers, lakes and streams, with remarks on denudation and the causes of the great changes of climate which occurred just prior to the historical period. *Geol. Mag.* **2**, 433–73.

U.N.E.S.C.O. 1969: Discharge of selected rivers of the world. *UNESCO Studies and Reports in Hyd.* **5**.

URSIC, S. J. and DENDY, F. E. 1965: Sediment yields from small watersheds under various land uses and forest covers. *Proc. Federal. Inter-Agency Sedimentation Conference, U.S.D.A. Misc. Pub.* **970**, 47–52.

U.S. DEPARTMENT OF AGRICULTURE 1964: Stream-gaging stations for research on small watersheds. *U.S.D.A. Forest Service Agricultural Handbook* **268**.

U.S. GEOLOGICAL SURVEY 1967 Continuing: *Techniques of Water Resources Investigations.*

U.S. GEOLOGICAL SURVEY 1967 Continuing: Surface Water techniques. *Techniques of Water-Resources Investigations of the United States Geological Survey* Book **3**, Section A.

U.S. GEOLOGICAL SURVEY 1969: Laboratory theory and methods for sediment analysis. *Techniques of Water-Resources Investigations of the United States Geological Survey* Book **5**, Chapter C1.

U.S. GEOLOGICAL SURVEY 1970: Methods for collection and analysis of water samples for dissolved minerals and gases. *Techniques of Water-Resources Investigations of the United States Geological Survey* Book **5**, Chapter A1.

VAN BAVEL, C. H. M. 1965: Neutron scattering measurement of soil moisture: development and current status. In A. WEXLER (ed.), *Humidity and moisture measurement and control in science and industry*, New York, 171–84.

VAN DENBURGH, A. S. and FETH, J. H. 1965: Solute erosion and chloride balance in selected river basins in the Western conterminous United States. *Water Resources Res.* **1**, 537–41.

VAN DOREN, C. A., STAUFFER, R. S. and KIDDER, E. H. 1950: Effect of contour farming on soil loss and runoff. *Soil Sci. Soc. Amer. Proc.* **15**, 413–17.

VANONI, V. A., BROOKS, N. H. and KENNEDY, J. F. 1960: Lecture notes on sediment transportation and channel stability. *Calif. Inst. Technology Rept.* **KH-R1**.

VAN STRAATEN, L. M. J. U. (ed.) 1964: *Deltaic and shallow marine deposits*, Amsterdam. (464pp.)

VEMURI, V. and VEMURI, N. 1970: On the systems approach in hydrology. *Internat. Assoc. Sci. Hyd. Bull.* **15**, 17–38.

VICE, R. B. *et al.* 1969: Sediment movement in an area of suburban highway construction, Scott Run Basin, Fairfax County, Virginia, 1961–4. *U.S. Geol. Survey Water Supply Paper* **1591E**. (41pp.)

VITA-FINZI, C. 1969: *The Mediterranean Valleys*, Cambridge.

VORONKOV, P. P. 1963: Hydrochemical bases of the separation of local runoff and a method of separating its discharge hydrograph. *Meteorologiya i Gidrologiya* **8**, 21–8.

WADELL, H. 1935: Volume, shape and roundness of quartz particles. *J. Geol.* **43**, 250–80.

WALLING, D. E. 1971: Streamflow from instrumented catchments in south-east Devon. In K. J. GREGORY and W. L. D. RAVENHILL (eds.), *Exeter Essays in Geography*, Exeter, 55–81.

WALLING, D. E. 1971: Sediment dynamics of small instrumented catchments in south-east Devon. *Trans. Devonshire Assoc.* **103**, 147–65.

WALLING, D. E. 1971: *Instrumented catchments in south-east Devon. Some relationships between catchment characteristics and catchment response.* Unpub. Ph.D. Thesis University of Exeter.

WALLING, D. E. and GREGORY, K. J. 1970: The measurement of the effects of building construction on drainage basin dynamics. *J. Hyd.* **11**, 129–44.

WALLING, D. E. and TEED, A. 1971: A simple pumping sampler for research into suspended sediment transport in small catchments. *J. Hyd.* **13**, 325–37.

WALTZ, J. P. 1969: Ground Water. In R. J. CHORLEY (ed.), *Water, Earth and Man*, London, 259–68.

WARD, R. C. 1967: *Principles of Hydrology*, London. (403pp.)

WARD, R. C. 1967: Water balance in a small catchment. *Nature* **213**, 123–5.

WARD, R. C. 1971: Measuring evapotranspiration: A review. *J. Hyd.* **13**, 1–21.

WARD, R. C. 1971: Small Watershed Experiments: An appraisal of concepts and research developments. *University of Hull Occasional Papers in Geography* **18**.

WARK, J. W. and KELLER, F. J. 1963: *Preliminary study of sediment sources and transport in the Potomac River Basin*, Interstate Commission on the Potomac River Basin.

WARNICK, C. C. and PENTON, U. E. 1971: New methods of measuring water equivalent of snow pack for automatic recording at remote mountain locations. *J. Hyd.* **13**, 201–5.

WARWICK, G. T. 1964: Dry valleys in the southern Pennines. *Erdkunde* **18**, 116–23.

WATERS, R. S. 1966: Dartmoor Excursion. *Biuletyn Peryglacjalny* **15**, 123–8.

WAUGH, J. R. 1970: The relationship between summer low flows and geology in Northland, New Zealand. *N.Z. Ministry of Works–Water and Soil Division Misc. Hyd. Pub.* **6**.

WEIBEL, S. R. *et al.* 1964: Urban land runoff as a factor in stream pollution. *J. Water Pollution Conf. Fed.* **36**, 914–24.

WENTWORTH, C. K. 1930: A simplified method of determining the Average Slope of land surfaces. *Amer. J. Sci.* **20**, 184–94.

WERNER, P. W. 1951: On the origin of river meanders. *Trans. Amer. Geophys. Un.* **32**, 898–902.

WEYMAN, D. R. 1970: Throughflow on slopes and its relation to the stream hydrograph. *Internat. Assoc. Sci. Hyd. Bull.* **15**, 25–33.

WHIPKEY, R. Z. 1965: Subsurface stormflow from forested slopes. *Internat. Assoc. Sci. Hyd. Bull.* **10**, 74–85.

WILLIAMS, P. J. 1957: Some investigations into solifluction features in Norway. *Geog. J.* **123**, 42–58.

WILLIAMS, R. E. and FOWLER, P. M. 1969: A preliminary report on an empirical analysis of drainage network adjustment to precipitation input. *J. Hyd.* **8**, 227–38.

WILGAT, T. 1966: Distance from water—an index of density of hydrographic system. *Przeglad Geograficzny* **38**, 371–80.

WILM, H. G., COTTON, J. S. and STOREY, H. C. 1938: Measurement of debris-laden streamflow with critical depth flumes. *Trans. A.S.C.E.* **103**, 1237–78.

WILM, H. G. and STOREY, H. C. 1944: Velocity head rod calibrated for measuring stream flow. *Civil Eng.* **14**, 475–6.

WILSON, K. V. 1967: A preliminary study of the effect of urbanization on Floods in Jackson Mississippi. *U.S. Geol. Survey Prof. Paper* **575D**, 259–61.

WINKLER, E. H. 1970: Errors in using modern stream-load data to estimate natural rates of denudation: discussion. *Geol. Soc. Amer. Bull.* **81**, 983–4.

WINSLOW, J. H. 1966: Raised submarine canyons. An exploratory hypothesis. *Ann. Ass. Amer. Geog.* **56**, 634–72.

WISCHMEIER, W. H. and SMITH, D. D. 1958: Rainfall energy and its relationship to soil loss. *Trans. Amer. Geophys. Union* **39**, 285–91.

WISLER, C. O. and BRATER, E. F. 1959: *Hydrology*, New York. (408pp.)

WITZIGMAN, F. S. 1965: A summary of the work of the Inter-Agency Sedimentation Project. *Proc. Federal Inter-Agency Sedimentation Conference, U.S.D.A. Misc. Pub.* **970**, 166–77.

WOLDENBERG, M. J. 1966: Horton's laws justified in terms of allometric growth and steady state in open systems. *Geol. Soc. Amer. Bull.* **77**, 431–4.

WOLDENBERG, M. J. 1969: Spatial order in fluvial systems: Horton's laws derived from mixed hexagonal hierarchies of drainage basin areas. *Geol. Soc. Amer. Bull.* **80**, 97–112.

WOLMAN, M. G. 1954: A method of sampling coarse bed material. *Trans. Amer. Geophys. Un.* **35**, 951–6.

WOLMAN, M. G. 1955: The natural channel of Brandywine Creek, Pennsylvania. *U.S. Geol. Survey Prof. Paper* **271**.

WOLMAN, M. G. 1967: A cycle of sedimentation and erosion in urban river channels. *Geografiska Annaler* **49A**, 385–95.

WOLMAN, M. G. 1967: Two problems involving river channel changes and background observations. In *Quantitative Geography Part II Physical and Cartographic Topics. Northwestern University Studies in Geography* **14**, 67–107.

WOLMAN, M. G. and LEOPOLD, L. B. 1957: River flood plains: Some observations on their formation. *U.S. Geol. Survey Prof. Paper* **282C**, 87–107.

WOLMAN, M. G. and MILLER, J. P. 1960: Magnitude and frequency of forces in geomorphic processes. *J. Geol.* **68**, 54–74.

WOLMAN, M. G. and SCHICK, A. P. 1967: Effects of construction on fluvial sediment: urban and suburban areas of Maryland. *Water Resources Res.* **3**, 2, 451–62.

WOOD, M. J. and SUTHERLAND, A. J. 1970: Evaluation of a digital catchment model on New Zealand catchments. *J. Hyd. (N.Z.)* **9**, 323–35.

WOODRUFF, J. F. 1964: A comparative analysis of selected drainage basins. *Prof. Geographer* **16**, 15–19.

WOODRUFF, J. F. and HEWLETT, J. D. 1970: Predicting and mapping the average hydrologic response for the eastern United States. *Water Resources Res.* **6**, 1312–26.

WOODRUFF, J. F. and PARIZEK, E. J. 1956: Influence of underlying rock structures on stream courses and valley profiles in the Georgia Piedmont. *Ann. Ass. Amer. Geog.* **46**, 129–39.

WOODYER, K. D. 1968: Bankfull frequency in rivers. *J. Hyd.* **6**, 114–42.

WOODYER, K. D. and BROOKFIELD, M. 1966: The land system and its stream net. *C.S.I.R.O. Div. of land resources, Tech. Mem.* **66/5**.

WOOLDRIDGE, D. D. 1965: Tracing soil particle movement with Fe-59. *Soil Sci. Soc. Amer. Proc.* **29**, 469–72.

WOOLDRIDGE, S. W. and LINTON, D. L. 1955: *Structure, surface and drainage in south-east England*, London. (176pp.)

WOOLHISER, D. A. 1971: Deterministic approach to watershed modelling. *Nordic Hydrology* **2**, 146–66.

WONG, S. T. 1963: A multivariate statistical model for predicting mean annual flood in New England. *Ann. Ass. Amer. Geog.* **53**, 298–311.

WORLD METEOROLOGICAL ORGANISATION 1965: *Guide to Hydrometeorological Practices* **WMO168**, TP 82.

WORLD METEOROLOGICAL ORGANISATION 1965: Hydrological Analysis. In *Guide to Hydrometeorological Practices* **WMO168**, TP 82, Annex A, A1–67.

WRIGHT, C. E. 1967: Glass-fibre trapezoidal flumes. *The Surveyor and Municipal Engineer*, 23–7.

WRIGHT, C. E. 1970: Catchment characteristics influencing low flows. *Water and Water Engineering* **74**, 468–71.

YANG, C. T. 1970: On river meanders. *J. Hyd.* **13**, 231–53.

YATSU, E. 1966: *Rock control in geomorphology*, Tokyo. (135pp.)

YEVDJEVICH, V. M. 1964: Statistical and probability analysis of hydrologic data, Part II, Regression and Correlation analysis. In V. T. CHOW (ed.), *Applied Hydrology*, New York, 8–43—8–65.

YI-FU TAN 1966: New Mexican gullies: A critical review and some recent observations. *Annals. Ass. Amer. Geog.* **56**, 573–97.

YOUNG, A. 1960: Soil movement by denudation processes on slopes. *Nature* **188**, 120–22.

YOUNG, A. 1969: Present rate of land erosion. *Nature* **224**, 851–2.

YU-SI FOK, 1971: Law of stream relief in Horton's stream morphological system. *Water Resources Res.* **7**, 201–3.

ZAKRZEWSKA, B. 1967: Trends and methods in land form geography. *Ann. Ass. Amer. Geog.* **57**, 128–65.

ZELLER, J. 1967: Meandering channels in Switzerland. *Internat. Assoc. Sci. Hyd. Pub.* **75**, 174–86.

ZERNITZ, E. R. 1932: Drainage patterns and their significance. *J. Geol.* **40**, 498–521.

ZEUNER, F. E. 1946: *Dating the past: an introduction to geochronology*, London. (444pp.)

ZIEGLER, C. A., PAPADOPOULOS, J. and SELLERS, B. 1967: Radioisotope gauge for monitoring suspended sediment in rivers and streams. *Int. J. appl. Radiat. Isotopes* **18**, 585.

ZIMMERMAN, R. C., GOODLETT, J. C. and COMER, G. H. 1967: The influence of vegetation on channel form of small streams. Symposium on River Morphology. *Int. Ass. Sci. Hydrol. Pub.* **75**, 255–75.

ZINGG, T. 1935: Beitrag zur Schotteranalyse. *Schwiz, Min u. Pet. Mitt.*, bd. **15**, 39–140.

Index

Subjects indexed below are restricted to major references and to relevant material in maps, diagrams and tables which is indicated by page references in italics. Locations are not included unless they are elaborated in detail in the text, and references to authors are not included in view of the bibliography (pp. 407–447).